MODIFICATION REACTIONS OF OLIGOMETHYLHYDRIDESILOXANES

MODIFICATION REACTIONS OF OLIGOMETHYLHYDRIDESILOXANES

OMARI MUKBANIANI,
TAMARA TATRISHVILI
AND
GENNADY ZAIKOV
EDITORS

Nova Science Publishers, Inc.
New York

LIBRARY OF CONGRESS CATALOGING-IN-PUBLICATION DATA

Mukbaniani, O. V. (Omar V.)
Modification reactions of oligomethylhydridesiloxanes / Omari Mukbaniani and Tamara Tatrishvili ; Gennady Zaikov, editor.
p. cm.
Includes index.
ISBN 13: 978-1-60021-365-6
ISBN 10: 1-60021-365-0
1. Polymers--Research. 2. Silicones--Research. 3. Hydrosilylation. 4. Chemical reactions. I. Tatrishvili, Tamara. II. Zaikov, Gennadii Efremovich. III. Title.
QD281.P6M79 2006
547'.7--dc22

2006022423

Published by Nova Science Publishers, Inc. ✢ New York

Contents

PREFACE

In 1969 at the Moscow State University, high-molecular compounds were discussed at the International Congress, which took place within the framework of the International Union of Pure and Applied Chemistry (IUPAC). On this congress the chief of Soviet delegation Prof. V.A. Kargin spoke at a meeting. He had declared, that "we should not expect making of large-tonnage manufactures of new polymers, and progress of Mankind in this field will go on ways of modification of already known polymers, makings polymer blends, polymer composites and filled polymers". These forecasts come true in practice, and modification of known polymers is the important factor in making new polymers and new large-tonnage manufactures.

The present monograph is devoted to modification reactions of oligomethylhydride-siloxanes. It is the fixed block in a building of modification processes, confirming above the told forecast.

To speak about polyorganosiloxanes it is necessary to recollect first of all the names of Profs. K.A. Andrianov (USSR) and E. Rochow (USA), who have introduced the major contributions to development of this class of high-molecular compounds: synthesis, studying of properties and application. Major interest to this class of compounds was in the works of Profs. H. Mark (USA), Robert and Bob West (USA), K. Matyjaszewski (USA), R.J.P. Corriu (France), J. Chojnowski (Poland), E. Henge (Austria), W. Uhlig (Austria), A.G. Brook (USA), H. Ringsdorf (Germany), H.R. Kricheldorf (Germany), L.M. Khananashvili (Georgia), B. Marciniec (Poland),

The given monograph should be of major interest to a wide range of readers: students, post-graduate students, the science officers specializing in the field of polymers, and as engineers and technician, the firms working in factory laboratories manufacturing polymers (first of all, polyorganosiloxanes).

Editors and authors of the given collection will be grateful to readers for design notes which they can take into account in the further operation.

Editors:
Prof. O.V. Mukbaniani (Javakhishvili Tbilisi State University, Tbilisi, Georgia)
Dr. T.N. Tatrishvili (Javakhishvili Tbilisi State University, Tbilisi, Georgia)
Prof. G.E. Zaikov (N.M. Emanuel Institute of Biochemical Physics, Moscow, Russia)

In: Modification Reactions of Oligomethylhydridesiloxanes ISBN: 1-60021-365-0
Editors: O.Mukbaniani, et al. pp. 1-19

Chapter 1

HYDROSILYLATION REACTIONS OF METHYLHYDRIDESILOXANE TO N-ALKENES

O. Mukbaniani, T. Tatrishvili, G. Titvinidze, and N. Mukbaniani
[1]I. Javakhishvili Tbilisi State University,
I. Chavchavadze Ave., 1, 0128 Tbilisi, Georgia

ABSTRACT

The hydrosilylation reaction of α,ω–bis(trimethylsiloxy)methylhydridesiloxane to alkenes in the presence of catalyst – platinum hydrochloric acid (0,1 M solution in tetrahydrofuran) at 1:35 ratio of initial compounds, at various temperatures (30-70^{0}C) were investigated and comb-type methyl-siloxane oligomers with different alkyl substituted groups in the side chain have been obtained. It was shown that complete hydrosilylation of all active ≡Si-H groups does not take place. The hydrosilylation reaction order, activation energies and rate constants were found. The synthesized oligomers were characterized by ^{1}H, ^{13}C, COSY NMR and IR spectra data. For full characterization of hydride addition of methylhydridesiloxane to n-alkenes, by quantum-chemical half empiric AM1 method, for all initial, intermediate and final products, in modelling hydrosilylation reaction of methyldimethoxysilane to hexene-1, the heats of formations (ΔH_f), change of energy (ΔH) of the system depending on the change of distance ($R_{C\text{-}Si}$) between ≡C-Si≡ bonds, also the charge values (q) on the atoms, dipole moments (μ) and bond orders (Pij) are calculated. It was concluded, that the course of modelling hydride addition of methyldimethoxysilane to hexene-1 energetically is more favourable by the Anti-Markovnikov rule. Comb-type oligomers were studied by gel-permeation chromatographic, differential scanning calorimetric, thermogravimetric and wide-angle X-ray methods.

Key words: methylhydridesiloxane, alkenes, hydrosilylation, hydride addition, comb-type oligomers.

INTRODUCTION

The properties of organosilicon polymers depend on the structure of the macromolecular chain and on the nature of organic groups surrounding the silicon atom [1]. In previous studies a great attention is paid to the comb-type copolymers, where different type, length and natural organic substituted groups are bonded on the methylsiloxane hydrophobic matrix. It is possible to utilize the above-mentioned copolymers in a wide range by variation of these substituted groups. Nowadays it is known that some siliconorganic copolymers containing non-classical mesogenic groups in the main chain or in the side chain are characterized by liquid-crystalline properties [2-4].

In polydialkylsiloxanes, independently of the main chain, a formation of mesomorphic state mainly in the presence of non-classical mesogenic alkyl substituents with definite values of the length ($R=C_nH_{2n+1}$, $n\geq2$) is observed.

For synthesis of comb-type polymers there are two principal ways – polymerization of unsaturated or cyclic monomers, where the side fragment with mesogenic properties already exists in initial mo-nomer molecule. The second one is polymer analogical transformation, when previously obtained polymeric matrix occurs while attaching side branching groups.

Because of that, to obtain new comb-type organosilicon oligomers by modification of methylhydridesiloxane oligomers by insertion of various side branching groups represents an actual problem.

EXPERIMENTAL PART

The starting materials for the synthesis of comb-type methylalkylsiloxanes oligomers with various lengths of alkyl substituted groups in the side chain were α,ω–bis(trimethylsiloxy)methylhydridesiloxane, n-hexene-1, n-octen-1, n-decene-1, n-dodecene-1 and n-octadecene.

The initial α,ω–bis(trimethylsiloxy)methylhydridesiloxane with the degree of polymerization $n\approx35$, alkenes with normal structure: hexene-1, octen-1, decene-1, dodecene-1 and octadecene-1 was received from Fluka. The organic solvents were cleaned by drying and distillation.

IR spectra all of samples have been taken on an UR-20 instrument from KBr pellets, while the ^1H ^{13}C NMR spectra on a "Perkin-Elmer" at operating frequency of 250 MHz. All spectra were obtained with the use of $CDCl_3$ as a solvent and internal standard. A Perkin – Elmer DSC-2 differential scanning calorimeter was used to determine TGA and the thermal transitions (T_g) were read at the maximum of the endothermic or exothermic peaks. Heating and cooling scanning rates were 10^0C/min.

Gel-permeation chromatography investigation was carried out with the use of Waters Model 6000A chromatograph with an R 401 differential refractometer detector. The column set comprised 10^3 and 10^4 Å Ultrastyragel columns. Sample concentration was approximately 3% by weight in toluene and typical injection volume for the siloxane was 5 μL. standardization of the GPC was accomplished by the use of styrene or polydimethylsiloxane standards with the known molecular weight.

Wide-angle X-ray diffractograms were taken on a "DRON-2" instrument. A-CuK_α was measured without a filter, the angular velocity of the motor $\omega \approx 2$ deg/min.

Hydrosilylation Reaction of α,ω–Bis(Trimethylsiloxy)Methylhydridesiloxane to N-Hexene-1

Hydrosilylation reaction was carried out in a two-necked flask equipped with a tube for catalyst inlet, reflux condenser. The reaction products 0.9862 g (0.44 mmole) α,ω–bis(trimethylsiloxy)methylhydridesiloxane and 1.2810 g (0.0154 mole) n-hexene-1 and 5.4 ml dry toluene were placed into the flask and thermostatted in an oil bath until constant temperature was set up in an argon atmosphere. Then the catalyst 0.1 M solution of platinum hydrochloric acid in tetrahydrofuran ($5 \div 9 x 10^{-5}$ g per 1.0 g of starting substance) was introduced. After the reaction has been finished the solvent partially was eliminated, the reaction product was precipitated from toluene solution by n-hexane, and 2.1 g (93%) oligomer I was obtained. The hydrosilylation reaction with other alkenes was carried out by the above-mentioned reaction using the catalyst – platinum hydrochloric acid (0.1 M solution in tetrahydrofuran).

Result and Discussion

In the literature there is much information about hydrosilylation reactions of organohydrosilanes to alkenes. It was shown that hydrosilylation of organohydrosilanes to alkenes in the presence of catalyst platinum hydrochloric acid proceeds according to Anti-Markovnikov and Markovnikov rule [5, 6].

For synthesis of comb-type siliconorganic oligomers with alkyl groups in the side chain the hyd-rosilylation reaction of α,ω–bis(trimethylsiloxy)methylhydridesiloxane to alkenes, in the presence of catalyst 0,1 M solution of platinum hydrochloric acid in tetrahydrofuran or platinum on the carbon (5%) at 1:35 ratio of initial compounds has been investigated.

Preliminary heating of initial compounds in the temperature range of $30 \div 70^0C$ in the presence of a catalyst showed that in these conditions polymerization of n-alkenes, or other changes due to a break in the siloxane backbone, or elimination of methane in methylhydridesiloxanenot occurring. Besides, there are no changes in the NMR and IR spectra of initial compounds. By gas-liquid chromatography it was established that the polymerization of n-alkenes in this condition does not proceed.

During the hydrosilylation reaction, the decrease of the active ≡Si-H group on the time was observed. The hydrosilylation reaction was investigated in dry toluene solution ($C \approx 6{,}8 x 10^{-2}$ mole/l). It was established, that during the reaction all active ≡Si-H groups do not take place completely in the hydrosilylation reaction. On the Figure 1 one can observe the dependence of changes of active ≡Si-H groups on the time during the hydrosilylation reaction of α,ω– bis(trimethylsiloxy)methylhydridesiloxane to n-hexene-1.

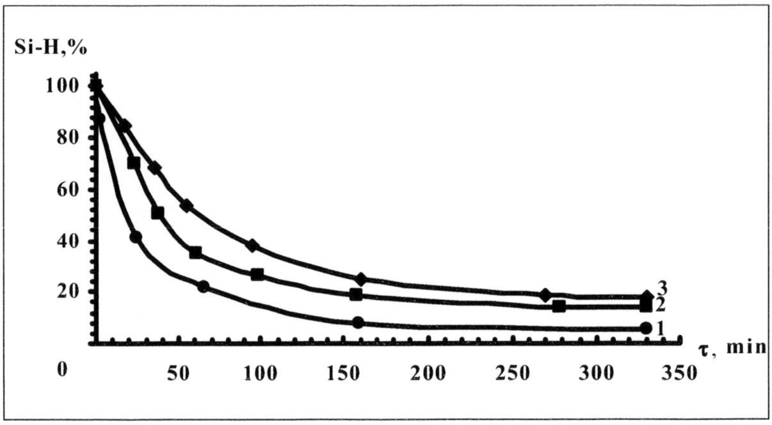

Figure 1. Dependence of changes of concentration of active ≡Si-H group on the time,upon hydrosilylation reaction of α,ω–bis(trimethylsiloxy)methylhydridesiloxane to hexene-1, where curve 1 is at 50^0C, curve 2 at 40^0C and curve 3 is at 30^0C.

From the Figure 1 it is evident that at 30^0C hydrosilylation reaction proceeds about ~80% conversion of active ≡Si-H, while at 50^0C hydrosilylation reaction proceeds about 92%. Consequently, with the rise of temperature the depth of hydrosilylation reaction increases.

So, the hydrosilylation reaction of α,ω–bis(trimethylsiloxy)methylhydridesiloxane to alkenes in ge-nerally proceeds according to the scheme 1. Where: [(a)+(b)](c)=m≈35; n=6-C_6H_{12} [I (50^0C); I^1 (40^0C); I^2(30^0C);]; n=8-C_8H_{16} [II (60^0C); II^1(50^0C); II^2(40^0C)]; n=10 - $C_{10}H_{20}$ (III); n=12 – $C_{12}H_{24}$ (IV); n=18 – $C_{18}H_{36}$ (V);

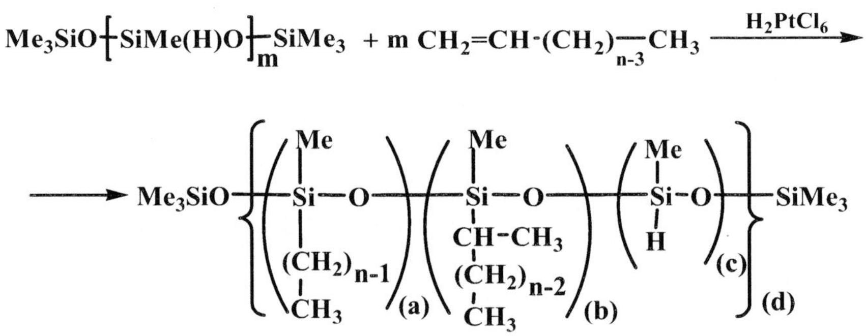

Scheme 1. Hydrosilylation reaction of α,ω–bis(trimethylsiloxy)methylhydridesiloxane to alkenes.

Hydrosilylation proceeds rapidly during the first hour, and then the reactions proceed at a low rate. Synthesised oligomers are vitreous liquid or solid products depending on the surrounding group well soluble in ordinary aromatic type organic solvents with η_{sp}≈0,07÷0,1. Structure and composition of oligomers was proven by elementary analysis, determination of molecular masses, IR, ^{1}H and ^{13}C NMR spectral data. Some physical-chemical properties of synthesized oligomers are presented in table 1.

Table 1. Some physical-chemical properties of synthesized oligomers.

№	Yield, %	η_{sp}*	Reaction temperature, ^0C	Conversion of active ≡Si-H % group	d_1, Å	T_g, ^0C	5% mass losses, ^0C	$\bar{M}_\omega x1 0^{-4}$
I	93	0,07	50	94	11,41	-112	270-300	2.86
I^1	92	0,06	40	86	-	-	-	-
I^2	92	0,06	30	82	-	-	-	-
II	91	0,09	60	89	13,6	-	-	2.95
II^1	90	0,08	50	79	-	-	-	-
II^2	90	0,08	40	66	-	-	-	-
III	88	0,09	70	86	16.07	-42	350-370	-
IV	86	0.09	70	82	18.64	-	-	4.50
V	85	0.10	70	78	20.54	+44	370-390	2.59

*In 1% toluene solution at 25^0C.

As it is seen from Figure 2 with the increase of the length of n-alkenes the hydrosilylation reaction rate, at the same temperature decreases.

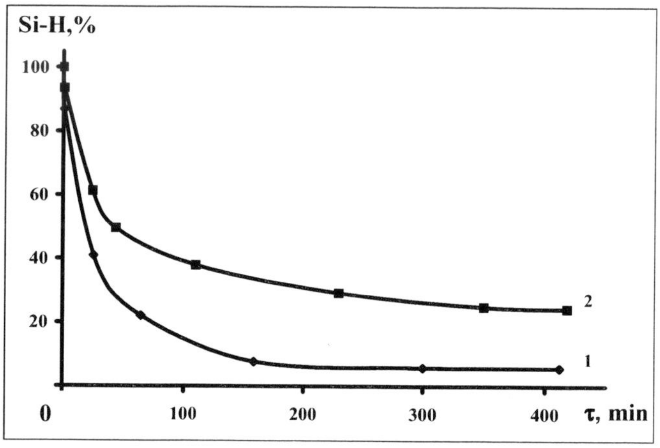

Figure 2. Dependence of changes of concentration of active ≡Si-H group on the time,upon hydrosilylation reaction of α,ω–bis(trimethylsiloxy)methylhydridesiloxane to hexene-1: curve – 1, octene-1: curve – 2, at 50^0C temperature.

In the IR spectra of synthesized oligomers one can observe absorption bands characteristic of asymmetric valence oscillation of linear ≡Si-O-Si≡ bonds 1025 cm^{-1}. In the region 840, 1200, 1275 and 2165 cm^{-1}, one can observe the absorption bands characteristic for $-SiMe_3$, $\equiv Si-CH_2$, ≡Si-Me and ≡Si-H bonds respectively. With the increase of the length of the side group's intensity of symmetric valence oscillations in the region of 2800-2900 cm^{-1}, the characteristic for CH_2 groups increases. In the ^1H NMR and COSY spectrum of oligomer I (Figure 3), one can observe singlet signals for protons in ≡Si-Me and $-SiMe_3$ group with chemical shift in the range δ≈0,1÷0,2 ppm. In the spectrum one can observe resonance signal for protons of methylene group in the fragment ≡Si-CH_2-, with a centre of chemical shifts δ≈0,5 ppm. Triplet signal with centre of chemical shift δ≈0,85 ppm corresponds to protons of methyl group in the fragment –CH_2-CH_3, also one can observe a signal with chemical shift in the range δ≈0,8÷0,95 ppm characteristic for protons in methylene group in fragments –C^2H_2- and –C^3H_2-. In the spectra these signals are overlapped to each other. The signals with a chemical shift δ≈1,1÷1,3 ppm characteristic for protons of methylene -C^4H_2- and -C^5H_2- group. Low intensity singlet signal with a chemical shift δ≈4,3 ppm is characteristic for protons in unreacted ≡Si-H group. In spectra one can observe a signal for protons of methine group in the fragment ≡Si-CH= with a chemical shift in the range 1,1÷1,3 ppm and for protons of methyl group in the fragment =CH-CH_3 with a chemical shift δ≈0,8 ppm, which indicate that the reaction proceeds as an anti-Markovnikov rule, as well as partially with the Markovnikov rule. Ratio of the products obtained by anti-Markovnikov and Markovnikov rule is equal to 5.4:1 accordingly.

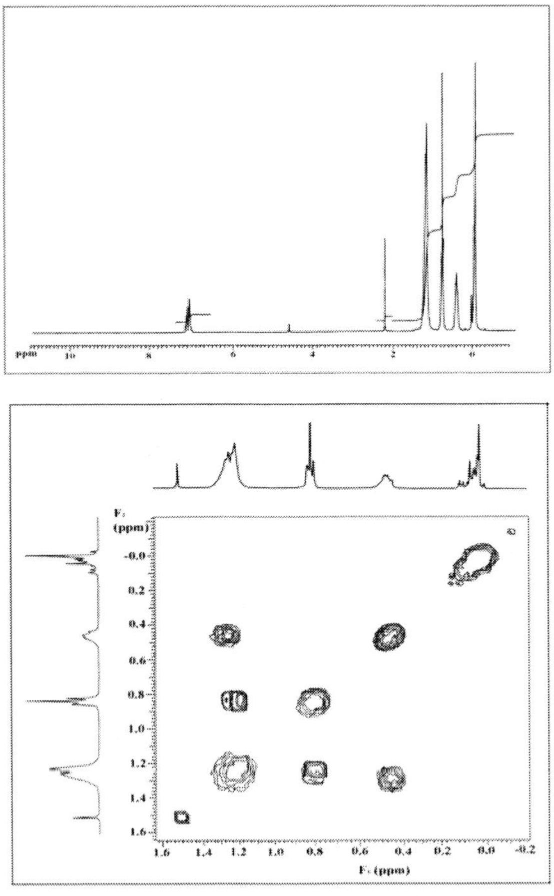

Figure 3. ^{1}H NMR and COSY spectrum of oligomer I.

^{13}C NMR spectrum of oligomer I (Figure 4) is incomplete agreement with ^{1}H NMR spectrum data and proved that the hydrosilylation reaction proceed both anti-Markovnikov rule and Markovnikov rule. In the spectrum one can observe a characteristic signal for carbon nucleus of -C^1H_2- group with chemical shift δ≈17.79 ppm, for -C^2H_2- group - δ≈17.99 ppm, for –C^3H_2- group - δ≈33.53 ppm, for –C^4H_2- group - δ≈32.06 ppm, for –C^5H_2- group - δ≈23.02 ppm. In the spectrum one can observe a resonance signal for carbon nucleus of methylene group in the fragment -CH_2-CH_3 with a chemical shift δ≈14.0 ppm; resonance signal for a carbon nucleus of methine group in the fragment ≡Si-CH with a chemical shift δ≈18.08 ppm and for carbon nucleus of methyl group in the fragment =CH-CH_3 with chemical shift δ≈2.15 ppm, which proved the proceeding of hydrosilylation reaction by two ways.

In the ^{1}HNMR and COSY spectrum of oligomer II (Figure 5) one can observe a singlet signal for protons of methyl group in the fragment ≡Si-Me and –$SiMe_3$ with a centre of chemical shift δ≈0.1 ppm, for protons of methylene group in the fragment ≡Si-CH_2-, with centre of chemical shift δ≈0.45 ppm. From spectrum it is seen triplet signal for protons of methyl group in the fragment CH_2-CH_3, with chemical shift δ≈0.8 ppm. One can also observe the resonance signal with chemical shift in the range δ≈ 075÷ 0.95 ppm characteristic for protons in methylene group in the fragments –C^2H_2- and –C^3H_2-. In the spectrum these signals overlap each other.

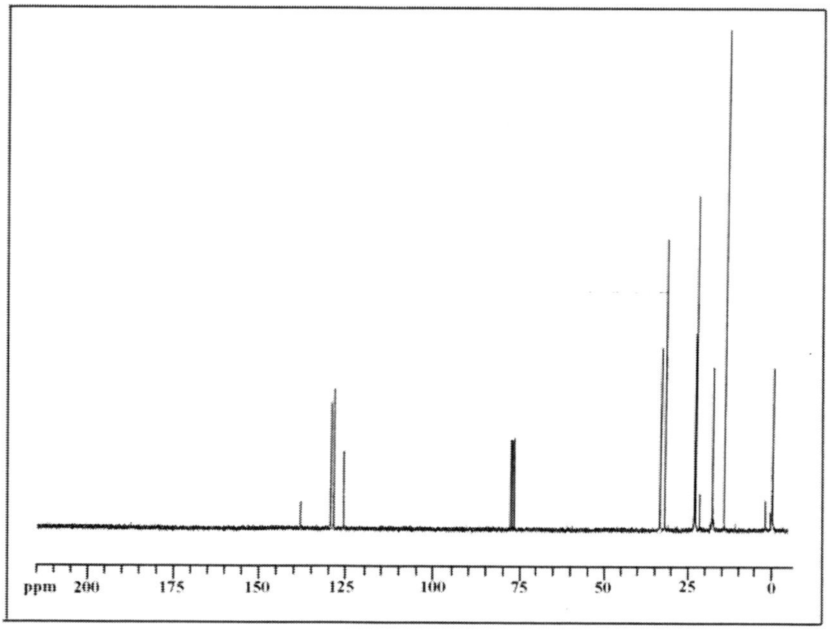

Figure 4. ^{13}C NMR spectrum of oligomer I (solvent and standard $CDCl_3$).

In the spectrum one can observe signals with a centre of chemical shift δ≈1,15÷1,3, characteristic for protons of methylene groups in the fragment -C^4H_2-, -C^5H_2-,–C^6H_2- and -C^7H_2-. A resonance signal with a chemical shift δ≈4.3 ppm corresponds to protons of an ≡Si-H group. A resonance signal with a chemical shift which corresponds to protons of methine group in the fragment ≡Si-CH, with a centre of chemical shift δ≈1.1-1.3 ppm; signal with the centre of chemical shift δ≈0.8 ppm, characteristic for protons of the methyl group in the fragment $CH-CH_3$. This proved that the hydrosilylation reaction of methylhydridesiloxane to n-octene-1, as it is mentioned in case of hexene-1, proceeds both by Anti-Markovnikov and Markovnikov rule, with the ratio of obtained product 7.8:1 accordingly.

^{13}C NMR spectrum of oligomer II (Figure 4) is in complete agreement with ^{1}H NMR spectrum data and proved that the hydrosilylation reaction proceed both Anti-Markovnikov rule and Markovnikov rule. In the spectra one can observe resonance signals: for $-C^1H_2-$ group with chemical shift $\delta \approx 18.1$ ppm, for $-C^2H_2-$ group with chemical shift $\delta \approx 17.9$ ppm; for $-C^3H_2-$ group with a chemical shift $\delta \approx 17.7$ ppm; for $-C^4H_2-$ group with a chemical shift $\delta \approx 33.8$ ppm; for $-C^5H_2-$ group with a chemical shift $\delta \approx 33.7$ ppm; for $-C^6H_2-$ group with chemical shift $\delta \approx 23.5$ ppm and for $-C^4H_2-$ group with a chemical shift $\delta \approx 23.3$ ppm.

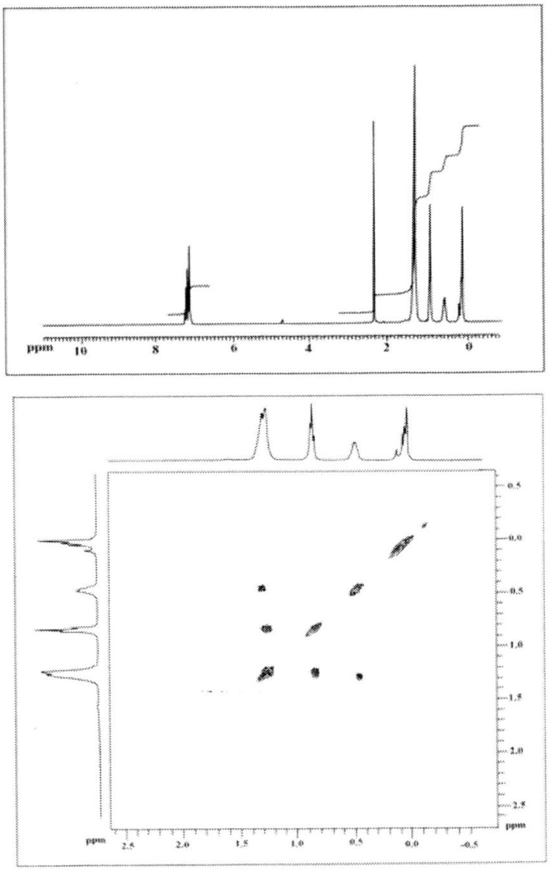

Figure 5. ^{1}H NMR and COSY spectrum of oligomer II (solvent and standard $CDCl_3$).

In the spectra one can observe signals of a carbon nucleus for methyl group in the fragment $-CH_2-CH_3$ with a chemical shift $\delta \approx 14.4$ ppm; for a carbon nucleus of methine group in the fragment $\equiv Si-CH-$ with a chemical shift $\delta \approx 18.0$ ppm and for methyl nucleus in the fragment $-CH-CH_3$, with a chemical shift $\delta \approx 2.17$ ppm. Once again ^{13}C NMR spectra proved proceeding of the hydrosilylation reaction by two directions.

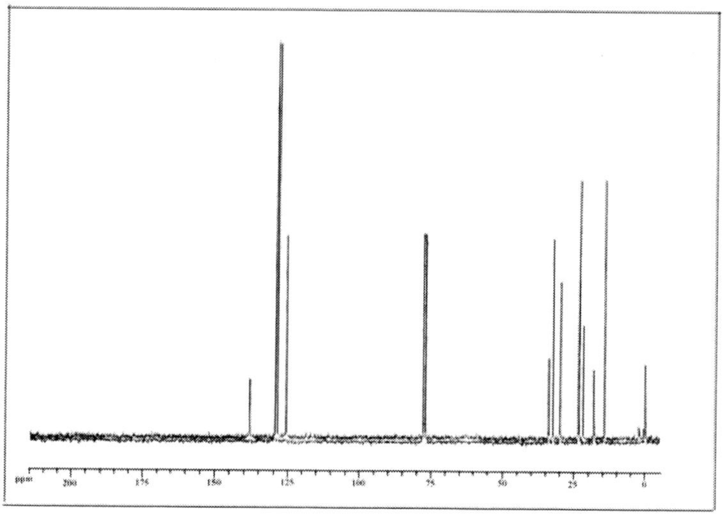

Figure 6. ^{13}C NMR spectrum of oligomer II (solvent and standard $CDCl_3$).

For full characterization of hydride addition of methylhydridesiloxane to hexene-1, by quantum-chemical semi empiric AM1 method [7], for all initial, intermediate and final products, in modelling the reaction of hydrosilylation of methyldimethoxysilane to hexene-1, the heats of formations (ΔH_f), changes of energy (ΔH) of the system depending on the change of distance ($R_{C\text{-}Si}$) between ≡C-Si≡ bonds, also the charges values (q) on the atoms, dipole moments (μ) and bonds orders (Pij) are calculated (see tables 3 and 4).

For the beginning the hydride addition of modelling compound methyldimethoxysilane [$Me(MeO)_2SiH$] to hexene-1 according to the anti-Markovnikov rule was considered. The model reaction proceeds according to the following scheme 2 with formation of compound VI:

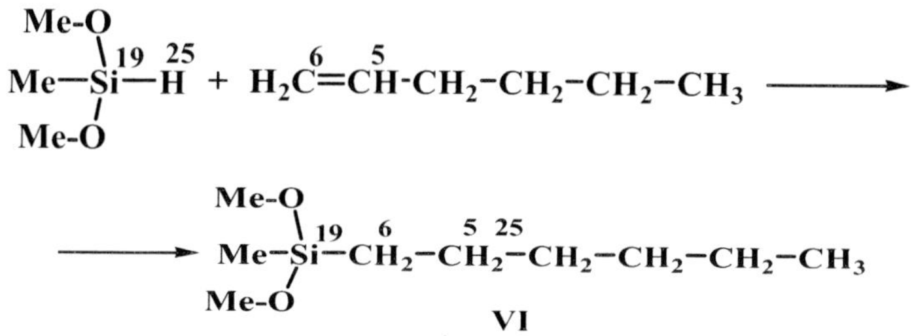

VI

Scheme 2. Modelling hydrosilylation reaction of methyldimethoxysilane with hexene-1 according to the anti-Markovnikov rule.

In hexene-1 the distance between the double bonded carbon atom (C_6) and silicon was taken about 1.0 Å more than the bond length, expected in a product. Figure 7 showed the starting position of compounds in hydrosilylation reaction of methyldimethoxysilane to hexene-1:

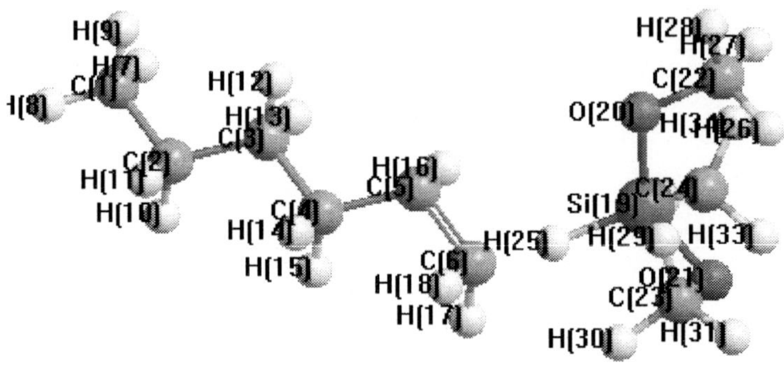

Figure 7. Starting position of compounds in modelling hydrosilylation reaction of methyldimethoxysilane to hexene-1 according to the anti-Markovnikov rule.

The change of step in distances (R_{C_6-Si}) between atom of silicon and double bonded carbon atom (C_6) was 0.05 Å. Dependence of change of energy (ΔH) of the system on the distances (R_{C-Si}) between silicon and carbon atoms is presented in Figure 8.

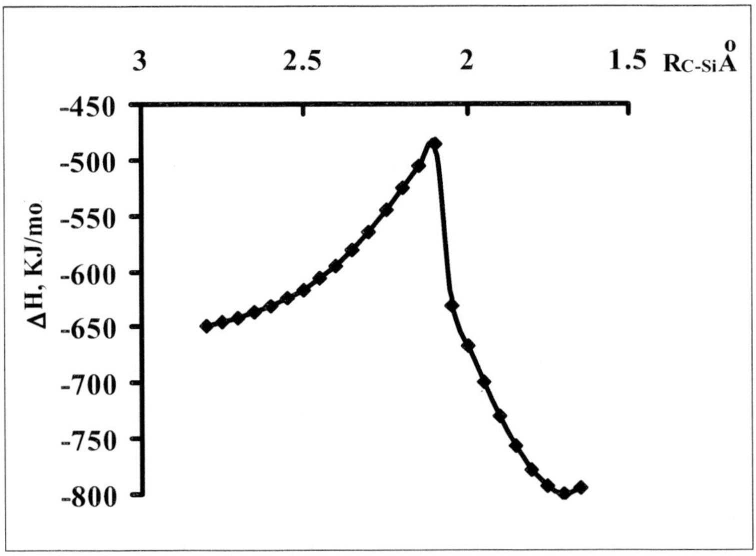

Figure 8. Dependence change of energy (ΔH) on distance (R_{C-Si}) between silicon and carbon atoms during modelling hydride addition reaction of methyldimethoxysilane to hexene-1 according to the anti-Markovnikov rule.

As it is seen from Figure 8, an approximation of silicon to carbon (C_6) atom up to R_{C-Si}=2.10 Å distances, the system energy rises. At the same time the bond order between carbon C_6 and C_5 atoms in the hexene-1 molecule decreases from 1.948 up to 1.471. The bond order between hydrogen and silicon atoms also decreases (R_{Si-H}=0.897-0.579) and in modelling compound methyldimethoxy-silane the process of formation of new bonds (PC^6-Si=0.016-0.219 and PC^5-H=0.0001-0.155) is observed. In case of distance up to 2.05 Å the system energy sharply decreases, the double C=C bond passes in ordinary C-C bond (PC^5-C^6=1.004), and the bond order between silicon and carbon atom C_6 reaches to 0.864. Thus the hydrogen is completely eliminated from silicon (P_{Si-H}=0.001) and attaches to carbon atom C_5 (P_{C_5-H}=0.959).

With consecutive reduction of distances the system energy monotonously decreases and with strengthening forms new bonds. The heat of formation of the hydrosilylation product (ΔH=-799.92 kJ/mol) is in good correlation with heat of formation (ΔH=-807.12 kJ/mol) of compound VI obtained by hydride addition of methyldimethoxysilane to hexene-1 according to the anti-Markovnikov rule. Activation energy of modelling reaction proceeding by anti-Markovnikov rule is equal to E_{act}=161.9 kJ/mol. In Figure 9 one can see final structure of modelling hydrosilylation reaction of methyldimethoxysilane to hexene-1 according to the Anti-Markovnikov rule.

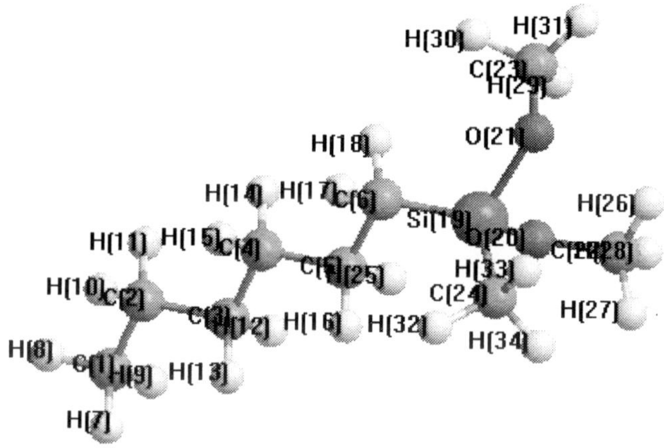

Figure 9. Final structure of the product obtained by modelling hydrosilylation reaction of methyldimethoxysilane to hexene-1 according to the anti-Markovnikov rule.

The modelling hydrosilylation reaction according to the Markovnikov rule of methyldimetho-xysilane to hexene-1 is also considered. The reaction proceeds according to the following scheme 3 with formation of compound VII:

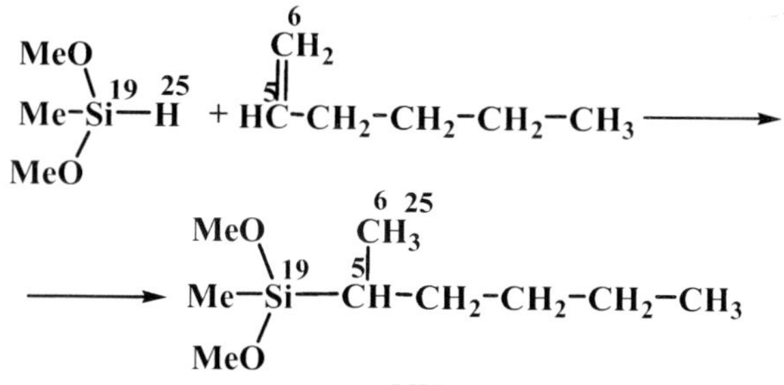

VII

Scheme 3. Modelling hydrosilylation reaction of methyldimethoxysilane with hexene-1 according to the Markovnikov rule.

Figure 10 shows the starting position of initial compounds in modelling the hydrosilylation reaction of methyldimethoxysilane to hexene-1 according to the Markovnikov rule:

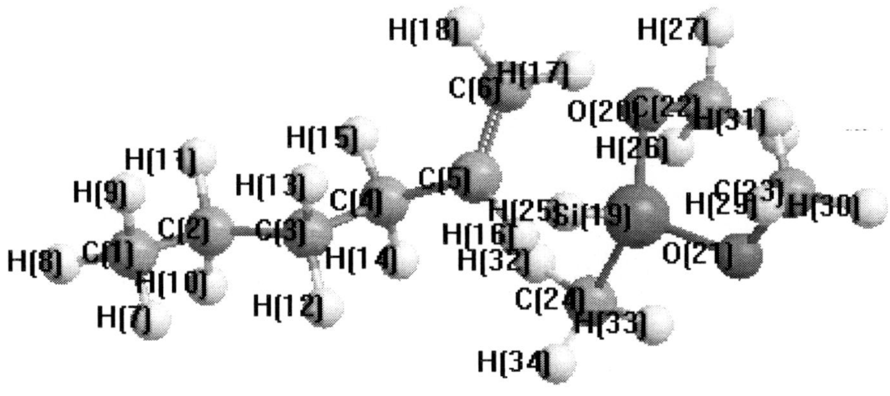

Figure 10. Starting position of compounds in modelling hydrosilylation reaction of methyldimethoxysilane to hexene-1 according to the Markovnikov rule.

In Figure 11 the dependence of change of energy (ΔH) of the system on the distances between silicon and carbon atoms ($R_{C_5\text{-}Si}$) is presented. Similarly the above-mentioned distances between the double bonded carbon atom (C_5) and silicon was taken about 1.0 Å more than the bond expected in a product. The change of step in distances ($R_{C_5\text{-}Si}$) between silicon atom and double bonded carbon atom (C_5) was about 0.05 Å.

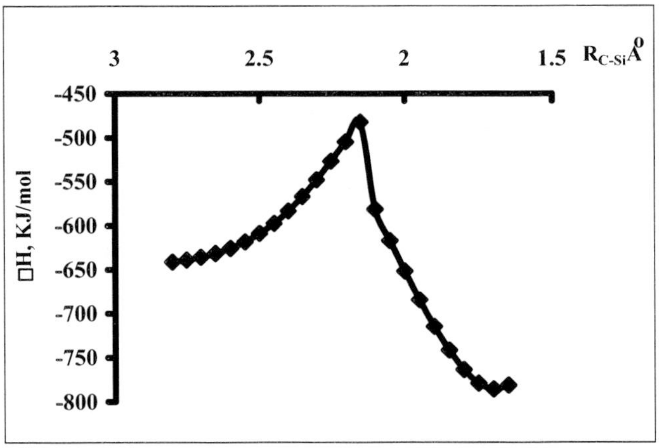

Figure 11. Dependence of change of energy (ΔH) on the distance (RC^5 -Si) between silicon and carbon atoms during modelling hydride addition of methyldimethoxysilane to hexene-1 according to the Markovnikov rule.

As it is seen in Figure 11, at approximation of silicon to carbon atom (C_6) up to $R_{C\text{-}Si}$=2.15 Å distances the system energy rises. At the same time the bond order between carbon atoms C_5 and C_6 in a molecule hexene-1 decreases from 1.934 up to 1.659. The bond order between hydrogen and silicon atoms also decreases ($R_{Si\text{-}H}$=0.885-0.700) and in methyldimethoxysilane the process of formation of new bonds ($P_{C_5\text{-}Si}$=0.023-0.147 and $P_{C_6\text{-}H}$=0.002-0.085) is observed.

In case of distance up to 2.10 Å the system energy sharply decreases, the double C=C bond passes in ordinary C-C bond ($P_{C_5\text{-}C_6}$=1.013), and the bond order between the silicon and carbon atoms C_5 reaches to 0.840. Thus the hydrogen is completely eliminated from silicon ($P_{Si\text{-}H}$=0.001) and attached to carbon atom C_6 ($P_{C_6\text{-}H}$=0.973).

With consecutive reduction of distances, the system energy of monotonously decreases and with strengthening of newly formed bonds. The heat of formation of a product of hydride addition (ΔH=-792.03 kJ/mol) is in good correlation with heat of formation (ΔH=-785.79 kJ/mol) of compound VII obtained by hydride addition of methyldimethoxysilane to hexene-1 according to the Markovnikov rule. Activation energy of modelling reaction proceeding by Markovnikov rule is equal to E_{act} =158.9 kJ/mol.

In Figure 12 one can see the final structure of modelling hydrosilylation reaction of methyldimethoxysilane to hexene-1 according to the Markovnikov rule.

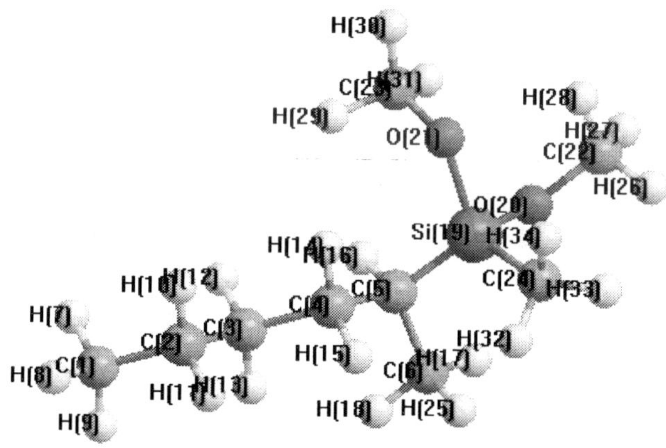

Figure 12. Final structure of the product obtained by modelling hydrosilylation reaction of methyldimethoxysilane to hexene-1 according to the Markovnikov rule.

Table 3. Change of energy (ΔH) of the system depending on the change of distance ($R_{C\text{-}Si}$) between ≡C-Si≡ bonds, the charges values (q) on the atoms, dipole moments (μ) and bonds orders (Pij), during modelling hydrosilylation reaction of methyldimethoxysilane to hexene-1 according to the anti-Markovnikov rule.

$R_{6\text{-}19}$	$R_{5\text{-}25}$	Δ H	q_5	q_6	q_{19}	q_{25}	μ	$P_{5\text{-}6}$	$P_{19\text{-}25}$	$P_{6\text{-}19}$	$P_{5\text{-}25}$
2.80	2.32	-649.33	-0.1239	-0.2848	1.3658	-0.2399	0.853	1.9480	0.8973	0.0161	0.0001
2.75	2.27	-645.84	-0.1191	-0.2919	1.3675	-0.2409	0.933	1.9451	0.8967	0.0187	0.0001
2.70	2.22	-641.63	-0.1133	-0.3006	1.3700	-0.2428	1.034	1.9412	0.8955	0.0219	0.0002
2.65	2.17	-636.62	-0.1068	-0.3107	1.3741	-0.2459	1.163	1.9354	0.8932	0.0261	0.0005
2.60	2.12	-630.79	-0.1001	-0.3217	1.3793	-0.2501	1.317	1.9273	0.8893	0.0315	0.0011
2.55	2.07	-624.04	-0.0936	-0.3329	1.3843	-0.2543	1.489	1.9168	0.8843	0.0381	0.0021
2.50	2.02	-616.03	-0.0855	-0.3455	1.3889	-0.2580	1.660	1.9054	0.8787	0.0450	0.0035
2.45	1.97	-606.32	-0.0743	-0.3611	1.3941	-0.2619	1.829	1.8929	0.8725	0.0523	0.0052
2.40	1.92	-594.58	-0.0600	-0.3801	1.4006	-0.2666	2.003	1.8780	0.8647	0.0606	0.0075
2.35	1.87	-580.62	-0.0417	-0.4034	1.4102	-0.2730	2.167	1.8593	0.8543	0.0701	0.0108
2.30	1.82	-564.35	-0.0199	-0.4309	1.4232	-0.2812	2.354	1.8345	0.8397	0.0818	0.0158
2.25	1.77	-545.89	0.0069	-0.4642	1.4414	-0.2919	2.552	1.8002	0.8184	0.0967	0.0238
2.20	1.72	-525.67	0.0402	-0.5059	1.4682	-0.3061	2.759	1.7495	0.7851	0.1169	0.0377
2.15	1.67	-504.78	0.0836	-0.5608	1.5104	-0.3252	2.972	1.6667	0.7272	0.1475	0.0655
2.10	1.62	-486.43	0.1516	-0.6507	1.5975	-0.3528	3.147	1.4713	0.5787	0.2187	0.1550
2.05	1.57	-631.66	-0.1655	-0.4422	1.3725	0.0716	0.528	1.0042	0.0067	0.8637	0.9320
2.00	1.52	-666.47	-0.1663	-0.4626	1.3938	0.0788	0.488	1.0018	0.0051	0.8618	0.9363
1.95	1.47	-699.39	-0.1658	-0.4843	1.4162	0.0846	0.467	0.9997	0.0040	0.8591	0.9401
1.90	1.42	-729.68	-0.1644	-0.5075	1.4391	0.0895	0.465	0.9977	0.0032	0.8558	0.9438
1.85	1.37	-756.33	-0.1620	-0.5321	1.4627	0.0935	0.501	0.9956	0.0025	0.8520	0.9471
1.80	1.32	-778.09	-0.1589	-0.5581	1.4869	0.0968	0.582	0.9934	0.0020	0.8478	0.9502
1.75	1.27	-793.31	-0.1548	-0.5855	1.5116	0.0993	0.703	0.9911	0.0016	0.8432	0.9532
1.70	1.22	-799.92	-0.1497	-0.6142	1.5368	0.1009	0.869	0.9886	0.0013	0.8383	0.9559
1.65	1.17	-795.37	-0.1437	-0.6441	1.5622	0.1016	1.071	0.9859	0.0011	0.8334	0.9586

Table 4. Change of energy (ΔH) of the system depending on the change of distance ($R_{C\text{-}Si}$) between ≡C-Si≡ bonds, the charges values (q) on the atoms, dipole moments (μ) and bonds orders (Pij), during modelling hydrosilylation reaction of methyldimethoxysilane to hexene-1 according to the Markovnikov rule.

$R_{5\text{-}19}$	$R_{6\text{-}25}$	ΔH	q_5	q_6	q_{19}	q_{25}	μ	$P_{5\text{-}6}$	$P_{19\text{-}25}$	$P_{5\text{-}19}$	$P_{6\text{-}25}$
2.80	2.31	-641.19	-0.1941	-0.2231	1.3844	-0.2516	1.381	1.9337	0.8849	0.0225	0.0015
2.75	2.26	-638.82	0.2001	0.2190	1.3857	-0.2523	1.433	1.9299	0.8832	0.0254	0.0018
2.70	2.21	-635.64	-0.2081	-0.2131	1.3874	-0.2534	1.478	1.9262	0.8811	0.0285	0.0023
2.65	2.16	-631.34	-0.2174	-0.2060	1.3892	-0.2548	1.530	1.9218	0.8787	0.0319	0.0028
2.60	2.11	-625.66	-0.2284	-0.1974	1.3914	-0.2565	1.587	1.9168	0.8757	0.0358	0.0037
2.55	2.06	-618.27	-0.2411	-0.1878	1.3942	-0.2585	1.657	1.9104	0.8720	0.0404	0.0047
2.50	2.01	-608.91	-0.2560	-0.1761	1.3976	-0.2610	1.737	1.9028	0.8674	0.0455	0.0062
2.45	1.96	-597.32	-0.2733	-0.1627	1.4021	-0.2641	1.836	1.8926	0.8614	0.0519	0.0083
2.40	1.91	-583.35	-0.2938	-0.1464	1.4082	-0.2681	1.935	1.8798	0.8533	0.0592	0.0111
2.35	1.86	-566.91	-0.3180	-0.1272	1.4167	-0.2732	2.051	1.8626	0.8423	0.0681	0.0153
2.30	1.81	-548.05	-0.3474	-0.1037	1.4287	-0.2801	2.181	1.8386	0.8266	0.0795	0.0216
2.25	1.76	-527.07	-0.3836	-0.0750	1.4462	-0.2891	2.336	1.8035	0.8033	0.0946	0.0318
2.20	1.71	-504.59	-0.4295	-0.0383	1.4732	-0.3017	2.477	1.7499	0.7659	0.1152	0.0493
2.15	1.66	-482.06	-0.4907	0.0099	1.5178	-0.3186	2.600	1.6593	0.7001	0.1470	0.0845
2.10	1.61	-581.21	-0.3771	-0.1729	1.3599	-0.0095	0.396	1.0134	0.0105	0.8399	0.9492
2.05	1.56	-616.97	-0.3934	-0.1779	1.3773	0.0025	0.430	1.0092	0.0077	0.8419	0.9541
2.00	1.51	-651.59	-0.4109	-0.1834	1.3957	0.0146	0.463	1.0059	0.0053	0.8428	0.9583
1.95	1.46	-684.47	-0.4303	-0.1881	1.4158	0.0255	0.511	1.0031	0.0037	0.8421	0.9615
1.90	1.41	-714.81	-0.4513	-0.1919	1.4372	0.0357	0.556	1.0007	0.0025	0.8403	0.9643
1.85	1.36	-741.64	-0.4737	-0.1947	1.4597	0.0451	0.620	0.9983	0.0014	0.8377	0.9667
1.80	1.31	-763.59	-0.4981	-0.1963	1.4838	0.0525	0.714	0.9959	0.0011	0.8339	0.9684
1.75	1.26	-779.01	-0.5237	-0.1967	1.5085	0.0594	0.815	0.9935	0.0007	0.8295	0.9699
1.70	1.21	-785.77	-0.5511	-0.1963	1.5339	0.0649	0.963	0.9910	0.0006	0.8246	0.9713
1.65	1.16	-781.29	-0.5797	-0.1946	1.5597	0.0695	1.137	0.9883	0.0005	0.8192	0.9727

Thus at comparison of the heat of formations of the compounds VI and VII, also of the Figures 8 and 11 (the character of curves of change of energy and the bonds orders values) it can be concluded that the course of modelling hydrosilylation reaction of methyldimethoxysilane to hexene-1 energetically is more favourable by the Anti-Markovnikov rule that is in agreement with NMR spectral data.

Figure 13 shows dependence of reverse concentration on the time during the hydrosilylation reaction of methylhydridesiloxane to hexene-1. One can see that at the initial stages the hydrosilylation reaction is of second order.

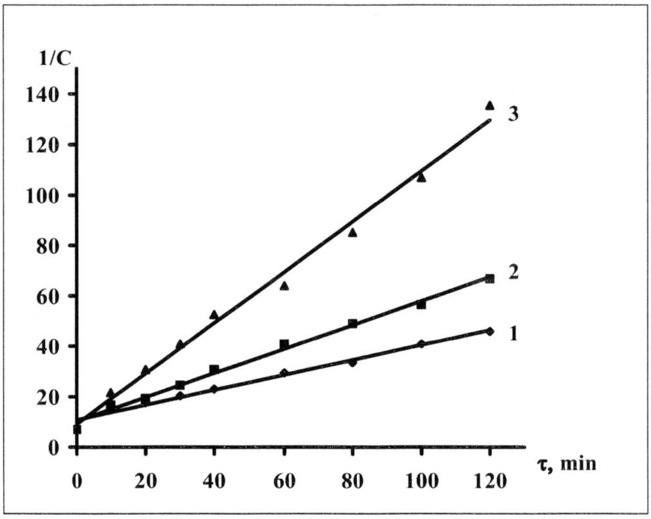

Figure 13. Dependence of inverse concentration on the time during the hydrosilylation reaction of methylhydridesiloxane to hexene-1. Where curve 1 is at 50^{0}C, curve – 2 at 40^{0}C and curve – 3 at 30^{0}C.

The reaction rate constants of hydrosilylation reactions of methylhydridesiloxane to hexene-1 and octene-1 at various temperatures were determined, the values of which are: $k_{30°C}$≈0.2225, $k_{40°C}$≈0.3948 l/mol·s and $k_{50°C}$ ≈ 0.7367 in case of hexene-1; $k_{40°C}$≈0.1337, $k_{50°C}$≈0,2373 and $k_{60°C}$≈0.4039 l/mol·s, in case of octene-1.

Figure 8 shows dependence of hydrosilylation reaction rate constants logarithm on the reverse temperature. From this data the activation energy of hydrosilylation reaction was calculated, which is equal to: $E_{act.}$≈46.3 kJ/mol (hexene-1) and $E_{act.}$≈44.7 kJ/mol (octene-1).

The synthesized oligomers were studied by GPC method. Figure 14 shows the molecular weight distribution of oligomer I. From which it is evident that oligomer I has trimodal molecular weight distribution. Its molecular weight is equal to: $\overline{M}_n \approx 7.23 x 10^3$, $\overline{M}_\omega \approx 2.86 x 10^4$, and polydispersity is - D≈3.96.

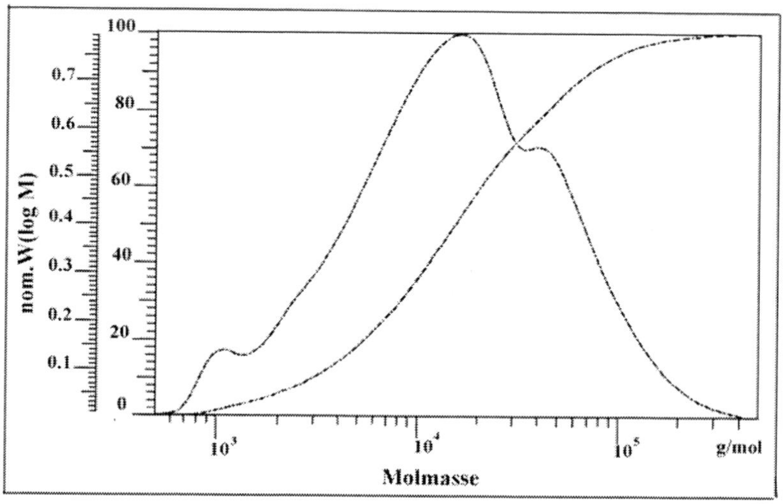

Figure 14. Gel-permeation chromatographic curves of oligomer I.

It is clear that the average molecular weights of synthesized oligomers exceed theoretical values several times of molecular weights calculated in the case of full hydrosilylation. It indicates that during hydro-silylation reaction branching process also takes place. In UV spectra of synthesized oligomers absorption maximums at 310 nm are observed, which characterize ≡Si-Si≡ bond and indicates proceeding of homodehydrocondensation reactions [8].

For oligomers wide-angle X-ray investigations have been carried out. As it is seen from Figure 15 the oligomers are one phase amorphous systems. With an increase of the length alkyl side groups the values of interchain distances d_l rises from $d_l \approx 11.41$ Å (oligomer I) up to $d_l \approx 20.54$ Å (oligomer V). The value of interchain distances d_l of synthesized comb-type oligomers is similar to the value of interchain distances of comb-type poly-1-alkylethylene polymers [9].

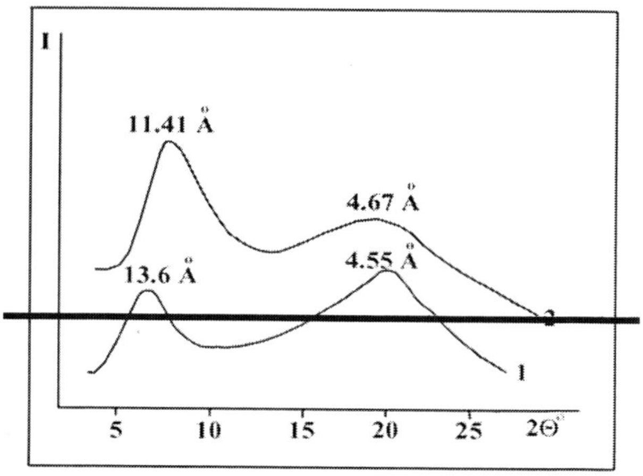

Figure 15. X-ray patterns of methylsiloxane comb-type oligomers. Where curve 1 corresponds to oligomer II and curve 2 corresponds to oligomer I.

Differential scanning calorimetric investigation for synthesized oligomers has been carried out. In Figure 16 DSC curves are presented. As it is seen from DSC curves in oligomer I only one endothermic peak is observed, with one temperature transition, which corresponds to glass transition temperature $T_g \approx -112^0C$. In contrast to literature data [10], where on methylsiloxane matrix are attached oxyhexyl groups, in oligomer I mesomorphic properties are not observed. Glass transition temperature of oligomer I exceed to glass transition temperature of methylsiloxane oligomers with oxyhexyl side gro-ups.

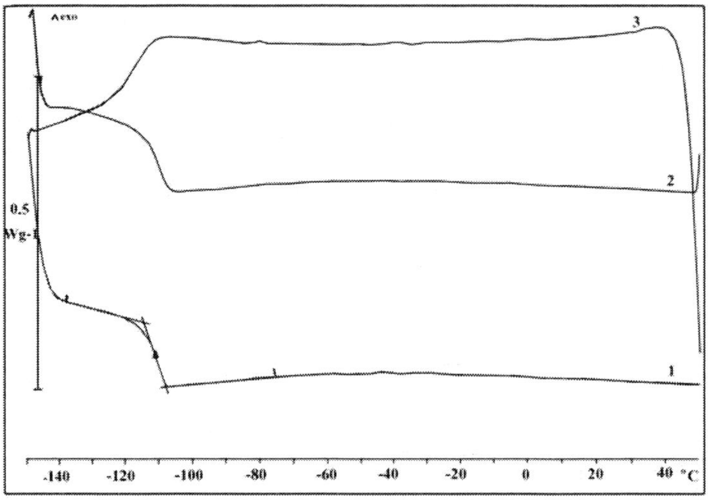

Figure 16. DSC curves of oligomer I, where 1 and 2 are heating scanning curves, curve 3 – cooling scanning.

So, by catalytic hydrosilylation reaction of α,ω-bis(trimethylsiloxy)methylhydride-siloxanes with n-alkenes methylalkylsiloxanes with various alkyl side groups in the chain were synthesized by us for the first time. It was shown that hydrosilylation reaction mainly proceeds according to the Anti-Markovnikov rule.

REFERENCES

[1] O.V.Mukbaniani,; L.M. Khananashvili, Intern. *J. Polymeric Materials,* 1994, 27, 31.

[2] Iu.K.Godovsky,; I.I Mamaeva,; N.N.Makarova,; V.S. Papkov,; N.N. Kuzmin. *Macromol. Chem., Rapid Commun.,* 1985, 6, 797.

[3] Iu.K.Godovsky,; N.N.Makarova,; V.S. Papkov,; N.N. Kuzmin, *Vysokomol. Soed.* 1985, 27, 164 (in Russian).

[4] Iu.K.Godovsky,; N.N.Makarova,; V.S.Papkov,; N.N. Kuzmin. *Macromol. Chem., Rapid Commun.,* 1985, 6, 443.

[5] G.N.Koroleva,; V.O. Reikhsfeld, *Zhurn. Obshch. Khim.,* 1967, 37, 2559 (in Russian).

[6] V.O. Reikhsfeld,; G.N. Koroleva. *Zhurn. Obshch. Khim.,* 1966, 36, 1474 (in Russian).

[7] M.I.S. Dewar,; E.G.Zeobish,; E.F.Healy,; J.S. Stewart, *J. Amer. Chem. Soc.,* 1985, 107, 3902

[8] B.A.Trofimov,; L.N.Parshina,; M.I.Khilko,; L.A. Oparina. Thesis of the All Russian Conference "Organosilicon Compounds: Synthesis, Characteristics and Application", Moscow, 2000, L-36.
[9] N.A. Plate,; V.P. Shibaev,; Comb-like Polymers and Liquid Crystals, Plenum Press, New York-London, 1987.
[10] G.N. Gurgenidze. Candidate Dissertation, *"Synthesis and Investigation of Properties of Comb-type Methylsiloxane Oligomers"*. Tbilisi State University, Tbilisi, Georgia, 2003.

In: Modification Reactions of Oligomethylhydridesiloxanes ISBN: 1-60021-365-0
Editors: O.Mukbaniani, et al. pp. 21-35

Chapter 2

HYDROSILYLATION REACTIONS OF METHYLHYDRIDESILOXANE TO STYRENE AND α-METHYLSTYRENE

O. Mukbaniani, T. Tatrishvili, G. Titvinidze, and N. Mukbaniani
I. Javakhishvili Tbilisi State University,
I. Chavchavadze Ave., 1, 0128 Tbilisi, Georgia

ABSTRACT

The hydrosilylation reaction of α,ω–bis(trimethylsiloxy)methylhydridesiloxane to styrene and α-methylstyrene in the presence of catalyst – platinum hydrochloric acid (0,1 M solution in tetrahydrofuran) at 1:35 ratio of initial compounds, at various temperatures (80-90^{0}C) were investigated. Methylsiloxane oligomers with aryl substituted groups in the side chain have been obtained. It was shown that complete hydrosilylation of all active ≡Si-H groups does not take place. The hydrosilylation reaction order, activation energies and rate constants were found. The synthesized oligomers were characterized by ^{1}H, ^{13}C, and IR spectra data. For full characterization of hydride addition of methylhydridesiloxane to styrene, by quantum-chemical half empiric AM1 method, for all initial, intermediate and final products, in modelling the hydro-silylation reaction of methyldimethoxysilane to styrene, the heats of formations (ΔH_f), the change of energy (ΔH) of the system depending on the change of distance ($R_{C\text{-}Si}$) between ≡C-Si≡ bonds, also the charge values (q) on the atoms, dipole moments (μ) and bond orders (Pij) are calculated. Synthesized oligomers were characterized by gel-permeation chromatographic, differential calorimetric scanning, thermogravimetric and wide-angle X-ray methods.

Key words: methylhydridesiloxane, styrene, α–methylstyrene, hydrosilylation, thermal stable.

INTRODUCTION

In the literature, there are known hydride addition reactions of organohydridechloro(flour)silanes with styrene at the presence of Speier catalyst [1-3] and Pt/C (on the carbon adsorbed Platinum) [4-6]. There are various opinions about the direction of hydride addition reactions of

organohydridechlorosilanes with styrene in which the reaction proceeds according to the Anti-Markovnikov rule and Markovnikov rule. The steric and inductive effects of displaced group considerably influences the reaction direction.

In the literature it is known that styrene is especially active in hydrosilylation reactions, the process frequently proceeds at room temperature, as catalysts are used compounds of Pt, Pd, Rh and Ni.

Taking into account the type of the used silanes, addition of hydrosilanes can occur both to α-, and to β-to the double bonded carbon [7, 8]:

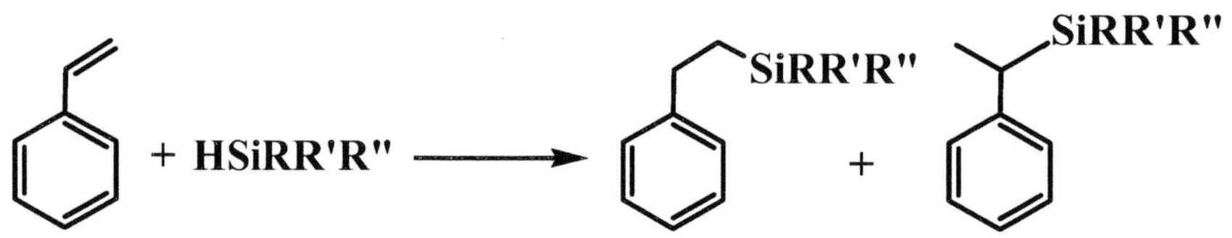

Where: R=R'=R", R≠R'≠R".

By Speier and coworkers [9] interaction of phenyl alkenes with general formula $Ph(CH_2)_nCH{=}CH_2$ or $PhCH{=}CH(CH_2)_nH$, (where: n=0÷4) with various hydrosilanes in the presence of H_2PtCl_6 have been investigated:

$$\equiv Si-H + Ph(CH_2)_nCH{=}CH_2 \xrightarrow{H_2PtCl_6} \underset{\textbf{A}}{\equiv Si(CH_2)_{n+2}Ph} + \underset{\textbf{B}}{\equiv SiCHPh(CH_2)_nMe}$$

n=0,1,2

$$\equiv Si-H + PhCH{=}CH(CH_2)_nH \xrightarrow{H_2PtCl_6} \underset{\textbf{A}}{\equiv Si(CH_2)_{n+2}Ph} + \underset{\textbf{B}}{\equiv SiCHPh(CH_2)_nMe}$$

n=0,1,2,4

In both cases reaction proceeds with formation of two products: A and B. Quantity ratio of these products depends on the nature of olefins and groups attached to silicon in hydrosilanes.

Hydrosilylation reaction of triethoxysilane to styrene in the presence of Karstedt's catalyst has been investigated [10]. It was established that the reaction proceeds according to two directions with formation α-and β-isomer. The ratio between products is about 39:61:

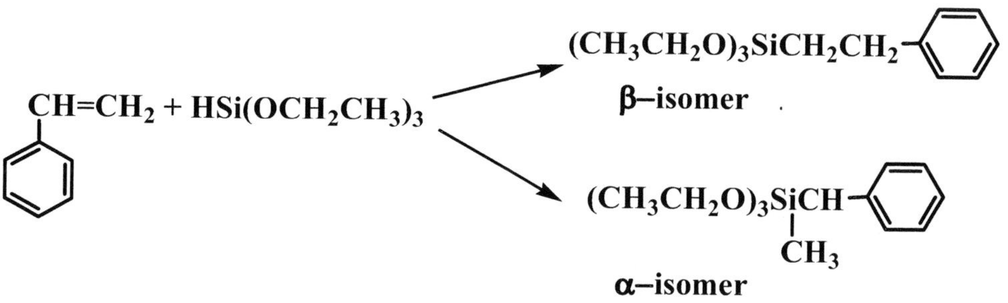

E. Chernishev and coworkers investigated the reaction ability of fluoroarylsilanes. It was established, that fluoroarylsilanes forms both products of hydrosilylation, basically α--isomers, and in the case of fluorodiphenylsilanes, α- and β-isomers are formed with the same quantity. During these processes field effects and steric effects play a significant role. Fluorohydrosilanes in comparison with chloro-derivatives have more reactionary – ability [11]. In case of m-nitrostyrene use, reduction of amino-styrene is possible. Replacement of alkyl substituted groups by aryl groups, causes slowing-down of process rate and reaction basically proceeds on direction of reduction. In the case of addition of triphenylsilane, reaction proceeds completely according to reduction direction [12]:

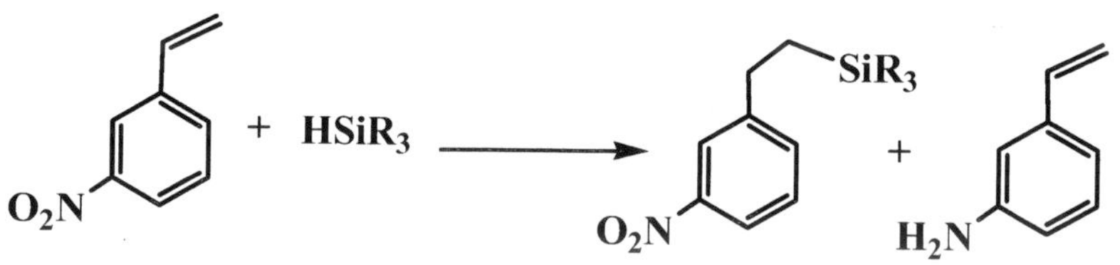

Significant works by A. Reikhsfield and employees were carried out [13, 14]. Kinetic regularity of hydrosilylation of different hydrosilanes to arylalkenes ($XC_6H_4CH=CH_2$) at the presence of platinum complexes was investigated.

The authors [15,16] carried out the reactions of hydrosilylation of chlorophenyl-, phenyldicyc-lohexyl-, butyl- and isoamilsilane to α-methylstyrene at the presence of $H_2PtCl_6{\cdot}6H_2O$ (0,1 M solution in tetrahydrofuran). In all cases addition occurs at the final carbon atom and mono-addition products are predominantly obtained. The kinetic investigation of the hydrosilylation reaction of phenylsilane and isoamilsilane with α-methylstyrene at various temperatures at the presence of $H_2PtCl_6{\cdot}6H_2O$ (0,1 M a solution in tetrahydrofuran) has been carried out. It was concluded, that the reaction proceeds faster, when positive charge on silicon atom and electronic density of double bond are greater.

Experimental Part

The starting materials for the synthesis of methylarylsiloxane oligomers were α,ω–bis(trimethyl-siloxy)methylhydridesiloxane styrene and α-methylstyrene.

The initial α,ω–bis(trimethylsiloxy)methylhydridesiloxane with the degree of polymerization n≈35, styrene and α-methylstyrene were received from Fluka. The organic solvents were cleaned by drying and distillation.

IR spectra of all samples have been taken on an UR-20 instrument from KBr pellets, while the ^{1}H and ^{13}C NMR spectra on a "Perkin-Elmer" at operating frequency of 250 MHz. All spectra were obtained with the use of $CDCl_3$ as solvent and internal standard. A Perkin – Elmer DSC-2 differential scanning calorimeter was used to determine TGA and the thermal transitions (T_g) were read at the maximum of the endothermic or exothermic peaks. Heating and cooling scanning rates were 10^0C/min.

Gel-permeation chromatography investigation was carried out with the use of Waters Model 6000A chromatograph with an R 401 differential refractometer detector. The column set comprised 103 and 104 Å Ultrastyragel columns. Sample concentration was approximately 3% by weight in toluene and typical injection volume for the siloxanes was 5 μL. Standardization of the GPC was accomplished by the use of styrene or polydimethylsiloxane standards with the known molecular weight.

Wide-angle X-ray difractograms were taken on a "DRON-2" instrument (Russian model). A-CuKα was measured without a filter, the angular velocity of the motor ω≈2 deg/min.

Hydrosilylation Reaction of α,ω–Bis(Trimethylsiloxy)Methylhydridesiloxane to Styrene

Hydrosilylation reaction was carried out in a three-necked flask equipped with a tube for catalyst inlet, reflux condenser, inert gas carrier tube and mechanical stirrer.

The reaction products 0,9768 g (0,49 mmole) α,ω–bis(trimethylsiloxy) methylhydridesiloxane and 1,5533 g (0,0149 mole) styrene and 5,4 ml dry toluene were placed into the flask and thermostated in an oil bath until constant temperature 90°C was set up. Then the catalyst 0,1 M solution of platinum hydrochloric acid in tetrahydrofuran ($5 \div 9x10^{-5}$ g per 1,0 g of starting sub-stance) was introduced. The (%) amount of active ≡Si-H was determined by Chugaev-Cereveti-nov method during reaction. After the reaction had finished the solvent was partially eliminated, the reaction product was precipitated from the toluene solution by n-hexane, and 2,36 g (91%) oligomer I^1 was obtained. The hydrosilylation reaction with α-methylstyrene and styrene at various temperatures (80, 85^0C) was carried out by the above-mentioned method.

Results and Discussions

From literature it is known, that by insertion of various heteroatom or cyclic fragments in the linear dimethylsiloxane backbone breaks the spiral structure of the polydimethylsiloxane chain and changes the physical-chemical indices and thermal-oxidative stability of copolymers [17]. The insertion of more rigid phenyl fragments in the main polydimethylsiloxane (PDMS) backbone hinder the depolymerization processes of the polymeric chain which is accompanied by release of cyclic D_n type dimethylcyclosiloxanes $[Me_2SiO]_n$. Such types of polymers are characterized by higher thermal-oxidative stability.

Increased thermal-oxidative stability of copolymers may be explained with the presence of high resistance phenyl groups to oxidation and with their inhibiting effect on oxidation of methyl groups [18, 19].

The aim of our work was, by means of modification reaction of methylhydridesiloxane, to synthesize new organosilicon oligomers with various aryl groups in a side chain. Starting materials α,ω-trimethylsiloxymethylhydridesiloxane, styrene and α-methylstyrene were used.

For the purpose of synthesis of linear organosilicon monomers with aryl side groups, the hydrosilylation reaction of α,ω-trimethylsiloxymethylhydridesiloxane (n≈35) to styrene and α-methylstyrene were studied.

By preliminary heating of initial compounds in the temperature range 80-90^{0}C, in the presence of catalyst, it was established that in these conditions, polymerization, elimination of hydrogen or other changes of initial compounds does not take place.

The hydrosilylation reactions at 1:35 molar ratio, in the presence of platinum hydrochloric acid as a catalyst, at various temperatures 80, 85 and 90^{0}C, in absolute dry toluene solution (C=0,15 mole/l) were studied. The course of the reaction was observed by a decrease of the amount of active ≡Si-H groups.

It was established, that during hydrosilylation reaction all active ≡Si-H groups do not react. In Figure 1 one can see the dependence of changes of concentration of active ≡Si-H group on the time, upon hydrosilylation reaction of α,ω–bis(trimethylsiloxy) methylhydridesiloxane to styrene (80, 85 and 90^{0}C) and α-methylstyrene (90^{0}C).

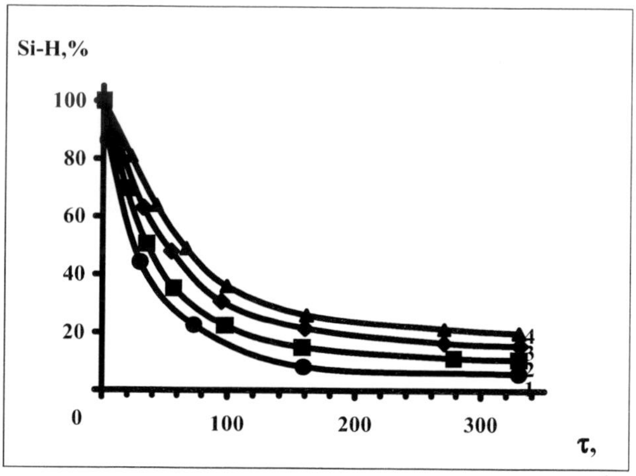

Figure 1. Dependence of changes of concentration of active ≡Si-H group on the time, in the hydrosilylation reaction of α,ω–bis(trimethylsiloxy)methylhydridesiloxane to styrene and α-methylstyrene, where curves 1,2,4 correspond to hydride addition of methylhydridesiloxane to styrene at 90, 85 and 800C accordingly and curve 3 corres-ponds corresponds to hydride addition of methylhydridesiloxane to α-methylstyreneat 90°C.

As it is seen from Figure 1, with the rise of temperature during hydrosilylation of methyl-hydride-siloxane to styrene the conversion of active ≡Si-H groups increases from 77% (curve 4) up to 95% (curve 1). In case of α-methylstyrene the conversion of active ≡Si-H groups at a temperature of 90^{0}C is 80%. The low depth of hydrosilylation reaction may be explained by steric effect evok-ed by methyl groups in α-methylstyrene.

The hydrosilylation reactions of α,ω–bis(trimethylsiloxy)methylhydridesiloxane to styrene and α-methylstyrene generally proceeds according to the scheme 1, in which [(a)+(b)](c)=m≈35; R=H (I^1 – 80^{0}C; I^2 – 85^{0}C; I – 90^{0}C); Me (II – 90^{0}C).

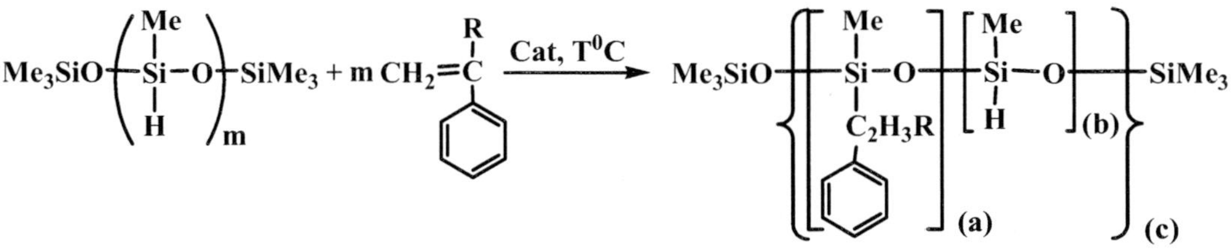

Scheme 1. Hydrosilylation reactions of α,ω–bis(trimethylsiloxy)methylhydridesiloxane to styrene and α-methylstyrene.

The hydrosilylation reaction mainly proceeds rapidly in the first hours, and then the reaction proceeds at a low rate. The results of the reaction the transparent oligomers with η_{sp}≈0.07-0.09, well soluble in aromatic type organic solvents were obtained. By functional and elementary analysis, also by IR, ^{1}H and ^{13}C NMR spectra data the composition and structure of obtained oligomers were established. Some physical-chemical properties and yields of synthesized oligomers are presented in table 1.

In the IR spectra of synthesized oligomers, at 2160-2165 cm^{-1} one can observe absorption bands, characteristic for un-reacted ≡Si-H bonds. This indicates that during hydrosilylation, conversion of all active ≡Si-H groups does not take place. In the IR spectra of oligomers one can see characteristic bands for asymmetric valence oscillation of the linear ≡Si-O-Si≡ bonds at 1020 cm^{-1}, characteristic bands for ≡Si-Me, -SiMe$_3$ bonds at 1275 and 840 cm^{-1} accordingly. In the spectra one can observe characteristic bands for mono substituted benzene rings in the range of 720, 1490-1600 and 3030-3070 cm^{-1}. Absorption bands at 1130 and 1435 cm^{-1} indicate the presence of phenyl groups.

In the ^{1}H NMR spectra of oligomer I, one can observe a singlet signals of protons in methyl group in ≡Si-Me, -SiMe$_3$ fragment with the chemical shift δ=0.1 и δ=0.3 ppm, broadened signal of protons in -CH-CH$_3$ fragment with the chemical shift δ=0.8 ppm. Also, the broadened signal of methyl protons in ≡Si-CH$_2$- fragment with the chemical shift δ=1.2 ppm, that points that that the hydride addition reaction also proceeds according to the Markovnikov rule. In the ^{1}H NMR spectra one can also observe a small intensity signal with the center of chemical shift at δ=4.3 ppm, which characterized un-reacted ≡Si-H bonds. The results of ^{1}H NMR spectra together with the IR spectra data indicates the various linked structure of synthesized oligomers.

Table 1. Some physical-chemical properties and yields of synthesized oligomers.

Oligomer #	Reaction Temperature ^0C	Yield, %	η_{sp}*	$\overline{M}_\omega x10^{-3}$, $\overline{M}_n x10^{-3}$, (D)	T_g, ^0C	d_1, Å	5% Mass losses
I^1	80	91	0.07	-	-65	-	-
I^2	85	92	0.07	-	-	-	-
I	90	94	0.09	15.94 5.10 (3.03)	-50	8.52	280
II	90	91	0.07	34.21 9.75 (3.51)	-56	8.48	275

*In 1% solution of Toluene at 25^{0}C.

In the ^{1}H NMR spectra one can also observe a broadened signal in a low field with the center of chemical shift δ=2.6 ppm characteristic for protons in methylene group in Ph-CH_2 fragment. In the spectra one can observe multiple signal characteristic for protons of the phenyl group in the range δ=6.7-7.4 ppm. The ratio of isomers obtained according to the anti-Markovnikov and Markovnikov rule is about ~1.2:1.

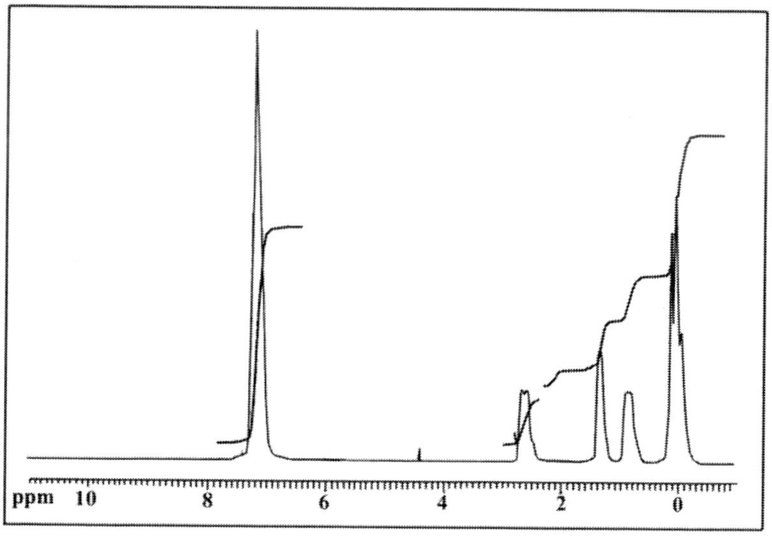

Figure 2. The ^{1}H NMR spectrum of oligomer I.

In the ^{13}C NMR spectra of oligomer I one can observe the signal with chemical shift at δ=21.9 ppm characteristic for carbon of methylene group attached with silicon (-Si-CH_2-), signal with chemical shift δ=30.9 ppm characteristic for carbon of methylene group in Ph-CH_2 fragment, the signal at δ=29.4 ppm characteristic for methine (CH) groups in -CH-CH_3 fragment and the signal at δ=15.2 ppm, which is referred to carbon of methyl group in CH-CH_3 fragment.

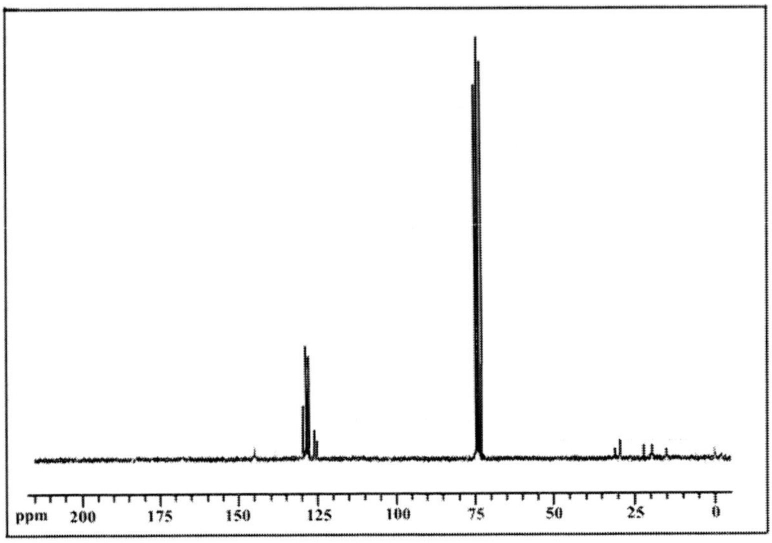

Figure 3. The ^{13}C NMR spectrum of oligomer I.

^{1}H NMR spectra of oligomer II contains singlet signals of protons of methyl group in ≡Si-Me, -$SiMe_3$ fragments with the chemical shift δ=0.1 and δ=0.3 ppm, broadened signal of protons of methylene group in ≡C-CH_3 fragment, with the chemical shift δ=1.0 ppm. In spectrum one can also observe broadened signal of methyl protons in -Si-CH_2- fragment with the chemical shift δ=1.3 ppm.

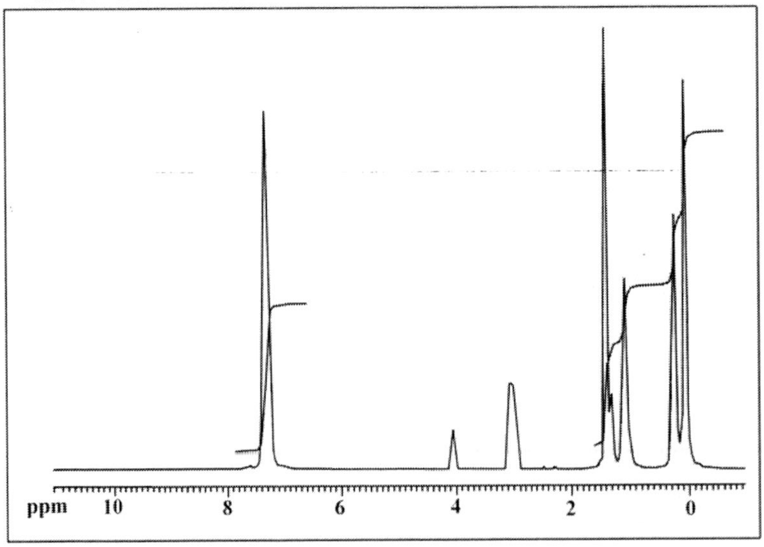

Figure 4. The ^{1}H NMR spectrum of oligomer II.

Also one can observe a wide signal of methyl protons in $=CH\text{-}CH_3$ fragment with the chemical shift $\delta=1.3$ ppm and broadened signal with the center $\delta=3.0$ ppm characteristic for protons of a CH group in the same fragment. The results of the ^1H NMR spectra indicate that the hydride addition reaction proceeds according to anti-Markovnikov and Markovnikov rule.

^{13}C NMR spectra of oligomer II (Figure 5) is in complete agreement with ^1H NMR spectra data. In spectra one can see resonance signal for carbon nucleus of methylene group ($-Si\text{-}CH_2-$) at silicon atom with center of chemical shift at $\delta=34.5$ ppm, resonance signal with center of chemical shift at $\delta=26.8$ ppm, characteristic for carbon nucleus of methine groups in the fragment Ph-CH, signal with center at $\delta=25.1$ ppm characteristic for methyl groups in the fragment $CH\text{-}CH_3$ and signal at $\delta=19.6$ ppm characteristic for carbon nucleus of methyl group in the fragment $=C(CH_3)_2$. This indicates that hydride addition of methylhydridesiloxane to α-methylstyrene proceeds both by anti-Markovnikov and Markovnikov rule.

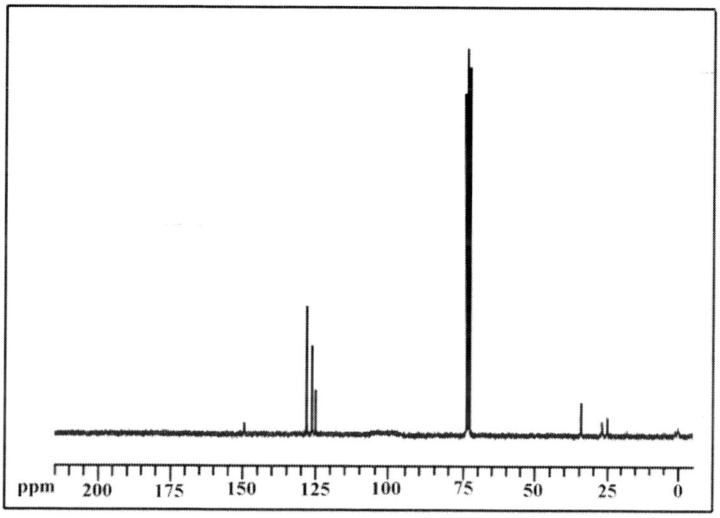

Figure 5. The ^{13}C NMR spectrum of oligomer II.

For full characterization of hydride addition of methylhydridesiloxane to styrene, by quantum-chemical semi empiric AM1 method [20], in modelling reaction of hydrosilylation of methyl-dimethoxysilane to styrene, for all initial, intermediate and final products the heats of formations (ΔH_f), changes of energy (ΔH) of the system depending on the change of distance ($R_{C\text{-}Si}$) between ≡C-Si≡ bonds, also the charges values (q) on the atoms, dipole moments (μ) and bonds orders (Pij) are calculated (see tables 2 and 3). Catalyst and solvent effects are not provided for calculations.

For the beginning the hydride addition of modelling compound $[Me(MeO)_2SiH]$ methyldimetho-xysilane to styrene according to the anti-Markovnikov rule was considered. The reaction of hydride addition by anti-Markovnikov rule proceeds according to the scheme 2 with formation of compound III.

In styrene the distance between the double bonded carbon atom (C_8) and silicon was taken about 1.0 Å more than the bond length expected in a product.

The change step of distances ($R_{C_8\text{-}Si}$) between silicon and double bonded carbon atom (C_8) was about 0,05 Å. Dependence of change of energy (ΔH) of the system on the distances ($R_{C\text{-}Si}$) between silicon and carbon atoms is presented on the Figure 6.

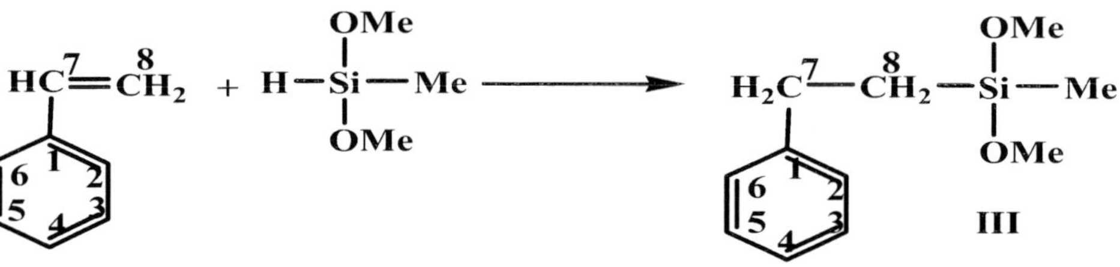

Scheme 2 The modelling reaction of hydride addition of methyldimethoxysilane to styrene according to the anti-Markovnikov rule.

As it is seen from Figure 6, at approximation of silicon to carbon (C_8) atom up to $R_{C\text{-}Si}$=2.25 Å distances the system energy rises. At the same time the bond order between carbon C_8 and C_7 atoms in the styrene molecule decreases from 1.922 up to 1.640. The bond order between hydrogen and silicon atoms also decreases ($R_{Si\text{-}H}$=0.888-0.693) and in modelling compound methyl-dimethoxysilane the process of formation of new bonds ($P_{C\text{-}Si}$=0.017-0.127 and $P_{C_7\text{-}H}$=0.002-0.099) is observed. In the case of distance up to 2.20 Å the system energy sharply decreases, the double C=C bond passes in ordinary C-C bond ($P_{C_7\text{-}C_8}$=1.004), and the bond order between silicon and carbon atom C_8 reaches to 0.861. Thus the hydrogen completely eliminated from si-licon ($P_{Si\text{-}H}$=0.011) and attaches to carbon atom C_7 ($P_{C_7\text{-}H}$=0.918).

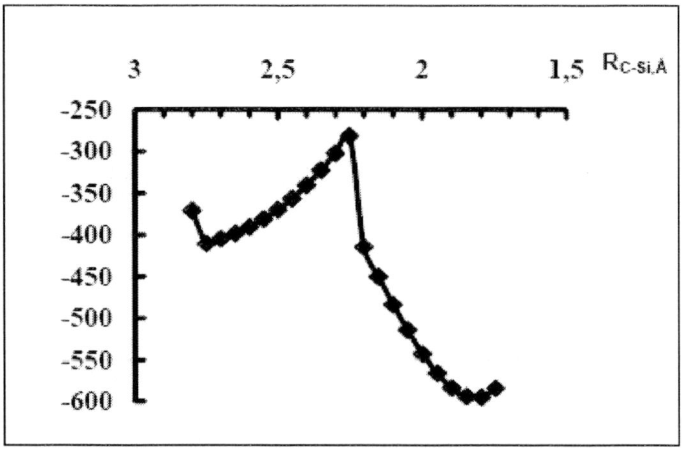

Figure 6. Dependence change of energy (ΔH) on distance ($R_{C\text{-}Si}$) between silicon and carbon atoms during modelling hydride addition reaction of methyldimethoxysilane to styrene according to the anti-Markovnikov rule.

With consecutive reduction of distances the system energy monotonously decreases and form new bonds. The heat of formation of hydrosilylation product (ΔH= -595.4 kJ/mol) is in good correlation with heat of formation (ΔH= -601.2 kJ/mol) of compound III obtained by hydride addition of methyldimethoxysilane to styrene according to the anti-Markovnikov rule. Activation energy of modelling reaction proceeding by anti-Markovnikov rule is equal to E_{act}=89.6 kJ/mol.

The modelling hydrosilylation reaction of methyldimethoxysilane to styrene according to the Markovnikov rule is also considered. The modelling reaction proceeds according to the scheme 3 with formation of compound IV.

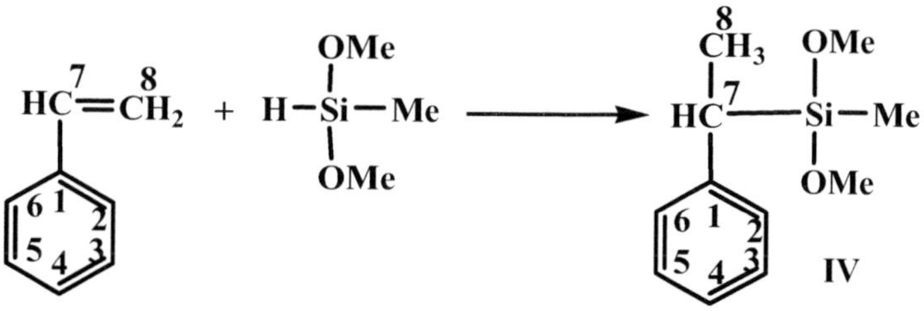

IV

Scheme 3. The modelling reaction of hydride addition of methyldimethoxysilane to styrene according to the Markovnikov rule.

In Figure 7 the dependence of change of energy (ΔH) of the system on the distances between silicon and carbon atoms ($R_{C_7\text{-}Si}$) is presented. Similarly the above-mentioned distances between the double bonded carbon atom (C_7) and silicon were taken about 1.0 Å more than the bond, expected in a product. The change in step of distances ($R_{C_7\text{-}Si}$) between silicon atom and double bonded carbon atom (C_7) was about 0.05 Å.

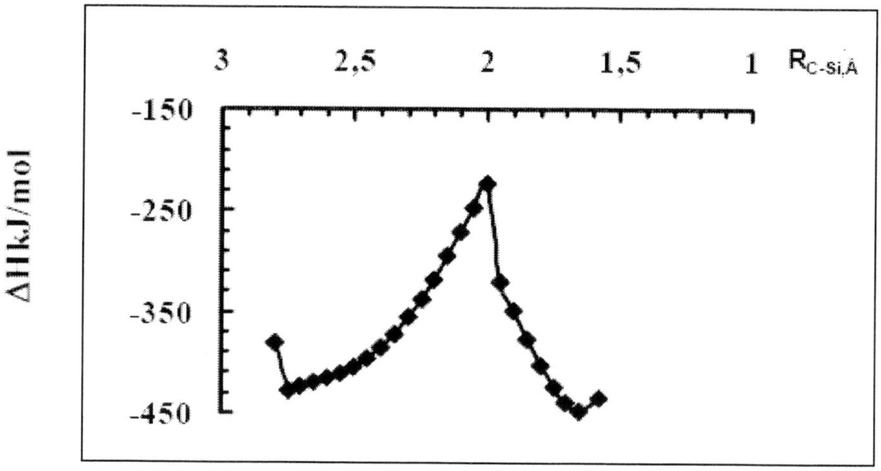

Figure 7. Dependence of change of energy (ΔH) on the distance ($R_{C_7\text{-}Si}$) between silicon and carbon atoms during modelling hydride addition of methyldimethoxysilane to styrene according to the Markovnikov rule.

As it is seen from Figure 7, at approximation of silicon to carbon (C_7) atom up to R_{C-Si}=2.00 Å distances the system energy rises. At the same time the bond order between carbon C_7 and C_8 atoms in a molecule styrene decreases from 1.894 up to 1.556. The bond order between hydrogen and silicon atoms also decreases (R_{Si-H}=0.877-0.669) and in methyldimethoxysilane the process of formation of new bonds (P_{C-Si}=0.016-0.197 and P_{C_8-H}= 0.011-0.065) is observed. In case of distan ce up to 1.95

Å the system energy sharply decreases, the double C=C bond passes in ordinary C-C bond ($P_{C_7-C_8}$=1.006), and the bond order between silicon and carbon atom C_7 reaches to 0.794. Thus the hydrogen completely eliminated from silicon (P_{Si-H}=0.019) and attaches to carbon atom C_8 (P_{C_8-H}= 0.932).

With consecutive reduction of distances, the system energy monotonously decreases and forms new bonds. The heat of formation of a product of hydride addition (ΔH = -448,4 kJ/mol) is in good correlation with the heat of formation (ΔH=-454.2 kJ/mol) of compound IV obtained by hydride addition of methyldimethoxysilane to styrene according to the Markovnikov rule. Activation energy of modelling reaction proceeding by Markovnikov rule is equal to E_{act} =156.4 kJ/mol.

Thus at comparison of the heat of formations of compound III and IV activation energies and also of the Figures 6 and 7 (the character of curves of change of energy and the bond orders value) one can conclude, that the course of modelling hydrosilylation reaction of methyldimethoxysilane to styrene energetically is more favorable by Anti-Markovnikov rule that is in agreement with NMR spectral data, but the difference is closely approximated (anti-Markovnikov rule: Markovnikov rule= 1.2:1).

In Figure 8 the dependence of reverse concentration on the time of the hydrosilylation reaction of methylhydridesiloxane to styrene is presented. One can see that hydrosilylation reaction at the initial stages is of second order. The hydrosilylation reaction rate orders at various temperatures were calculated: $k_{90^0C} \approx 3.0797\text{x}10^{-2}$, $k_{85^0C} \approx 2.3007\text{x}10^{-2}$ and $k_{80^0C} \approx 1.6781\text{x}10^{-2}$ l/mol·s.

From the dependence of reverse temperature on the logarithm of rate constants the activation energy of the hydrosilylation reaction was found and it is equal to $E_{act.} \approx$ 66.7 kJ/mol.

The synthesized oligomers were studied by GPC method (Figure 8,9). The oligomers I and II have bimodal molecular weight distribution. For the synthesized oligomers average molecular masses were determined. For oligomer I it's equal to: $\overline{M}_\omega \approx 1.59\text{x}10^4$, $\overline{M}_n \approx 5.17\text{x}10^3$, $\overline{M}_z \approx 2.913\text{x}10^4$, and polydispersity is equal to D=3.08. For oligomer II - $\overline{M}_\omega \approx 3.42\text{x}10^4$, $\overline{M}_n \approx 9.75\text{x}10^3$, $\overline{M}_z \approx 1.18\text{x}10^5$ and polydispersity is equal to D=3.51.

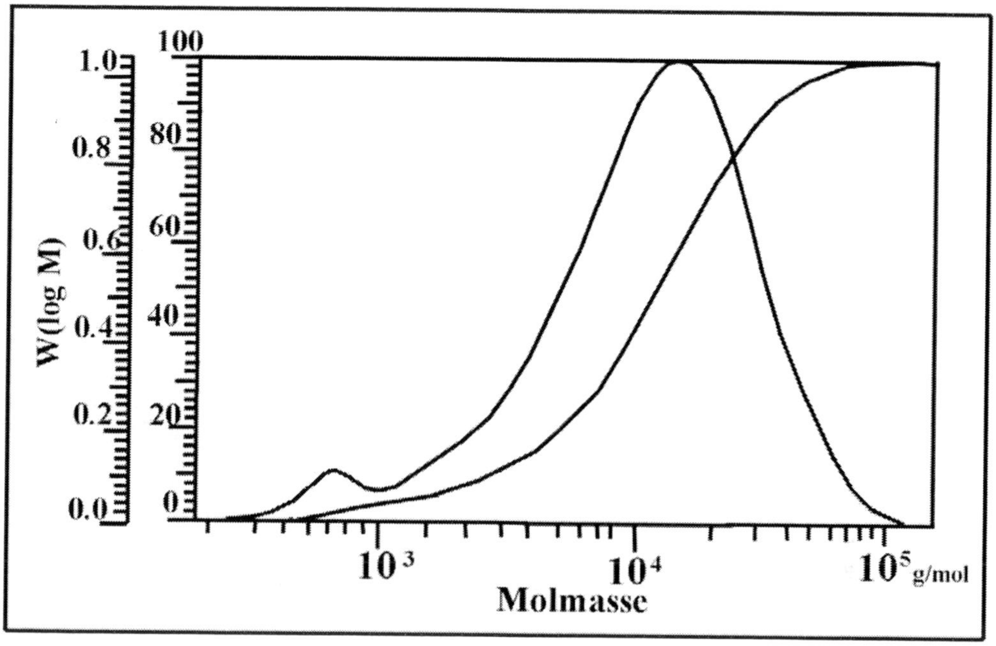

Figure 8. Gel permeation chromatographic curves of oligomer I.

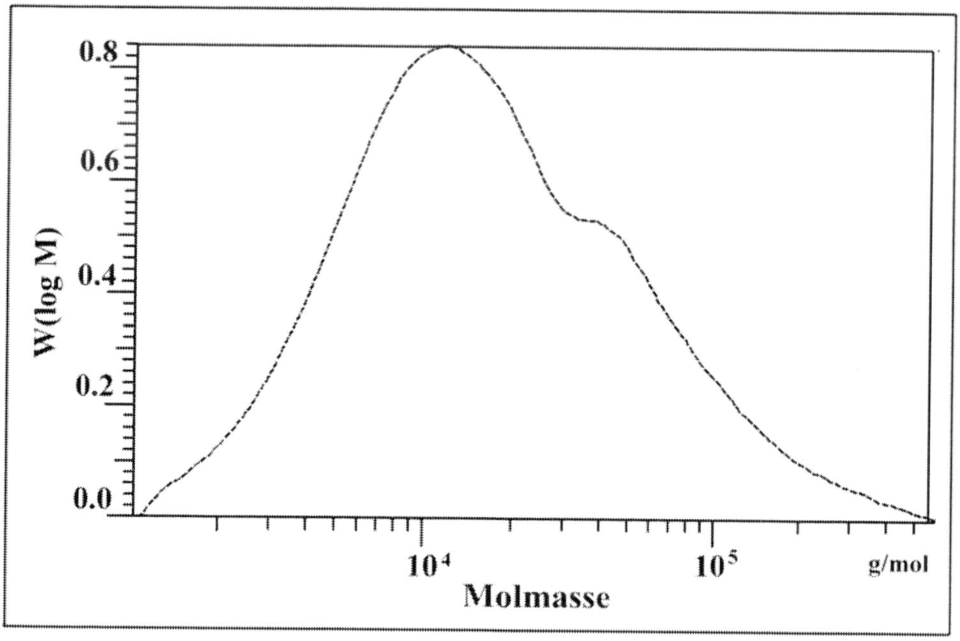

Figure 9. Gel permeation chromatographic curves of oligomer II.

As it's seen from Table 1, molecular masses exceeds calculated molecular masses 3-6 times. Increased molecular masses proved that during hydrosilylation reaction branched oligomers are obtained. Obtaining of branching systems may proceed with participation of residual ≡Si-H groups in intermolecular cross-linking reactions with obtainment of ≡Si-O-Si≡ and ≡Si-Si≡ bonds, which was earlier discussed [21,22].

By DSC investigation it was shown, that oligomers are characterized with one endothermic peak, which corresponds to glass transition temperature. For oligomer I glass transition temperature is equal to $T_g \approx -50^0C$ and for oligomer II - $T_g \approx -56^0C$.

Thermogravimetric investigations of synthesized oligomers were carried out. For oligomers the mass losses at 250^0C temperature is not observed, while unblocked polydimethylsiloxane polymer is completely destroyed in the temperature range $270\text{-}280^0C$ [7]. 5% mass losses are observed at $\sim 350^0C$, the main destruction process proceeds from 400^0C. It is seen that the modification effect, induced by insertion of aryl fragments in dimethylsiloxane chain increased thermal-oxidative stability of synthesized oligomers.

REFERENCES

[1] M.E. Dolgaia, E.A. Chernishev, Li Guanlian. *"Chemistry and Practical Application of Orga-nosilicon Compounds"*, 1958 Part 1, Leningrad, 99 (in Russian).

[2] A.D. Petrov, E.A. Chernishev, M.E. Dolgaia, Yu.P. Egorov, L.A. Leites. Zhurn. Obshch. Khim., 1960, 30, 376 (in Russian).

[3] J.W. Ryan, J.L. Speier. *J. Org. Chem.*, 1959, 24, 2052.

[4] J.W. Speier, J.A. Webster, G.H. Barnes. *J. Amer. Chem. Soc.*, 1957, 79, 974.

[5] R.A. Pike, R.C. Bochert. U.S. Patent 2.954.390, 1960, (Union Carbide Corp.).

[6] G.H. Wagner, D.L. Bailey, A.N. Pines, M.L. Dunham, D.B. McIntire. *Ind. Eng. Chem.*, 1953, 45, 367.

[7] V.P. Yuriev, I.M. Salimgareva. "Hydrosilylation Reactions of Olefins", *Nauka,* 1982, 223 (in Russian).

[8] J.S. Saam, J.L. Speier. *J. Amer. Chem. Soc.*, 1961, 83, 1351.

[9] M.C. Musolf, J.W. Speier. *J. Org. Chem.*, 1964, 29, 2519.

[10] N. Larry, J. Lews. *Amer. Chem. Soc.*, 1990, 112, 59998.

[11] E.A. Chernishev, M.E. Dolgaia, E.D. Lubuzh. Izv. Acad. Nauk USSR, *Ser. Khim.*, 1965, 650 (in Russian).

[12] K.A., Andrianov, V.E. Sidorov, L.M. Tartakovskaia. Izv. Acad. Nauk USSR, *Ser. Khim.,* 1971, 1118 (in Russian).

[13] V.O. Reikhsfeld, M.I. Astrakhanov. *Zhurn. Obshch. Khim.*, 1977, 47, 1494 (in Russian).

[14] V.O. Reikhsfeld, M.I. Astrakhanov. *Zhurn. Obshch. Khim.*, 1977, 47, 1497 (in Russian).

[15] G.N. Koroleva, V.O. Reikhsfeld. *Zhurn. Obshch. Khim.*, 1967, 37, 2559 (in Russian).

[16] V.O. Reikhsfeld, G.N. Koroleva. *Zhurn. Obshch. Khim.*, 1966, 36, 1474 (in Russian).

[17] V.P. Mileshkevich. *Kauchuk and Rezina*, 1978, 6, 4 (in Russian).

[18] T.H. Thomas, T.C. Kendrick. *J. Polymer Sci.*, 1969, 7, 537.

[19] A.F. Bulkina, V.S. Papkov, A.A. Zhdanov, K.A. Andrianov. *Vysokomol. Soed.*, 1978, 20A, 70 (in Russian).

[20] M.I.S. Dewar, E.G., Zeobish, E.F. Healy, J.S. Stewart. *J. Amer. Chem. Soc.*, 1985, 107, 3902.

[21] T.A.C. Flipsen, R. Derks, H. Van Der Vegt, A.J. Pennings, G. Handzzionnou. *J. Polymer Science,* 1997, 35, 41.

[22] B.A. Trofimov, L.N. Parshina, M.I. Khilko, L.A. Oparina. *Thesis of the All-Russian Confe-rence "Organosilicon Compounds: Synthesis, Characteristics and Application",* Moscow, 2000, L-36 (in Russian).

In: Modification Reactions of Oligomethylhydridesiloxanes ISBN: 1-60021-365-0
Editors: O.Mukbaniani, et al. pp. 37-56

Chapter 3

HYDROSILYLATION REACTIONS OF METHYLHYDRIDESILOXANE TO TRIMETHYLSILYLATED AND TRIETHOXYSILYLATED ESTER OF ACRYLIC AND METHACRYLIC ACIDS

O. Mukbaniani, N. Pirtskheliani and T. Tatrishvili
I. Javakhishvili Tbilisi State University, I.
Chavchavadze Ave., 1, Tbilisi 380028, Georgia

ABSTRACT

The reaction of hydrosilylation of α,ω–bis(trimethylsiloxy)methylhydridesiloxane (n≈35) to trimethylsilylated and triethoxysilylated esters of acrylic and methacrylic acids at various ratios of initial compounds, in the presence of platinum hydrochloric acid, or platinum on the carbon Pt/C has been investigated. It was established that in the presence of platinum catalysts, hydrosilylation of all active ≡Si-H groups do not take place and oligomers with various chain links are obtained. During hydride addition the reaction order, reaction rate constants and activation energy were found.

By IR and NMR spectra data the structures of synthesized oligomers were determined. On the basis of NMR spectra data was offered, that the hydrosilylation proceeds mainly with formation of intermediate product, by regrouping of this transition complex the oligomers with various structures of molecules are obtained. Gel chromatographic permeation, differential calorimetric scanning and wide-angle X-ray analysis of synthesized oligomers were carried out.

Key words: trimethylsilylated ester, triethoxysilylated ester, hydrosilylation and methylhydridesiloxane.

INTRODUCTION

In the literature, there is much information about hydrosilylation of acrylate and methacrylate [1-4]. In a study by A.D. Petrov the reaction of hydride addition of triethylsilane to methyl acrylate in the presence of Spier catalyst has been investigated and it was shown that in hydrosilylation reaction 1,2- or 1,4-addition depend on the nature of organic group attached at the silicon and on the structure of carbonyl compounds.

For example hydrosilylation of triethylsilane to methyl acrylate proceeds with 1,4-addition, while hydrosilylation of trichlorosilane to methyl acrylate proceeds with 1,2-addition according to Anti-Markovnikov rule. Consequently it was established, that hydrosilylation ability in triorganosilanes gradually changes with replacement of electro donor alkyl groups by electronegative halogens [1-3]:

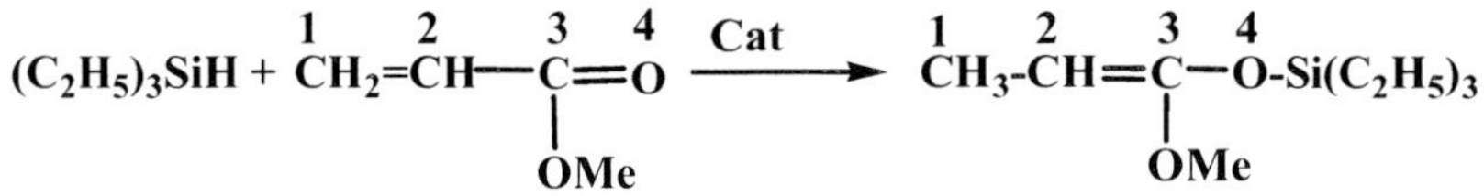

1,4 -Addition

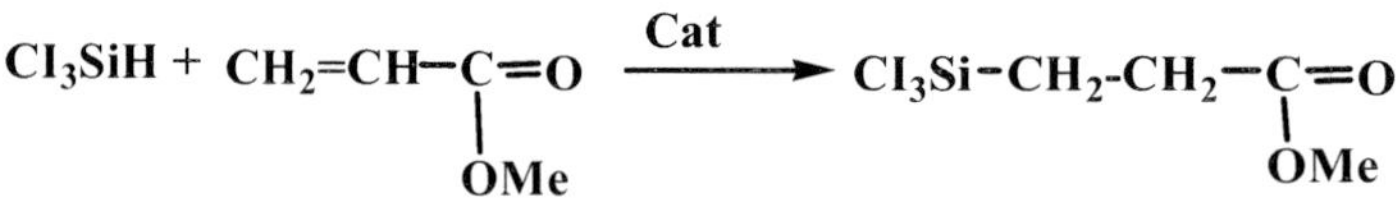

During hydride addition of trichlorosilane to methacrylate obtaining α-isomer, it may be explained that the electrophilic trichlorosilyl group attaches to the carbon atom which has more electronic density. In methyl acrylate electro donor methyl group in α-position appears nucleophilic properties in C=C fragment of β-carbon atoms [4,5].

Hydrosilylation of methyldichlorosilane to methyl acrylate in the presence of platinum on the carbon proceeds by 1,2- addition according to Anti-Markovnikov or Markovnikov rule and by 1,4-addition as conjugated system [1]:

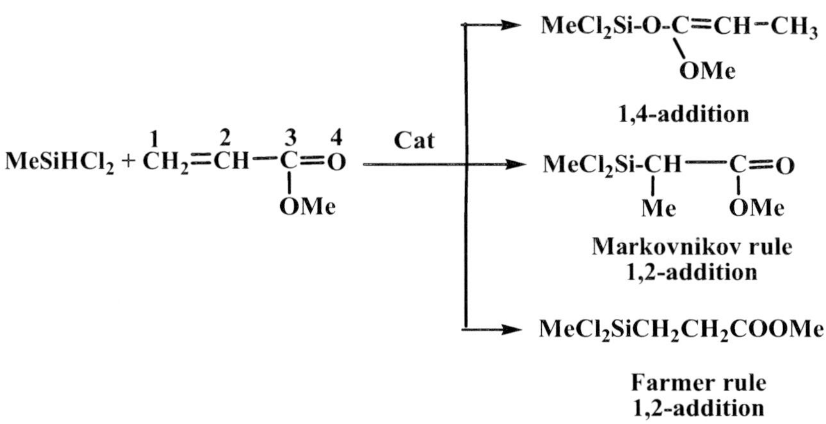

Hydrosilylation of triethylsilane to methyl methacrylate in the presence of catalyst Pt/C proceeds according to the Anti-Markovnikov rule (1,2-addition), with formation of triethyl(2-carbometoxypropyl)silane [1]:

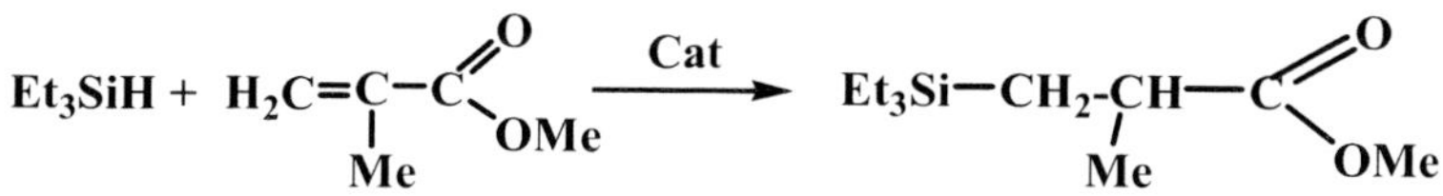

Hydrosilylation of methyldichlorosilane to acrylic esters in the presence of catalysts Pt/C or Pt/Al_2O_3 proceed with formation of two isomeric products [6,7]:

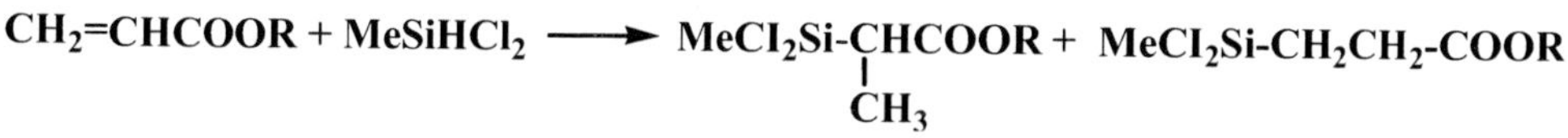

In contrast to authors [1-7], hydrosilylation reactions of acrylate proceeds with formation of α, β and O-products, according to the following scheme [8,9]:

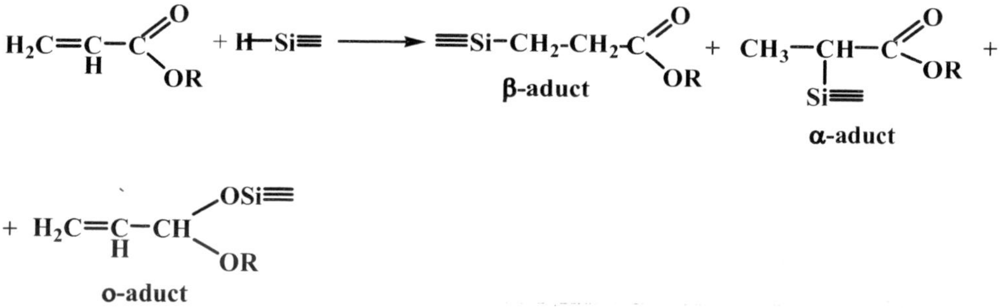

For methyl acrylate the yield of β-addition product is equal to -10,3%, α-addition product – 57,8% and o-addition product – 1,9%. Substituted groups create a definite effect on formation of structure of hydrosilylation product.

The reaction of hydrosilylation of acrylate and methacrylate with 1,4-bis(dimethylsilyl)benzene in the presence of hydroquinone and Spier catalyst or platinum on the carbon was studied by the authors [10] and it was shown that the reaction proceeds with formation of corresponding acyloxysilanes, according to the following scheme:

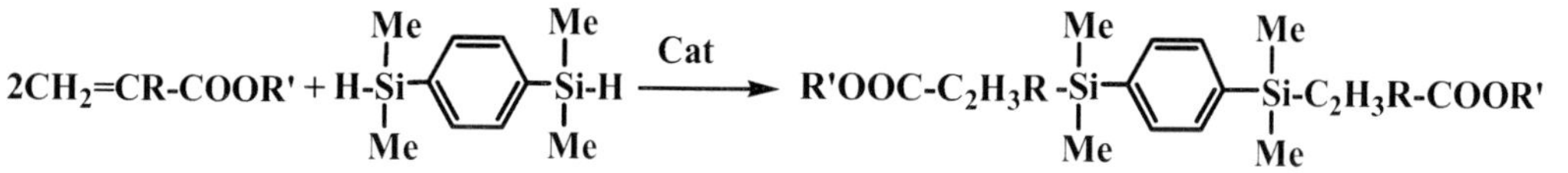

\Where: R=H, Me; R'=Me, Et.

In hydrosilylation reaction, the hydroquinone was used for preventing the polymerization reactions.

The reaction of hydrosilylation of methylhydridesiloxane to unsaturated vinyl derivatives (alkenes, acrylic acid, acrylic esters) in the presence of Spier catalyst investigated by the authors [11] and methylsiloxane oligomers with carboxyl, alkyl and ester side-groups has been obtained, according to the following scheme:

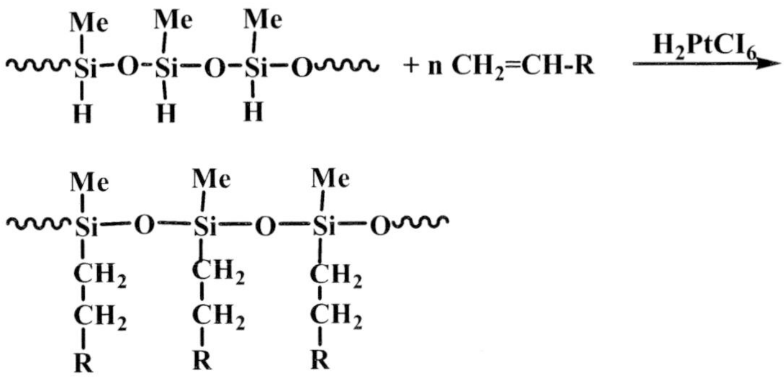

Where: R=-COOH, -COOAlk.

During investigation of hydrosilylation reaction of oligomethylhydridesiloxane to acrylic and methacrylic acid by the authors established, that along with the hydrosilylation reaction, the dehydrocondensation reaction takes place [12]:

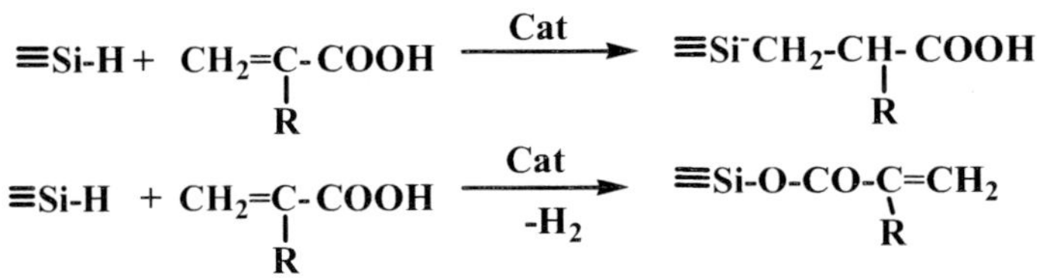

Therefore, in the literature there is contrasting information about course and mechanism of the hydrosilylation reaction of hydrideorganosilanes to acrylate and methacrylate.

EXPERIMENTAL PART

The starting materials for the synthesis of comb-type methylsiloxane oligomers were used α,ω–bis-(trimethylsiloxy)methylhydridesiloxane and trimethylsilylated esters of acrylic and methacrylic acids.

The initial α,ω–bis(trimethylsiloxy)methylhydridesiloxane with the degree of polymerization n≈35, acrylic and methacrylic acids were received from Fluka. Before reaction, the acrylic and methacrylic acids were distilled. By drying and distillation the organic solvents used in reaction hydrosilylation were cleaned. Trimethylsilylated esters of acrylic and methacrylic acid were obtained according to the literature data [13,14]. Triethoxysilylated ester of methacrylic acid obtained according to the literature data [15,16].

IR spectra all of samples have been taken on an UR-20 instrument from KBr pellets, while the ^{1}H NMR spectra on a "Perkin-Elmer" at an operating frequency of 250 MHz. All spectra were obtained with the use of $CDCl_3$ as a solvent and internal standard. A Perkin – Elmer DSC-2 differential scanning calorimeter was used to determine TGA and the thermal transitions (T_g) were read at the maximum of the endothermic or exothermic peaks. Heating and cooling scanning rates were 10^{0}C/min.

Gel-permeation chromatography investigation was carried out with the use of Waters Model 6000A chromatograph with an R 401 differential refractometer detector. The column set comprised 10^3 and 10^4 Å Ultrastyragel columns. Sample concentration was approximately

3% by weight in toluene and typical injection volume for the siloxane was 5 μL. Standardization of the GPC was accomplished by the use of styrene or polydimethylsiloxane standards with the known molecular weight.

Wide-angle X-ray diffractograms were taken on a "DRON-2" instrument. A-CuK_α was measured without a filter, the angular velocity of the motor $\omega \approx 2$ deg/min.

Hydride Addition of α,ω–Bis(Trimethylsiloxy)Methylhydridesiloxane to Trimethylsilylated Ester of Acrylic Acid

The hydrosilylation reaction was carried out in a three-necked flask equipped with a mechanical stirrer, an argon inlet, reflux condenser and calcium chloride drying tube. The initial reagents were placed in the flask and thermostated in an oil bath until constant temperature was achieved. 0,1 M solution of Platinum hydrochloric acid in tetrahydrofuran ($5\div9x10^{-5}$ g per 1,0 g of starting substance) was used as a catalyst. The reaction was carried out in an argon atmosphere in anhydrous toluene solution, at temperature range 60^0C. Then 5 ml anhydrous toluene was added to the reaction mixture, was filtered, connected to the vacuum and the toluene and unreacted trimethylsilylated ester of acrylic acid were removed at $30 \div 40^0$C.

The hydrosilylation reaction in the presence of platinum on the carbon and with trimethylsilylated ester of methacrylic acid and triethoxysilylated acrylic and methacrylic acids has been carried out in the same manner.

Results and Discussion

For the purpose of synthesis of comb-type methylsiloxane oligomers, containing trimethylsylilated and triethoxysilylated esters of propion- and isobutyric acids in the side chain, hydrosilylation reaction of α,ω–bis(trimethylsiloxy)methylhydridesiloxane ($n \approx 35$) to trimethylsilylated and triethoxysilylated esters of acrylic and methacrylic acids, at 1:35 ratio of initial compounds, in the presence of catalyst platinum hydrochloric acid or Pt/C (0.1% weight) were investigated.

The reaction of hydrosilylation of α,ω–bis(trimethylsiloxy)methylhydridesiloxane to trimethyl- and triethoxysilylated acrylic and methacrylic acids was examined at different temperatures in dilute solutions of anhydrous toluene ($C \approx 5.3x10^{-2}$ mol/l). Preliminary heating of the initial compounds during 5-6 h in the presence of catalyst platinum hydrochloric acid or platinum on the carbon in the temperature range 60-90^0C showed that polymerization and polycondensation of initial compounds, break of siloxane backbone, or elimination of methane and hydrogen in methylhydridesiloxane, or other changes do not take place. It has been shown, that conversion of all active ≡Si-H groups does not proceed completely and various linked oligomers are obtained.

Hydrosilylation reaction of α,ω–bis(trimethylsiloxy)methylhydridesiloxane to trimethylsilylated esters of acrylic and methacrylic acids carried out at 40-60^0C temperatures, in the presence of platinum on the carbon, where the catalyst Pt/C- 5% was taken ~ 0.1% of the initial compounds.

During hydrosilylation reaction of α,ω–bis(trimethylsiloxy)methylhydridesiloxane to trimethylsilylated ester of methacrylic acid the conversion of active ≡Si-H groups at 60^0C temperature does not exceed 62% (during 4 h). Because of that, hydrosilylation reaction was carried out at higher temperatures (80-90^0C).

Hydrosilylation reaction of α,ω–bis(trimethylsiloxy)methylhydridesiloxane to triethoxysilylated esters of acrylic and methacrylic acids was carried out at 60-80^0C temperatures, in the presence of 0.1M solution of platinum hydrochloric acid in tetrahydrofuran. It was established that during hydrosilylation to triethoxysilylated esters of acrylic acid at 80^0C temperature, the conversion of ≡Si-H groups composed 95%, whereas in case of triethoxysilylated esters of methacrylic acid is about 84%. This fact can be explained by the steric effect of methyl groups that it caused in methacrylic derivatives.

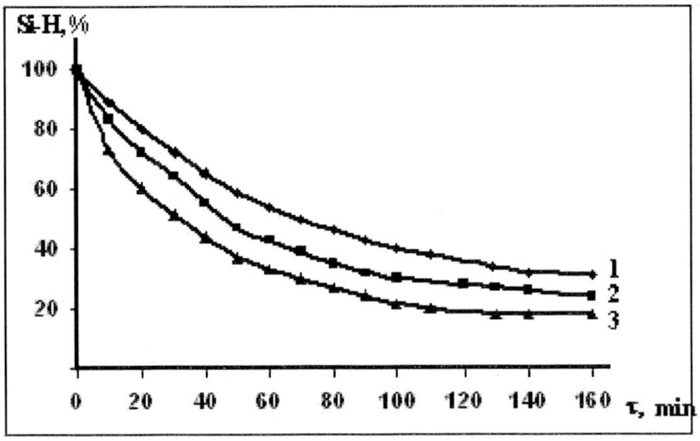

Figure 1. Dependence of changes of concentration of active ≡Si-H group on the time, during hydride addition of α,ω–bis(trimethylsiloxy)methylhydridesiloxane to trimethyl- silylated ester of acrylic acid, where curve 1 is at 40^0C temperature, curve 2 at 50^0C and curve 3 is at 60^0C.

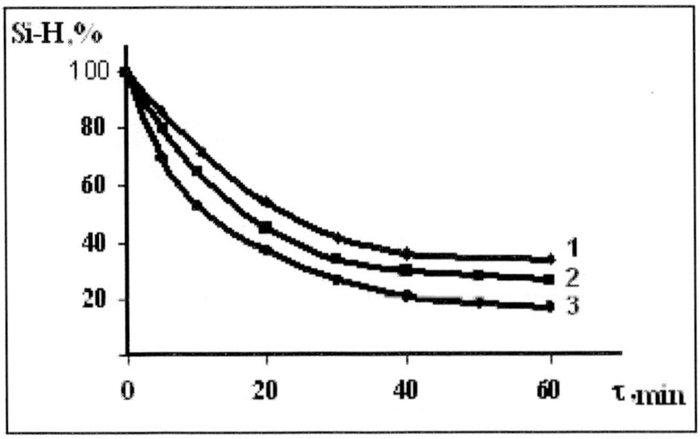

Figure 2. Dependence of changes of concentration of active ≡Si-H group on the time, during hydride addition of α,ω–bis(trimethylsiloxy)methylhydridesiloxane to trimethyl- silylated ester of methacrylic acid, where curve 1 is at 80^0C temperature, curve 2 at 85^0C and curve 3 is at 90^0C.

In Figures 1 and 2 the dependence of conversion of active ≡Si-H groups on the time, during hhydrosilylation reaction of α,ω–bis(trimethylsiloxy)methylhydridesiloxane to trimethylsilylated ester of methacrylic acid is presented. With the rise of temperature, the conversion of active ≡Si-H groups (the depth of hydrosilylation) and the rate of hydrosilylation reaction increases.

Hydrosilylation reaction of α,ω–bis(trimethylsiloxy)methylhydridesiloxane to trimethylsilylated esters of acrylic and methacrylic acids proceeds according to the following scheme:

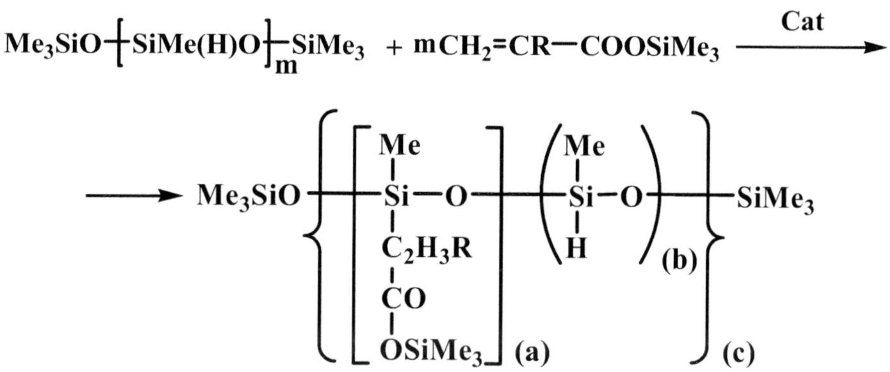

Where: m≈35, [(a)+(b)](c)=m≈35. R=H, R'=Me: I[1] (40^{0}C), I[2] (50^{0}C), I (60^{0}C). R=R'=Me: II[1] (80^{0}C), II[2] (85^{0}C), II (90^{0}C).

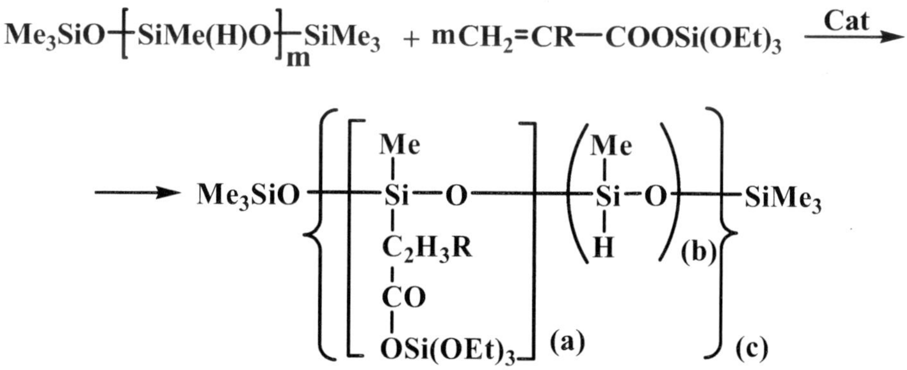

Where: m≈35, [(a)+(b)](c)=35. R=H, R'=OEt: III[1] (60^{0}C), III[2] (70^{0}C), III (80^{0}C). R=Me, R'=OEt: IV[1] (60^{0}C), IV[2] (70^{0}C), IV (80^{0}C).

The synthesized oligomers are vitreous liquid or solid products, depending on the surrounding group, well soluble in aromatic type ordinary organic solvents with η_{sp}≈0,03÷0,05. The structure and composition of synthesized oligomers are, proven by elementary analysis, determination of molecular masses, IR, ^{1}H and ^{13}C NMR spectral data. Some physical-chemical properties of synthesized oligomers are presented in table 1÷4.

Table 1. Same physical-chemical properties and yields of comb-type methylsiloxane oligomers.

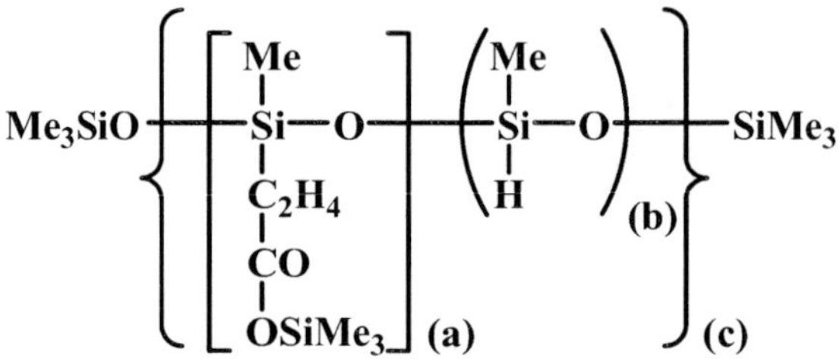

#	Yield, %	η_{sp}^{*}	d_1, Å	$\bar{M}_{\omega}x10^{-3**}$	T_g, ^0C	Elementary composition, %		
						C	H	Si
I[1]	77	0.03	8.65	_	-80	38.77 38.72	7.80 7.50	29.92 29.53
I[2]	84	0.04	_	_	_	38.95 39.16	7.81 7.46	29.75 29.02
I	89	0.05	8.65	4.42	-72	39.76 40.03	7.84 7.43	29.00 28.15

* In 1% toluene solution at 25^0C.

**Average molecular masses have been determined by GPS method.

*** Here and in other cases – over line are calculated values, under line experimental values.

Table 2. Same physical-chemical properties and yields of comb-type methylsiloxane oligomers.

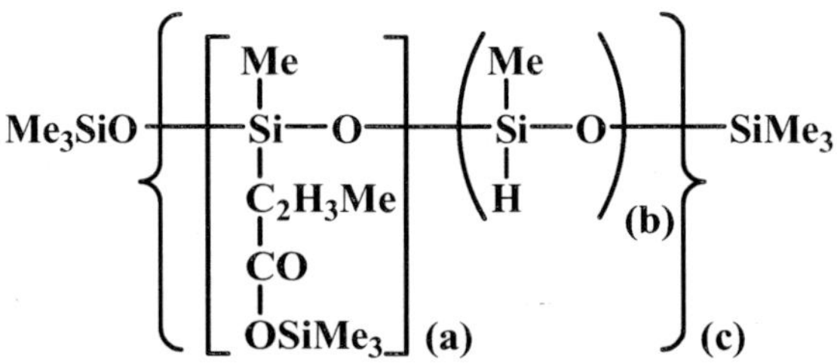

Oligomer №,	Yield, %	$\eta_{sp.}$	d_1,Å	Depth of hydrosilylation, %	$T_{g.}$, ^0C	Elementary composition, %		
						C	***H***	**Si**
II[1]	79	0.04	-	66	-	41.12 40.72	8.14 7.70	28.49 28.53
II[2]	87	0.04	-	72	-	41.79 41.16	8.18 7.96	27.89 28.02
II	92	0.05	9.56	78	-65	42.31 41.93	8.21 7.93	27.42 27.15

Table 3. Same physical-chemical properties and yields of comb-type methylsiloxane oligomers.

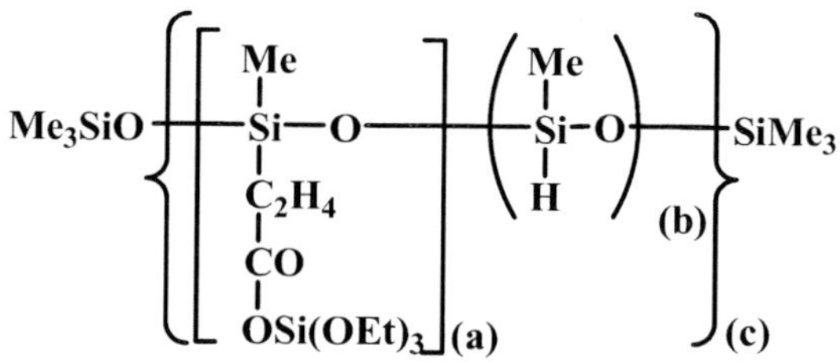

Oligomer №	Yield, %	T,^{0}C	Depth of hydrosilylation, %	η_{sp}***	d_1, Å	k, mol/l·s
III[1]	80	60	90	0.03	-	3.2006
III[2]	85	70	93	0.05	-	5.2102
III	89	80	95	0.05	8.63	1,2890

Table 4. Same physical-chemical properties and yields of comb-type methylsiloxane oligomers.

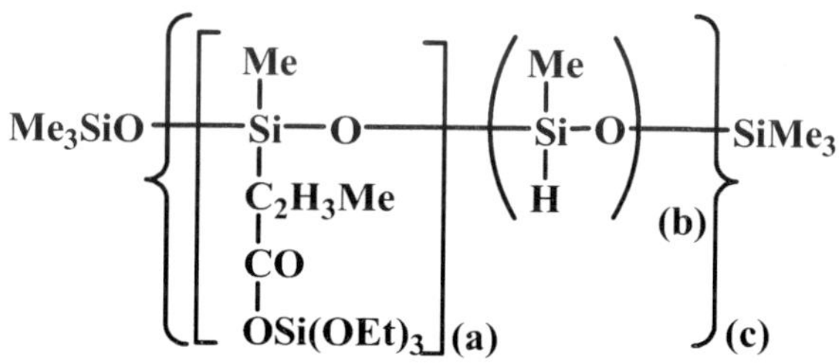

Oligomer №	Yield, %	T,^{0}C	Depth of hydrosilylation, %	η_{sp}	d_1, Å	k, mol/l·s
IV[1]	78	60	63	0.03	-	0.7678
IV[2]	79	70	74	0.03	_	1.2498
IV	80	80	84	004	8.63	2. 3008

In the IR spectra of synthesized oligomers, one can observe an absorption band characteristic for asymmetric valence oscillation of ≡Si-O-Si≡ and ≡CO-O-C≡ bonds in the region 1020 and 1155 cm^{-1} respectively. Absorption bands of ≡SiMe and -$SiMe_3$ groups at 1260 cm^{-1} and 840 $cм^{-1}$ accordingly; for valence oscillation of ≡C-H bonds at 2950-3100 cm^{-1}

region. In the spectra, one can observe absorption bands characteristic for unreacted ≡Si-H bonds at 2160-2165 cm^{-1}.

In ^1H NMR spectra of oligomers I (Figure 3) one can observe a singlet signal with chemical shift δ≈0.1 ppm, characteristic for protons in $-SiMe_3$ and ≡SiMe groups. A wide signal with chemical shift δ≈0.8 ppm, is characteristic for protons in $\equiv Si\text{-}CH_2\text{-}$ group; triplet signal with chemical shift δ≈2.3 is characteristic for methylene protons in the $-CH_2\text{-}CO\text{-}$ group. It means that hydrosilylation reaction mainly proceeds according to the anti-Markovnikov rule. In spectra one can observe light intensive signal with chemical shift δ≈4.2 ppm, characteristic for unreacted ≡Si-H groups, which has been proven through obtaining various liked oligomers.

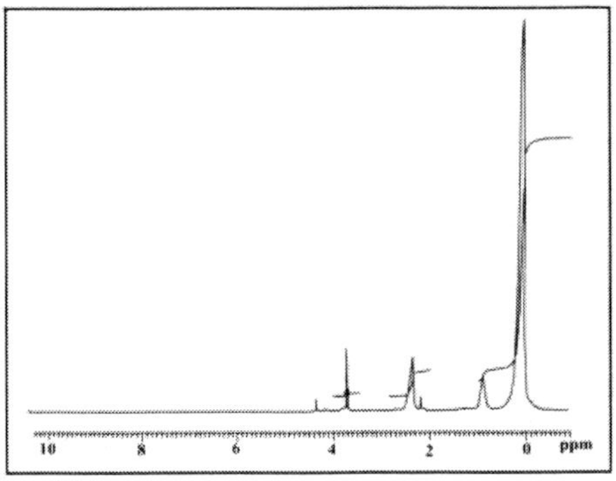

Figure 1. ^1H NMR spectrum of oligomer I.

In 13C NMR spectra of oligomer I (Figure 4) one can observe resonance signal characteristic for carbon: in carboxyl group at δ≈178.3 ppm, in ≡SiMe- group at δ≈7.5 ppm, in -SiMe3 group at δ≈7.28 ppm. Resonance signal at δ≈25.7 ppm is characteristic for carbon atoms in the group -CH2- attached at silicon atom; signal at δ≈56.7 ppm is characteristic for carbon atoms in the –CH2-CO- group.

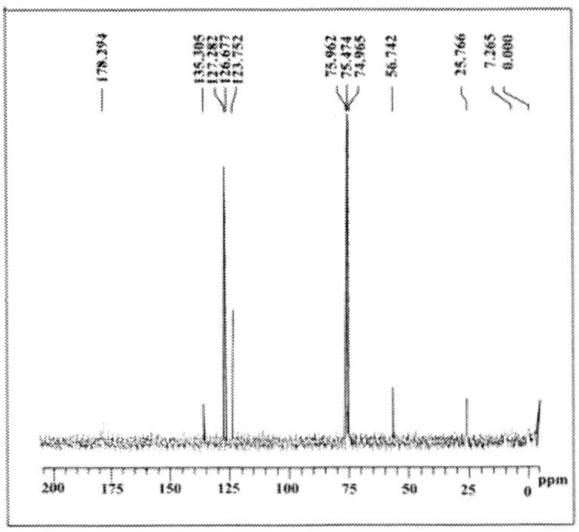

Figure 2. ^{13}C NMR spectrum of oligomer **I**.

As for[1]H NMR spectra of oligomer - II, here processing the reaction according to the Markovnikov rule has the least steric effect caused by methyl groups and the greatest size of reactant siloxane chain. About a least probability of processing of the reaction under this direction confirmed that in spectra, intensity of singlet signal of methyl group is smaller, in comparison with intensity of doublet signal and equal to 1:2, i.e. Hydrosilylation by Markovnikov rule proceeds ~ 33% and by anti-Markovnikov rule ~ 67%.

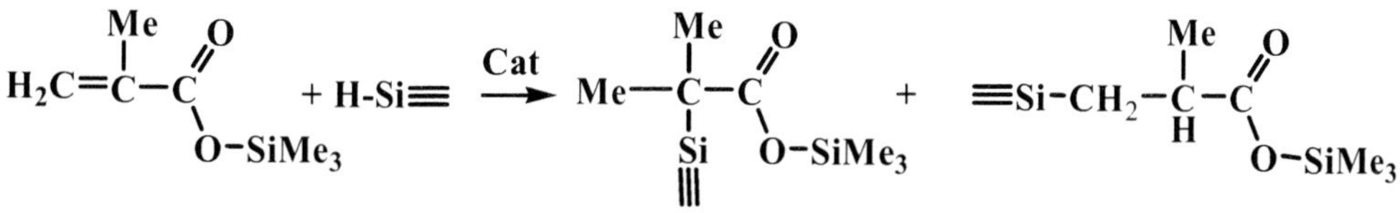

I.e. one of the possible variants is hydrosilylation against Markovnikov rule.

However, as in the previous case, here 1,4-addition to silylated esters with formation of transition complex most likely to take place. The obtained transition complex then undergoes regrouping according to the Eltekov rule. Because of that, ^{1}H NMR spectra of mentioned adduct, is characterized by following data (Figure 5); in structure XIII^a for methyl protons of ≡Si-Me and ≡SiMe$_3$ groups one can observe singlet signals with chemical shift δ≈0.07 and δ≈0.15 ppm accordingly. The singlet signal with chemical shift δ≈1.27 ppm characterized methyl protons in Me-C≡ group. In the case of the structure XIII^b one can observe the singlet signals characteristic for methyl groups at silicon with chemical shift δ≈0.07-0.1 ppm; for methyl protons (Me-C) doublet signal with chemical shift δ≈1.23 ppm (J≈5.28 Hz); for ≡Si-CH$_2$ group doublet signal at δ≈0.97 ppm and for CH group multiplet signal at δ≈2.6 ppm.

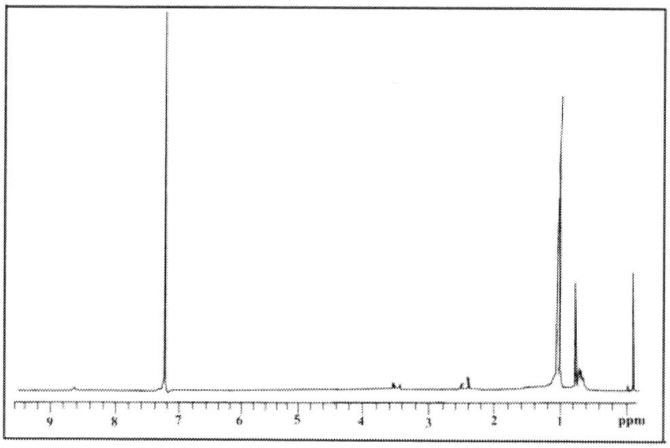

Figure 3. ^{1}H NMR spectra of oligomer II.

In ^{13}C NMR spectra of this oligomer (Figure 6) one can observe two group of signals in case of structure XIII^{a}for carbon: –COO– resonance signal at 171.57 ppm, Me – 19.02 ppm, for quaternary carbon ≡C- - 388 ppm, ≡Si-Me – 1.3 ppm and -SiMe$_3$ – 14.4 ppm. In case of structure XIII^b one can see resonance signal for carbon: –COO–167.6 ppm, Me – 18.2 ppm, ≡Si-CH$_2$– 19.4 ppm, =CH – 33.9 ppm, ≡Si-Me 1.29 ppm and for -SiMe$_3$ – 13.98 ppm.

Consequently, the NMR spectra data unambiguously confirmed that in case of trimethylsilylated ester of methacrylic acid the reaction proceeds in both directions.

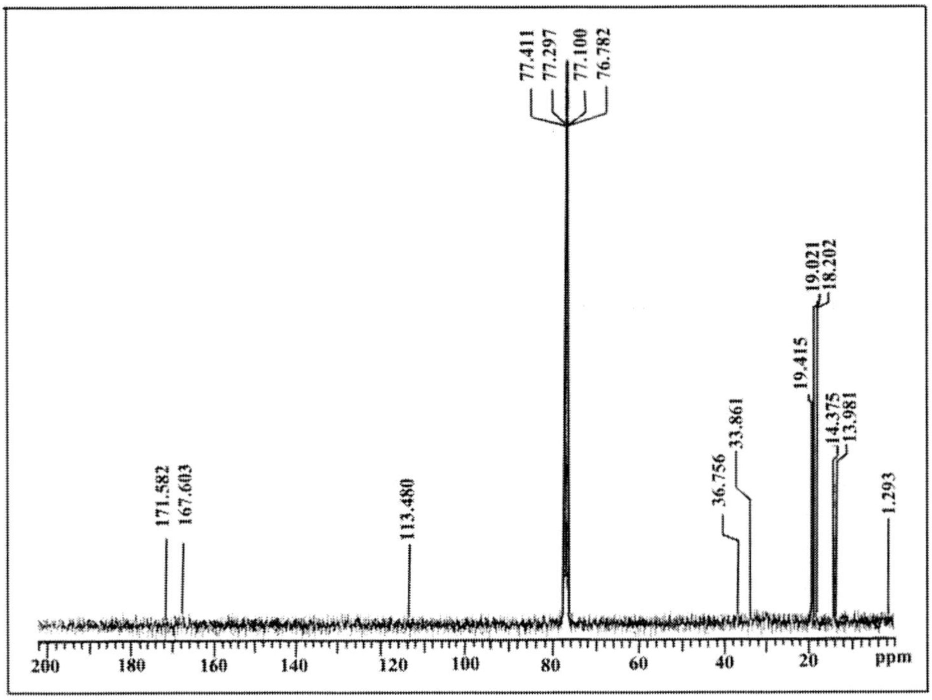

Figure 4. ^{13}C NMR spectra of oligomer II.

^{1}H and ^{13}C NMR spectral investigation for oligomers III and IV analogically has been performed (Figure 7 and 8). During hydrosilylation of methylhydridesiloxane to triethoxysilylated esters of acrylic and methacrylic acids obtaining of 1.2- and 1.4-addition products is possible:

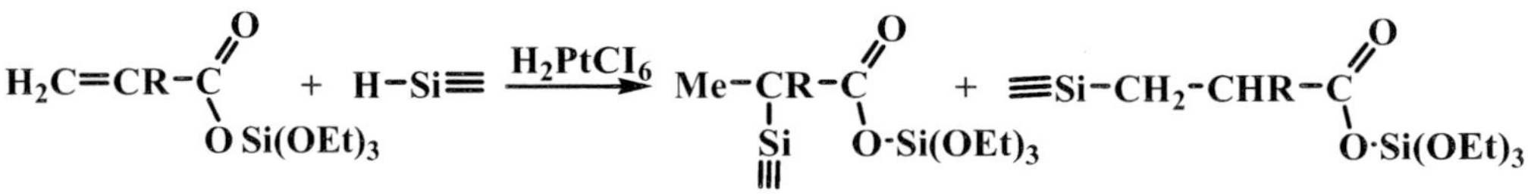

III^{a} - 33% III^{б} - 67%
IV^{a} - 26% IV^{б} - 74%

According to the spectra (Figure 7) really we have two mixtures of adduct. For methyl groups of structure III^{a} one can observe a signal with a constant of spin-spin interaction $J \approx 7.62$ Hz - $\delta \approx 1.1$ ppm.

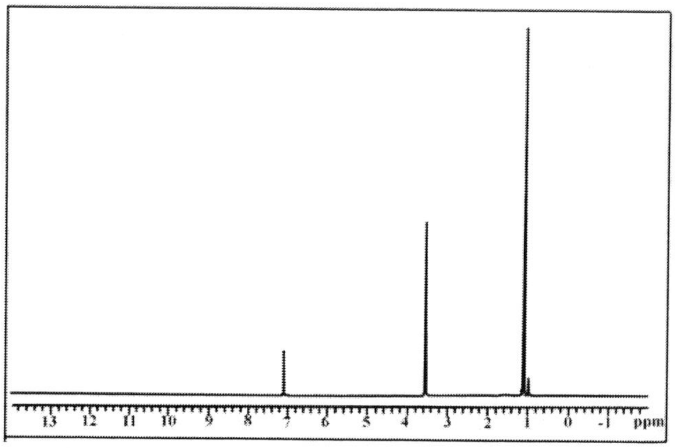

Figure 5. ^{1}H NMR spectrum of oligomer III.

Methine proton (CH) on which three protons of methyl groups influence, have quartet character with the same value of constant spin-spin interaction and resounds in δ≈2.16 ppm. Also in a spectrum are available methylene and methyl protons for the corresponding ethoxy-group with chemical shift at δ≈3.58 ppm. (quartet J=7.03 Hz) and δ≈1,1 ppm. (triplet J=7.03 Hz).

^{13}C NMR spectra of oligomer III (for structure III[a]), corresponds to a proton spectrum and contains five singlet signals. In the weaker field at 174.9 ppm the carbon atom of -COO- carboxyl groups resounds. The atom of tertiary carbon of methine groups, connected with atom of silicon, is characterized by chemical shift δ≈29.9 ppm. From two methyl groups that group which is part of the ethoxy-group, resounds in a stronger field δ≈15.5 ppm, than the methyl group connected to tertiary carbon (20,19 ppm). The atom of carbon of methylene groups, connected with the atom of electro negative oxygen, resounds in rather weaker fields with chemical shift δ≈60.4 ppm.

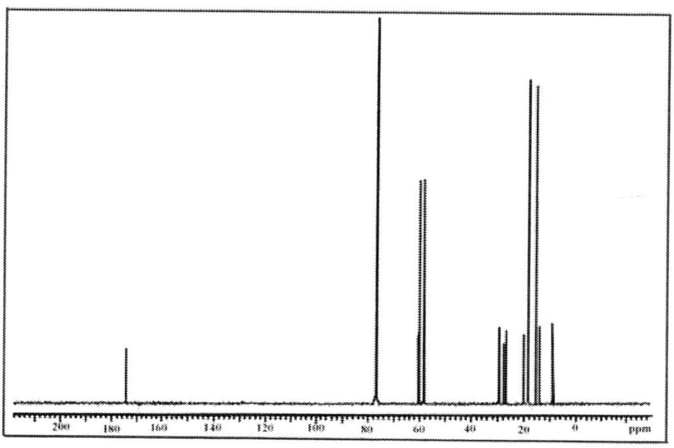

Figure 6. ^{13}C NMR spectra of oligomer III. (Solvent and standard - $CDCl_3$).

In ^{1}H NMR spectra of structures II, as opposed to structure II[a], its spectra does not contain signals of methyl group and is characterized by signals for the protons which are included in structure with the following chemical shifts: –Si-CH_2- δ≈1.24 ppm (triplet *J*=7.03 Hz), –CH_2-COO– - δ≈2.2 ppm (triplet *J*=7.03 Hz); Chemical shifts ethoxy-group are similar to chemical shifts of ethoxy-group, included in II[a] structure.

^{13}C NMR spectra of oligomer II with the structures II[b], is also submitted from five singlet signals. From them in the stronger field the primary carbon of methyl group included in the ethoxy-group, with chemical shift δ≈14.5 ppm resounds. The atom of secondary carbon connected to atom of silicon, resounds at δ≈23.3 ppm, and for methylene group connected with carboxyl -COO- group, having anelectronegative mesomeric effect, the resonance signal is observed at δ≈27.4 ppm. Chemical shift of carbon atom of carboxyl groups corresponds to structure II[a]. As for trimethylsilyl- and Me-Si≡ groups, primary carbon atoms of methyl groups are shown in the stronger field 9.4 ppm. In tables 5 and 6 the NMR spectral data for II[a] and II[b] structures are presented.

Table 5.[1] H NMR spectral data for oligomer II.

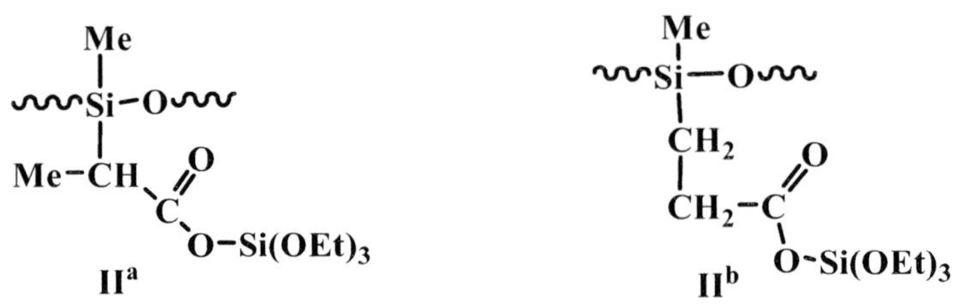

Structure	-Me	CH-Si≡	OCH_2-	≡SiCH_2-	-Me	-CH_2-	Me-Si	%
II[a]	1.1 d J=7.62	2.16 q J=7.62	3.58 q J=7.03	-	1.1 t J=7.03	-	0.07s	72.3
II[b]	-	-	3.58 q J=7.03	1.24 t J=7.03	1.57d J=7.27	2.2 t J=7.03	0.07s	27.7

Table 6.[13] C NMR spectral data for oligomer II.

Structure	Me	CH-Si≡	OCH_2	-COO-	≡SiCH_2-	-CH_2-	Me	$SiMe_3$	SiMe
III[a]	20.19	29.9	60.04	174.9	-	-	15.5	9.4	9.2
III[b]	-	-	-	174.9	23.2	27.4	14.5	9.4	9.2

The product of hydrosilylation of methylhydridesiloxane to triethoxysilylated ester of methacrylic acid by Markovnikov rule is formed in a small amount (1:2.8), in comparison with a product of 1,4-addition by anti-Markovnikov rule. This fact is explained by steric effect of the methyl group, which are included in a molecule triethoxymethacryloxysilane. Thus, the hydrosilylation reaction proceeds on two directions:

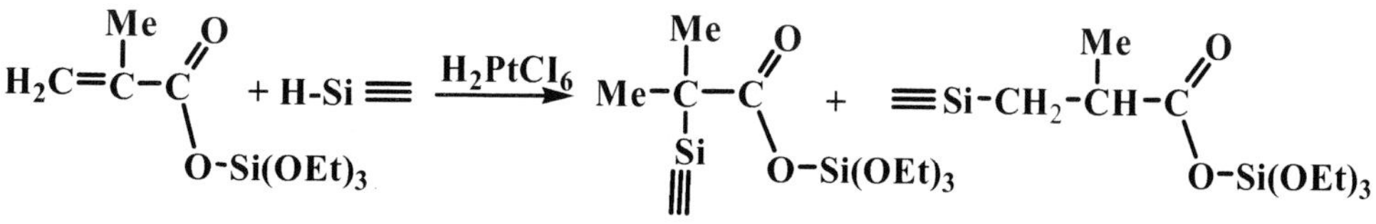

III[a] III[b]

In ^1H NMR spectra of compound III (structure III[a]) protons of methyl groups connected with a quaternary carbon atom gave signals with chemical shift δ≈1.05 ppm. The methylene and methyl protons included in ethoxy-group, give signals with chemical shifts δ≈3.96 ppm, (quartet, *J*=7.23 Hz) and δ≈1.08 ppm, (a triplet, *J*=7.23 Hz) accordingly (table 7, Figure 9). ^{13}C NMR spectra of the same oligomer (structure III[a]) proved the existence of five atoms. From these the atom of the carbon which is included in carboxyl group resounds in a maximum weaker field with chemical shift at δ≈172.0 ppm.

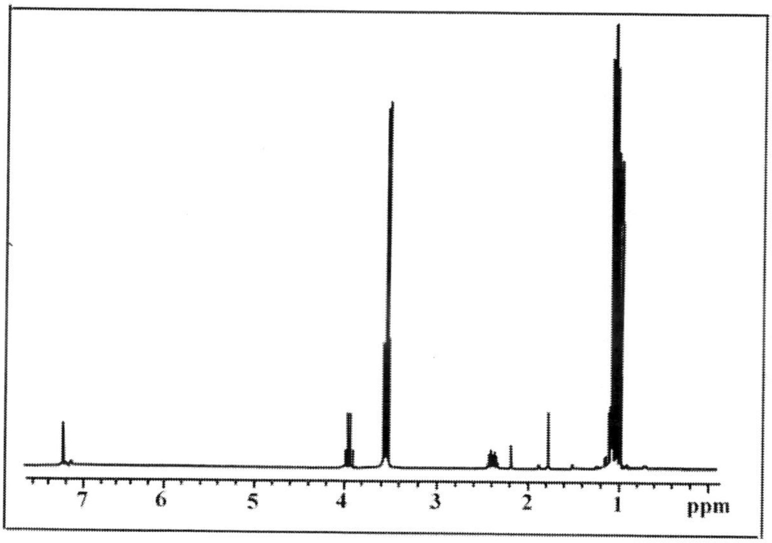

Figure 7. ^1H NMR spectra of oligomer IV. (Solvent and standard $CDCl_3$).

The chemical shift of the carbon in methylene group connected to the atom of electronegative oxygen, was δ≈60.4 ppm and the chemical shift of the quaternary carbon atom was δ≈34.2 ppm. For atoms of primary carbon of methyl groups, connected with a quaternary carbon atom, the chemical shift observed was δ≈19.26 ppm, and chemical shift for carbon atoms of methyl groups included in ethoxy-group - δ≈18.3 ppm (table 10).

In ^1H NMR spectra of a 1,4-addition (structure III[b]) proton of methylene groups, connected with an electropositive atom of silicon, give a doublet signal with chemical shift δ≈1.11 ppm, with a constant of spin-spin interaction *J*=7.04 Hz. The signal of a proton methine group has a sextet character with the center of chemical shift δ≈2.4 ppm, and thetriplet character had the protons of methyl groups connected with methine group, with chemical shift δ≈1.08 ppm. In ^{13}C NMR spectra of 1,4-addition of oligomer III (structure III[b]) the carbon atom of carboxyl-groups resounds in maximum weak area of a field with chemical shift δ≈178.0 ppm. For a carbon atom of methylene group connected to an electronegative atom of oxygen, chemical shift at δ≈58.5 ppm is observed.

For an atom of carbon of methylene group connected to electropositive atom of silicon, chemical shift δ≈29.2 ppm. For tertiary atom of the carbon which is included in methine group, chemical shift at δ≈33.9 ppm is observed. The methyl group connected to the tertiary carbon atom of a methyl group, included in ethoxy-group, results in a weaker field (δ≈19.1 ppm and δ≈18.5 ppm). The atoms of carbon which are included in trimethylsiloxy- and ≡Si-Me groups, the chemical shift at δ≈9.1 ppm is observed. In tables 7 and 8 ^{1}H and ^{13}C NMR spectra data of all structural isomers, included in IV oligomer are presented.

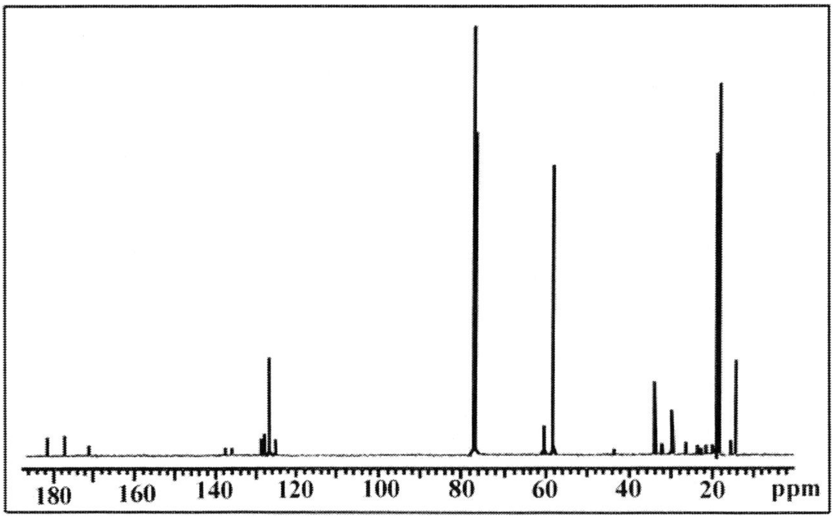

Figure 8. ^{13}C NMR spectra of oligomer IV.

Table 7. ^{1}H NMR spectral data for oligomer III

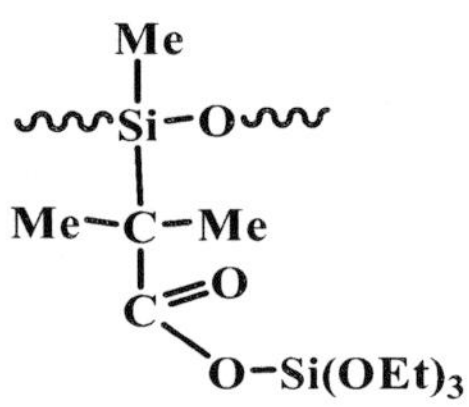

III^{a}

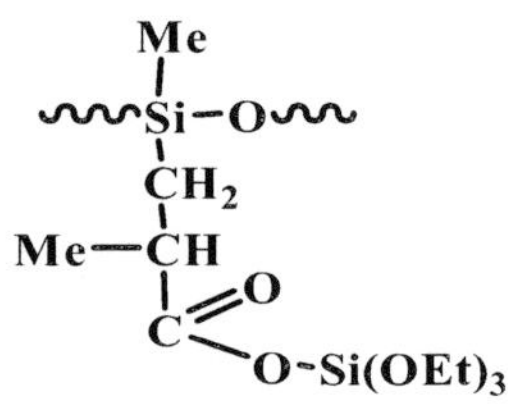

III^{b}

Structure	C-Me	Me	OCH_2-	$\equiv SiCH_2$-	Me	-CH-	Me-Si	%
III^{a}	1.05 s J=7,62	1.08 t J=7.03	3.96 q J=7.23	-	-	-	0.07s	26.4
$III^{б}$	-	1.08 t J=7.03	3.56 q J=7.03	1.11 d J=7.04	1.0 J=7.03	2.4 sex J=7.03	0.07s	73.6

Table 8.^{13}C NMR spectral data for oligomer III

Structure.	Me-C	-C≡	OCH_2	-COO-	$\equiv SiCH_2$-	Me	Me	$SiMe_3$	SiMe
III^a	19.26	34.2	60.4	172.0	-	18.3	-	14.2	9.1
III^b	-	-	58.5	178	22.2	18.5	19.1	14.2	9.1

In Figures 11 and 12 the curve of the dependences of inverse concentration on temperature are presented. It is established, that hydrosilylation reaction of methylhydridesiloxane to trimethylsilylated esters of acrylic and methacrylic acids has the second order. For hydrosilylation reactions the reaction rate constants and activation energy have been calculated: $k_{40}{}^{0}{}_{C}\approx 2.2217\times 10^{-1}$, $k_{50}{}^{0}{}_{C}\approx$ 2.3528×10^{-1}, $k_{60}{}^{0}{}_{C}\approx 2.3528\times 10^{-1}$ l/mol·s (trimethylacryloxysilane); $k_{80}{}^{0}{}_{C}\approx 1.0330$, $k_{85}{}^{0}{}_{C}\approx 2.3528$, $k_{90}{}^{0}{}_{C}$ $\approx$ 1.9484 l/mol·s (trimethylmethacryloxysilane).

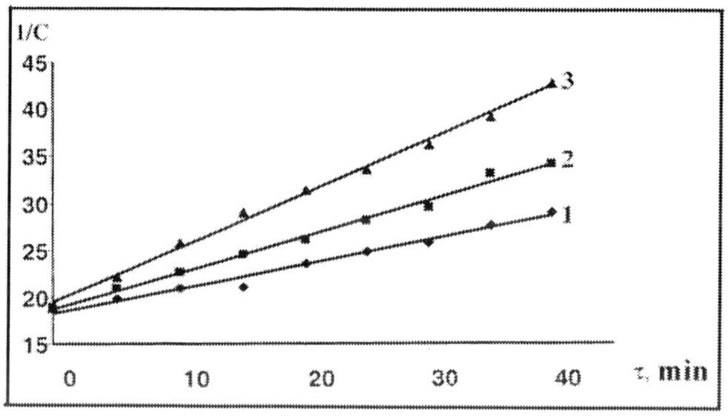

Figure 9. Dependence of inverse concentration on the time, during hydrosilylation reaction of methylhydridesiloxane to trimethylsilylated ester of acrylic acid. Where curve 1 corresponds to 400C, curve 2 - 500C and curve 3 - 600C.

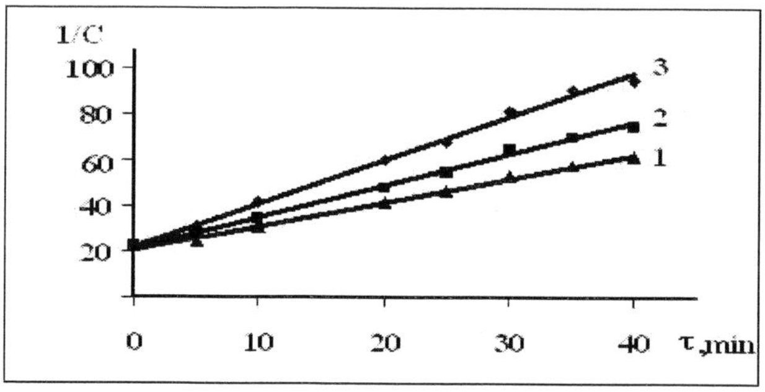

Figure 10. Dependence of inverse concentration on the time, during hydrosilylation reaction of methylhydridesiloxane to trimethylsilylated ester of methacrylic acid. Where curve 1 corresponds to 800C, curve 2 - 850C and curve 3 - 900C.

In Figures 13 and 14 dependence of the logarithm of rate constants on the inverse temperature during hydrosilylation reaction of methylhydridesiloxane to trimethylsilylated esters of acrylic and methacrylic acids are presented. From these reactions, the activation energies are calculated $E\approx 39{,}9$ kJ/mol (for trimethylacryloxysilane) and $E\approx 69{,}0$ kJ/mol (for trimethylmethacryloxysilane). Increased value of activation energy of hydrosilylation, in the case of trimethylmethacryloxysilane, is explained by steric factors.

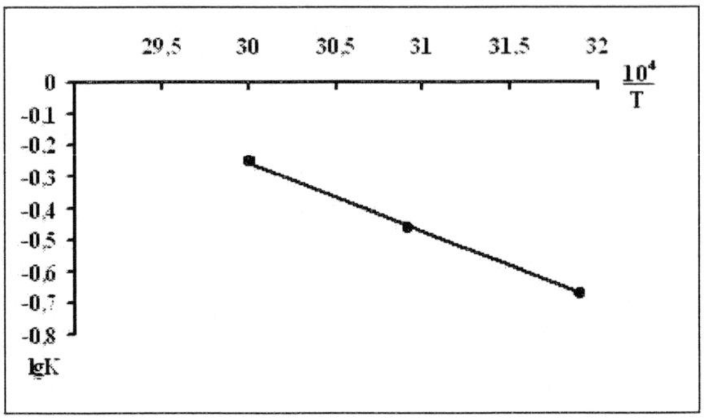

Figure 11. Dependence of logarithm of rate constants on the inverse temperature, during hydrosilylation reaction of methylhydridesiloxane to trimethylsilylated ester of acrylic acid.

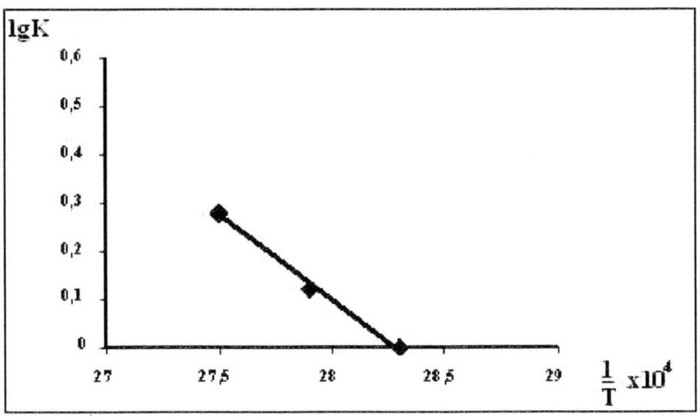

Figure 12. Dependence of logarithm of rate constants on the inverse temperature, during hydrosilylation reaction of methylhydridesiloxane to trimethylsilylated ester of methacrylic acid.

For synthesized oligomers the GPC investigations have been carried out and the average weight of molecular masses have been calculated. Their molecular masses are in the range $\overline{M}_\omega \approx 4.42 x 10^3 \div 4.55 x 10^3$

By thermomechanical investigations, the glass transition temperature for oligomer I and II have been determined with changes in the range $T_{gl} \approx -50 \div -72^0C$.

Wide angle X-ray investigation showed that synthesized oligomers are one-phase amorphous systems with value of interchain distance $d_I \approx 8.63$-8.65 Å.

For synthesized oligomers with triethoxy side groups hydrolytic condensation has been carried out in the presence of ammonium hydroxide or ammonium fluoride. Transparent silica hydrogels are obtained. The hardening process proceeds during 12 h. Synthesized silica hydro gels are perspective products for reception of silica aerogels.

Thus, for the first time hydrosilylation reactions of methylhydridesiloxane to trimethyl- and triethoxysilylated esters of acrylic and methacrylic acids in the presence of Pt/C or platinum hydrochloric acids had been investigated and methylsiloxane oligomers, containing trimethyl- and triethoxysilylated esters of propionic- and isobutyric acids in the side chain have been synthesized.

REFERENCES

[1] A.D. Petrov, S.I.Sadikh-Zade, E.I. Filatova. *Zhurn. Obshch., Khimii*, 1959, 29, 2936 (in Russian).

[2] S.I. Sadikh-Zade, A.D. Petrov. *Zhurn., Obshch., Khimii,* Chemistry and practical application siliconorganic compounds, part 1, Leningrad, 1958, 212 (in Russian).

[3] S.I. Sadikh-Zade, A.D. Petrov. Zhurn., Obshch. Khimii, 1959, 29 (10), 3194 (in Russian).

[4] R. Calas, N. Duffaut. Bull. Mens. Inform. *ITRG,* 1953, 7, 438.

[5] L.M. Sommer, F.P. Makray, D.W. Stevard, P.G. Gampbeh. *J. Amer. Chem. Soc.*, 1957, 79, 2764.

[6] J.I. Spier, J.A. Webster, G.H. Bernes. *J. Amer. chem.., Soc.*, 1957, 79, 974.

[7] B.P. 882101 (1961). (Union Carbide Corpor.). *"Chlorosilanes and derivatives from Acrylates".* Pike R.A., Black W.Th.

[8] Chengyou Kan, Xiangzheng Kong, Deshan Liu. *Polymer Preprints*, 2000, 41 (2),1243.

[9] V.P.Yuriev, I.M. Salimgariev. *"Hydrosilylation Reactions of Olefines"*, Printed By "Nauka", Moscow, 1982, 1-232 (in Russian).

[10] I.M. Gverdciteli, T.P. Doksopulo, E.I Chikovani. *Zhurnal Obshch. Khimii,* 1975, 47 (7), 352.

[11] S. Nitzsche. *Angew. Chem.,* 1960, 72 (3), 112 (in Russian).

[12] N. Pirtskheliani. Candidate dissertation, "Hydrosilylation reactions of α,ω-bis(trimethylsiloxy)methylhydridesiloxane to acrylic and methacrylic acids and their esthers", Tbilisi State University, 2004, Tbilisi Georgia.

[13] G.A. Razuvaev, L.M. Terman, B.A. Chudakova, I.N. Kostrikov. Vysokomol. Soed., 1978, 20A, (6), 1282 (in Russian).

[14] D.N. Andreev, E.V. Kukharskaia. *Zhurn. Obshch. Khimii,* 1960, 30, 2782 (in Russian).

[15] B.N. Dolgov, E.V. Kukharskaia, D.N. Andreev. *Vysokomol. Soed.*, 1960, 2(10), 1463 (in Russian).

[16] Patent USA, #2397287, 1946.

In: Modification Reactions of Oligomethylhydridesiloxanes ISBN: 1-60021-365-0
Editors: O.Mukbaniani, et al. pp. 57-75

Chapter 4

HYDROSILYLATION REACTIONS OF METHYLHYDRIDESILOXANE TO ACRYLATE AND METHACRYLATE

***O. Mukbaniani*[1]*, G. Zaikov*[2]*, N. Pirtskheliani*[1]*, T. Tatrishvili*[1] *and N. Mukbaniani*[1]**

[1]I. Javakhishvili Tbilisi State University, I. Chavchavadze Ave., 1, 380028 Tbilisi, Georgia

[2]Institute of Chemical Physics, Russian Academy of Sciences, Kosygin Str.4, 1117334 Moscow, Russia

ABSTRACT

The reaction of hydrosilylation of α,ω–bis(trimethylsiloxy)methylhydridesiloxane (n≈35) with acrylate and methacrylate at various ratios of initial compounds, in the presence of platinum hydrochloric acid, or platinum on the carbon Pt/C has been investigated. It was established that in the presence of platinum catalysts, hydrosilylation of all active ≡Si-H groups do not take place. During hydride addition the reaction order, reaction rate constants and activation energy were found.

IR and NMR spectra data determined the structures of synthesized oligomers. On the basis of NMR spectra data it was offered, that the hydrosilylation proceeds mainly with formation intermediate product, by regrouping of this transition complex the oligomers with various structures of molecules are obtained. It was shown that at high lengths of side-substituted groups, the NMR spectra proved the existence of conformers. Gel chromatographic permeation, differential calorimetric scanning and wide-angle X-ray analysis of synthesized oligomers were carried out.

Key words: acrylate, methacrylate, hydrosilylation and methylhydridesiloxane.

INTRODUCTION

In the literature, there is much information about hydrosilylation of acrylate and methacrylate [1-4]. A.D. Petrov investigated the reaction of hydride addition of triethylsilane to methyl acrylate in the presence of Spier catalyst and it was shown that in hydrosilylation reaction 1,2- or 1,4-addition depend on the nature of organic group attached at silicon and on the structure of carbonyl compounds.

For example hydrosilylation of triethylsilane to methyl acrylate proceeds with 1,4-addition, while hydrosilylation of trichlorosilane to methyl acrylate proceeds with 1,2-addition. Consequently it was established, that hydrosilylation ability in triorganosilanes gradually changes with replacement of alkyl groups by electronegative halogens [1-3]:

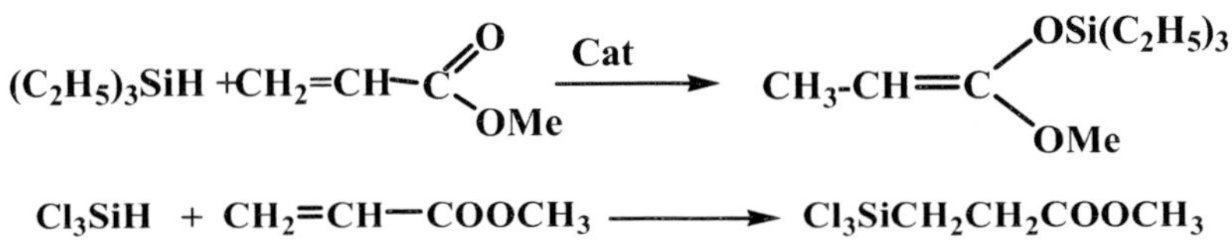

The hydride addition of trichlorosilane to methacrylate produces an α-isomer, which may explain why the electrophilic trichlorosilyl group attaches to the carbon atom which has more electronic density. In methyl acrylate electro donor methyl group in α-position appear nucleophilic properties in C=C fragment of β-carbon atoms [4,5].

Hydrosilylation of methyldichlorosilane to methyl acrylate in the presence of platinum on the carbon proceeds by 1,2- and 1,4-addition [1-3]:

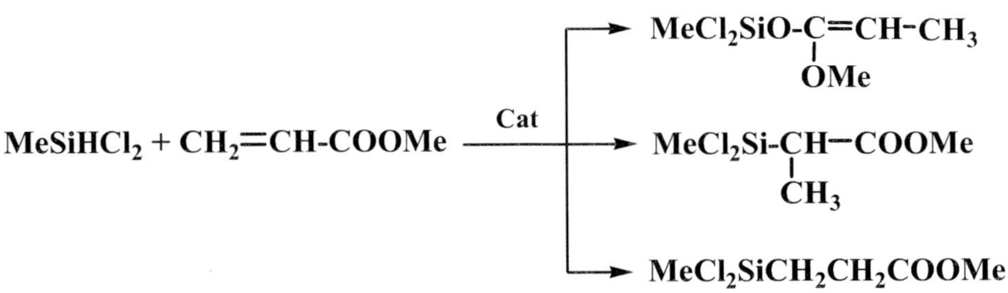

Hydrosilylation of triethylsilane to methyl methacrylate in the presence of catalyst Pt/C proceeds by 1,2-addition, with formation of triethyl(2-carbometoxypropyl)silane [1]:

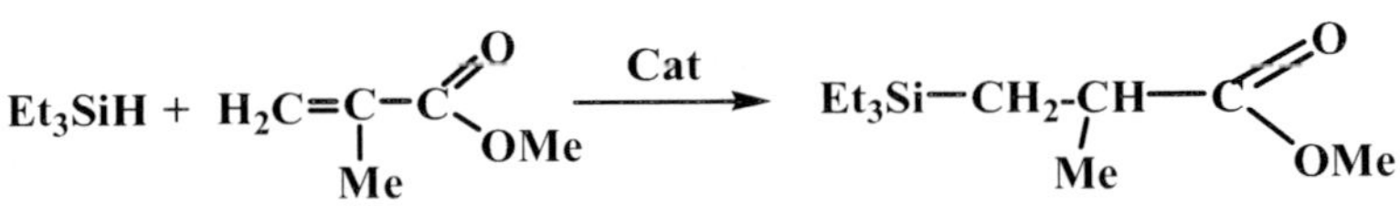

Hydrosilylation of methyldichlorosilane to acrylic esters in the presence of catalysts Pt/C or Pt/Al_2O_3 proceed with formation two isomeric products [6,7]:

$$CH_2{=}CHCOOR + MeSiHCl_2 \longrightarrow MeCl_2Si\text{-}\underset{CH_3}{\underset{|}{C}}HCOOR + MeCl_2Si\text{-}CH_2CH_2\text{-}COOR$$

In contrast to authors [1-7], hydrosilylation reactions of acrylate proceeds with formation of α, β and O-products according to the following scheme [8]:

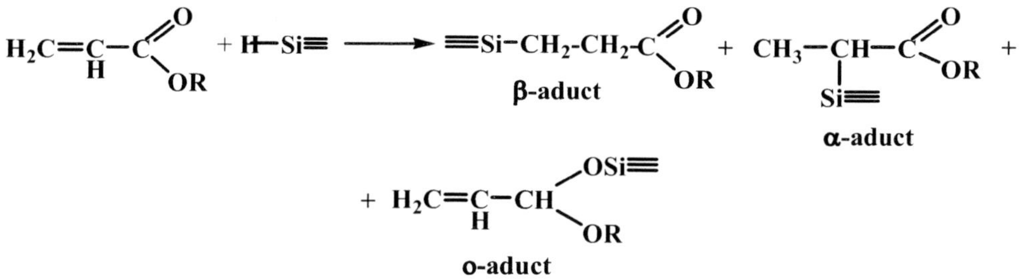

For methyl acrylate the yield of β-addition product is equal to -10.3 %, α-addition product – 57.8 % and o-addition product – 1.9 %. Substituted groups create a definite effect on formation of structure of hydrosilylation product.

The reaction of hydrosilylation of acrylate and methacrylate with 1,4-bis(dimethylsilyl)benzene in the presence of hydroquinone and Spier catalyst or platinum on the carbon was studied by the authors [9] and it was shown that the reaction proceeds with the formation of corresponding acyloxysilanes, according to the following scheme:

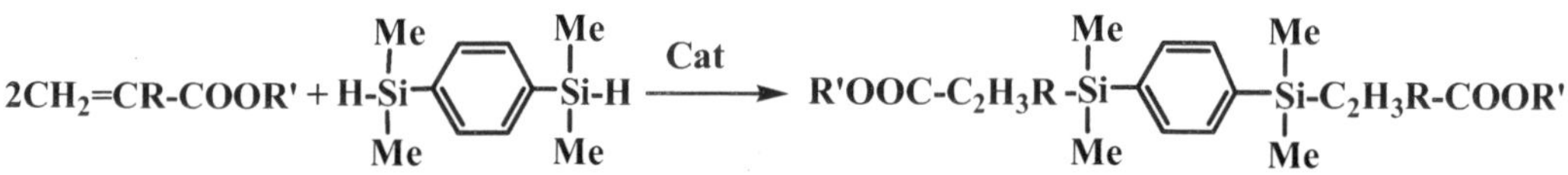

Where: R=H, Me; R'=Me, Et.

In hydrosilylation reaction the hydroquinone was used for preventing the polymerization reactions.

The reaction of hydrosilylation of methylhydridesiloxane to unsaturated vinyl derivatives (alkenes, acrylic acid, acrylic esters) in the presence of Spier catalyst has been investigated by the authors [10] and methylsiloxane oligomers with carboxyl, alkyl and ester side-groups have been obtained, according to the following scheme:

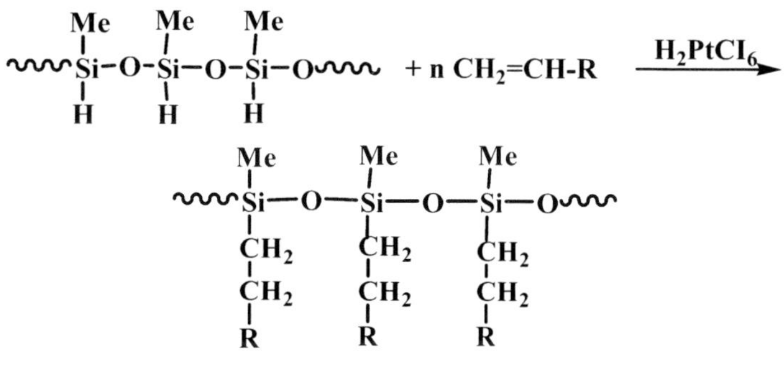

Where: R=C_nH_{2n}, -COOH, -COOAlk.

Therefore, in the literature there is contrasting information about the course and mechanism of hydrosilylation reaction of hydrideorganosilanes to acrylate and methacrylate.

Experimental Part

The starting materials for the synthesis of comb-type methylsiloxane oligomers were α,ω–bis-(trimethylsiloxy)methylhydridesiloxane, acrylate and methacrylates.

The initial α,ω–bis(trimethylsiloxy)methylhydridesiloxane with the degree of polymerization $n \approx 35$, acrylate and methacrylate were received from Fluka. Before reaction the acrylate and methacrylate were distilled or recrystallized. By drying and distillation the organic solvents used in reaction hydrosilylation were cleaned.

IR spectra of all the samples have been taken on an UR-20 instrument from KBr pellets, while the ^1H NMR spectra were taken on a "Perkin-Elmer" at an operating frequency of 250 MHz. All spectra were obtained with the use of $CDCl_3$ as solvent and internal standard. A Perkin – Elmer DSC-2 differential scanning calorimeter was used to determine TGA and the thermal transitions (T_g) were read at the maximum of the endothermic or exothermic peaks. Heating and cooling scanning rates were 10^0C/min.

Gel-permeation chromatography investigation was carried out with the use of Waters Model 6000A chromatograph with an R 401 differential refractometer detector. The column set comprised 10^3 and 10^4 Å Ultrastyragel columns. Sample concentration was approximately 3 % by weight in toluene and typical injection volume for the siloxane was 5 μL. Standardization of the GPC was accomplished by the use of styrene or polydimethylsiloxane standards with the known molecular weight.

Wide-angle X-ray diffractograms were taken on a "DRON-2" instrument. A-CuK_α was measured without a filter, the angular velocity of the motor $\omega \approx 2$ deg/min.

Hydride Addition of α,ω–Bis(Trimethylsiloxy)Methylhydridesiloxane to Methyl Acrylate

The hydrosilylation reaction was carried out in a three-necked flask equipped with a mechanical stirrer, an argon inlet, reflux condenser and calcium chloride drying tube. The initial reagents were placed in the flask and thermostated in an oil bath until constant temperature was achieved. 0.1 M solution of Platinum hydrochloric acid in tetrahydrofuran ($5 \div 9 x 10^{-5}$ g per 1.0 g of starting substance) was used as a catalyst. The reaction was carried out in an argon atmosphere in anhydrous toluene solution at a temperature of 60^0C. Then 5 ml anhydrous toluene was added to the reaction mixture, was filtered, connected to the vacuum and the toluene and unreacted methylacrylate were removed at 30 $\div 40^0$C.

The hydrosilylation reaction in the presence of platinum on the carbon with other acrylate and methacrylate has been carried out in the same manner.

RESULTS AND DISCUSSION

In the literature there are known comb-type methylsiloxane polymers with various organic branching fragments containing mesogenic groups in the side chain [11-13]. These polymers are characterized by liquid-crystalline properties. Besides in literature there are known diorganosiloxane polymers with liquid-crystalline properties not containing classical mesogenic groups. Such polymers corresponds methylpropyl-, diethyl- and dipropylsiloxane polymers with linear structure [14-16].

The present paper deals with synthesis and investigation of properties of comb-type methylsiloxane oligomers with various branching organic fragments. For the purpose of synthesis of siliconorganic comb-type oligomers with various ester groups in the side chain, the reaction of hydride addition of α,ω–bis(tri-methylsiloxy)methylhydridesiloxane to acrylate and methacrylate in the presence of 0.1 M solution of Platinum hydrochloric acid (in tetrahydrofuran), or platinum on the carbon Pt/C has been investigated. The reaction carried out at 1:35 ratio of initial compounds, at various temperatures (30-80^{0}C), as in melt condition so in dilute solution of dry toluene.

During the hydride addition, the changes of concentration of active ≡Si-H group on the time were observed. The hydride addition was investigated in dry toluene solution (C≈5.3x10^{-2} mol/l). It was established, that during the reaction not all active ≡Si-H groups completely take place in hydride addition. On the Figure 1 and 2 one can observe dependence of changes of concentration of active ≡Si-H.

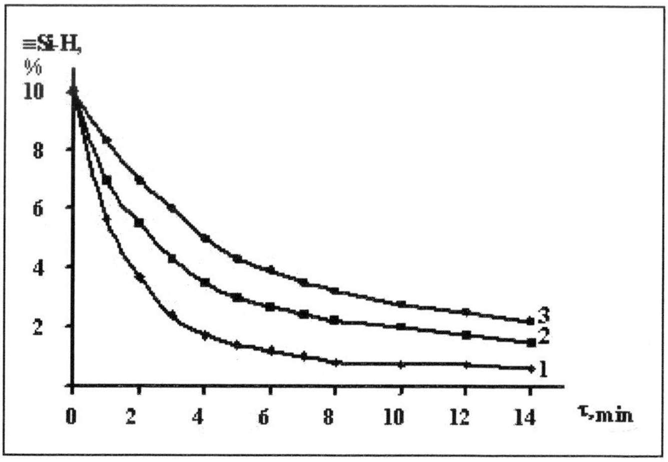

Figure 1. Dependence of changes of concentration of active ≡Si-H group on the time,upon hydride addition of α,ω–bis(trimethylsiloxy)methylhydridesiloxane to methyl acrylate, where curve 1 is at 50^{0}C temperature, curve 2 at 40^{0}C and curve 3 is at 30^{0}C.

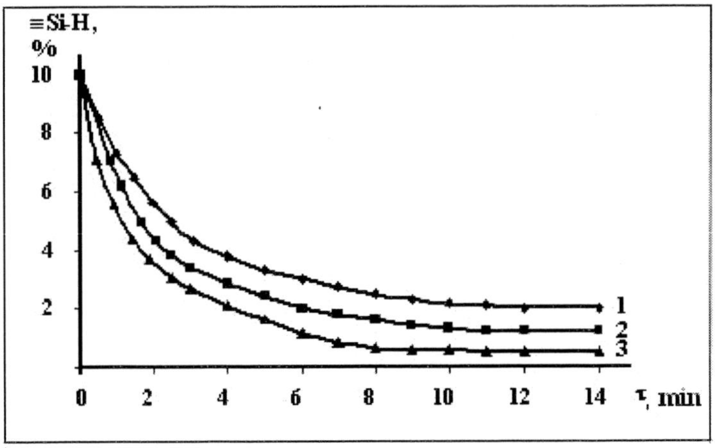

Figure 2. Dependence of changes of concentration of active ≡Si-H group on the time, upon hydride addition of α,ω–bis(trimethylsiloxy)methylhydridesiloxane to methyl methacrylate, where curve 1 is at 50^0C temperature, curve 2 at 55^0C and curve 3 is at 60^0C.

As it is seen from Figures 1 and 2 with an increase of temperature from 50^0C, to 60^0C the depth of hydride addition rises. At 50^0C temperature the depth of hydrosilylation in the time of methyl acrylate is equal to 95% and in the case of methyl methacrylate ~ 80%. Low rates of hydrosilylation in the case of methyl methacrylate may be explained by steric effect produced by isomeric structure. With an increase of the length of unsaturated esters the rate and the depth of hydrosilylation decreases. In case of butyl methacrylate at 60^0C hydrogen conversion is equal to ~78%.

So, hydride addition of methylhydridesiloxane to acrylate and methacrylate proceeds with formation of various-linked oligomers according to the following scheme:

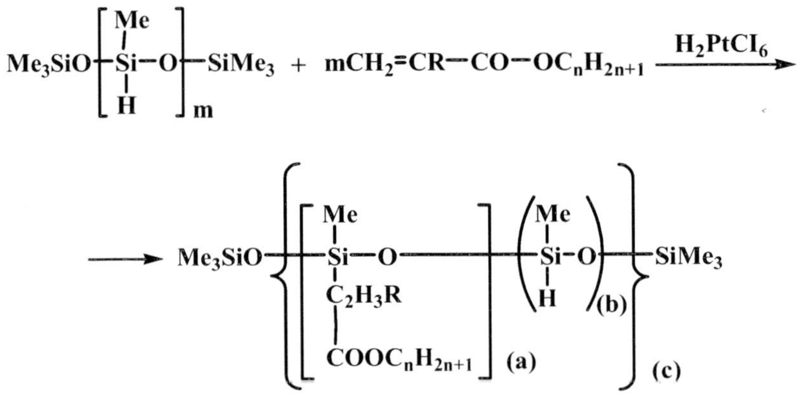

Where: [(a)+(b)](c)=m≈35; R=H; n=1 [Me - 30^0C-I^1, 40^0C - I^2, 50^0C - I]; n=2 (C_2H_5 - II); n=4 (n-C_4H_9 -III); n=6 (n-C_6H_{13} - IV); n=16 (n-$C_{16}H_{33}$ - V).

m≈35; R=Me, n=1 [Me - 50^0C(VI^1), 55^0C(VI^2), 60^0C(VI)]; n=4 [C_4H_9 - 60^0C(VII^1), 70^0C(VII^2), 80^0C (VII)].

Hydride addition proceeds vigorously during the first hour, then the reaction rate decreases. The synthesized oligomers are vitreous liquid or solid products, depending on the surrounding group, well soluble in ordinary aromatic type organic solvents with

η_{sp}≈0.03÷0.07. The structure and composition of oligomers, proven by elementary analysis, determination of molecular masses, IR, ^{1}H and ^{13}C NMR spectral data. Some physical-chemical properties of synthesized oligomers are presented in table 1 and 2.

Table 1. Some physical-chemical properties of synthesized comb-type oligomers.

#	Yield, %	R	η^*_{sp}	d_1, Å	T_g, ^{0}C	Elementary analysis, %**		
						C	H	Si
I^1	95	CH_3	0.05	8.63	-	*40.08* *39.79*	*6.98* *6.37*	*21.15* *20.87*
I^2	89	CH_3	0.04	-	-	*39.34* *39.00*	*6.62* *6.20*	*22.44* *22.08*
I	82	CH_3	0.03	-	-120	*38.51* *38.23*	*6.98* *6.60*	*23.26* *23.00*
II	95	C_2H_5	0.08	10.10	-93	*44.50* *44.12*	*7.61* *7.40*	*18.63* *18.28*
III	95	C_4H_9	0.05	-	-	*50.11* *49.79*	*8.54* *8.29*	*16.25* *15.68*
IV	93	C_6H_{13}	0,03	15,37	-87	*54,17* *53,76*	*9,23* *8,95*	*14,57* *14,21*
V	89	$C_{16}H_{33}$	0,05	27,1	-82	*66,02* *65,79*	*11,13* *10,86*	*9,14* *8,85*

Table 2. Some physical-chemical properties of synthesized comb-type oligomers.

#	Yield, %	R	η^*_{sp}	d_1, Å	T_g, ^{0}C	Elementary analysis, %**		
						C	H	Si
VI^1	76	CH_3	0.04	8,23	-	*43.25* *42.87*	*7.56* *7.21*	*20.05* *19.87*
VI^2	72	CH_3	0.04	-	-	*43.82* *43.24*	*7.57* *7.31*	*19.38* *19.07*
VI	80	CH_3	0.06	7.23	-110	*44.51* *44.11*	*7.59* *7.30*	*18.55* *18.70*
VII^1	73	C_4H_9	0.06	-	-	*51.38* *51.52*	*8.84* *8.60*	*16.21* *15.89*
VII^2	75	C_4H_9	0.06	-	-	*51.83* *51.74*	*8.87* *8.54*	*15.75* *15.79*
VII	70	C_4H_9	0.06	11.41	-88	*52.48* *52.32*	*8.91* *8.87*	*15.10* *14.78*

*In 1% toluene solution, at 25^{0}C.

**Over line calculated values, under line found values.

In the IR spectra of synthesized oligomers, one can observe an absorption band characteristic for asymmetric valence oscillation of ≡Si-O-Si≡ and ≡CO-O-C≡ bonds in the region 1020 and 1150 cm^{-1} respectively. Absorption bands of ≡Si-C≡ bonds at 1260 cm^{-1}, for valence oscillation of ≡C-H bonds at 2950-3100 cm^{-1} regions. It must be denoted that with an increase of the length of substituted organic side- groups the absorption intensity in this range rises. In the spectra, one can observe absorption bands characteristic for un-reacted ≡Si-H bonds at 2165 cm^{-1}.

For oligomers I-VII, ^{1}H and ^{13}C NMR investigation was carried out. As is seen from Figure 3, in ^{1}H NMR spectra of oligomer I, one can observe singlet signal for protons of methyl group in the fragment ≡SiMe and $-SiMe_3$ with chemical shift $\delta \approx 0.1$ ppm, for protons of methylene groups in the fragment $\equiv Si\text{-}CH_2\text{-}$, triplet signal with center of chemical shift $\delta \approx 0.85$ ppm; for protons of methylene group, in the fragment $-CH_2\text{-}CH_3$ with chemical shift $\delta \approx 1.05$ ppm. One can observe quartet signal for protons of methylene group in the fragment $-O\text{-}CO\text{-}CH_2\text{-}$, with chemical shift $\delta \approx 3.3$ ppm, this signals in the fragment $-CH_2\text{-}O\text{-}CO\text{-}$ undergo overlapping. Its displacement in low field is conditioned by influence of anisotropic cone of carbonyl group [17]. In spectra one can see low intensity singlet signal for protons of un-reacted ≡Si-H group with chemical shift $\delta \approx 4.2$ ppm.

In ^{1}H NMR spectra of oligomer II one can see singlet signals for methyl protons in the fragment $-SiMe_3$ and ≡Si-Me with center of chemical shifts $\delta \approx 0.1$ ppm. In spectra one can see triplet signal characteristic for methylene protons $-C^{1}H_2\text{-}$, $-C^{2}H_2\text{-}$ and $-C^{4}H_2\text{-}$ with center of chemical shift $\delta \approx 0.8$; $\delta \approx 3.6$ and $\delta \approx 3.95$ ppm accordingly (Figure 4).

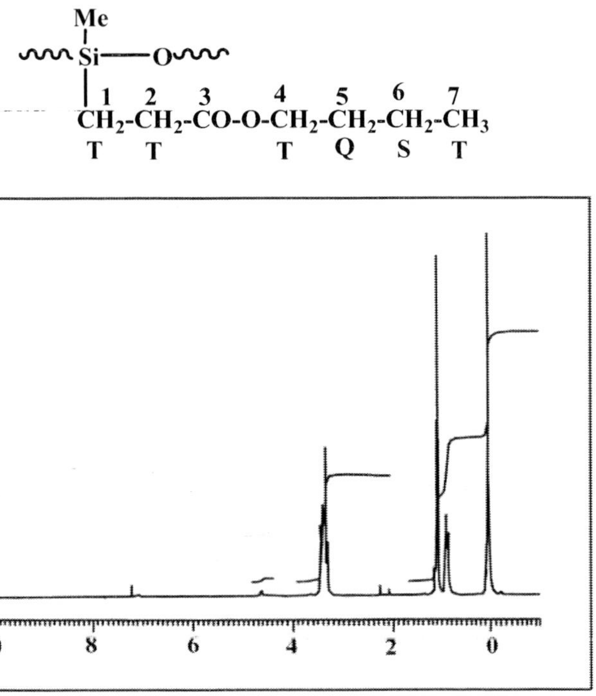

Figure 3. ^{1}H NMR spectrum of oligomer I.

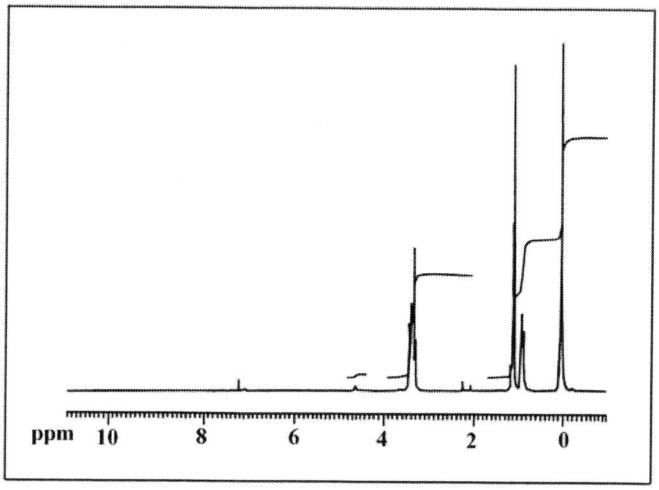

Figure 4. ^{1}H NMR spectrum of oligomer III.

In ^{1}H NMR spectrum of oligomer II one can see quintet signals characteristic for protons of methylene group in fragment -C^5H_2- with the center of chemical shift δ≈1.55 ppm; sextet signals for protons of methylene group -C^6H_2- with the center of chemical shift δ≈1.20 ppm. It must be denoted that quintet and sextet signals for protons of two methylene group -C^5H_2- and -C^6H_2- overlap each other and resonate in the range δ≈1.55-1.20 ppm. As is probably seen from the spectrum, there are conformational isomers, which appear because of lengthy organic substituted side-groups. In ^{1}H NMR spectrum of oligomer III one can see triplet signals with center of chemical shift δ≈1.0 ppm, characteristic for protons of methyl -C^7H_3 groups. Starting from ^{1}H NMR spectrum of oligomers hydride addition reaction mainly proceeds by anti-Markovnikov rule.

Presence of two groups of conformational isomers was verified by ^{13}C NMR spectral data (Figure 5), where signals appear as a pair. For example, for ≡Si-Me groups one can see the signals with chemical shifts δ≈10.17 ppm and δ≈10.56 ppm, for -C^1H_2- groups one can see the signals with chemical shifts δ≈14.73 ppm and δ≈14.89 ppm; for -C^7H_3 groups signal with chemical shift δ≈22.41 ppm; for -C^5H_2- - δ≈28.60; for –C^4H_2- signals with chemical shift δ≈65.11 and 68.29 ppm. It must be denoted that conformation of most was observed at this fragment.

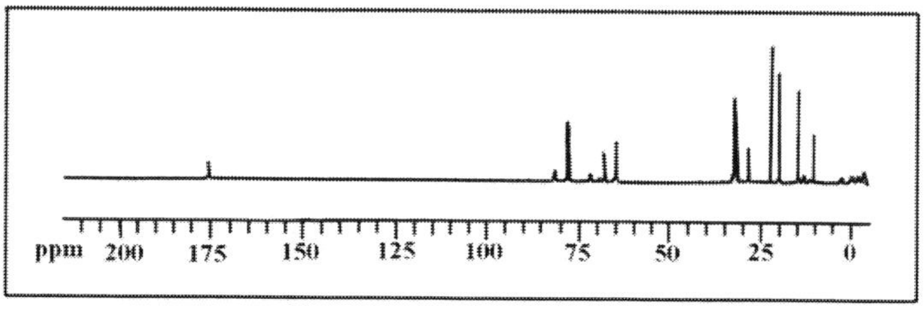

Figure 5. ^{13}C NMR spectrum of oligomer III .

In Figure 6 one can see ^{1}H NMR spectra of oligomer VI. From spectra it is evident that hydride addition of methylhydridesiloxane to methacrylate proceeds both by anti-Markovnikov rule and by Markovnikov rule, by 1,2- and 1,4-addition (with the ratio 1:1.3). Therefore, based on spectral data of oligomers, it was proposed that in general hydride addition of methylhydridesiloxane to methacrylate might proceed in several ways, see scheme 1:

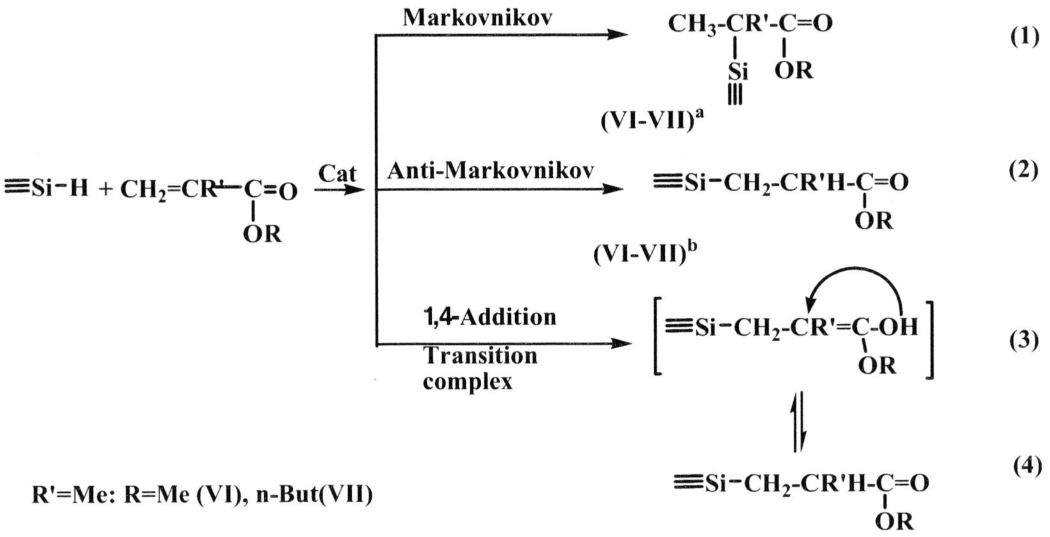

Scheme 1.The possible scheme of hydride addition of methylhydridesiloxane to acrylate and methacrylate.

As is seen from the possible scheme 1, in the case of 1,4-hydride addition reaction proceeds with ob-taining of intermediate-transition complex (3), by regrouping of this intermediate product according to the Eltekov rule (4), the product of 1,2-addition by anti-Markovnikov rule are obtained.

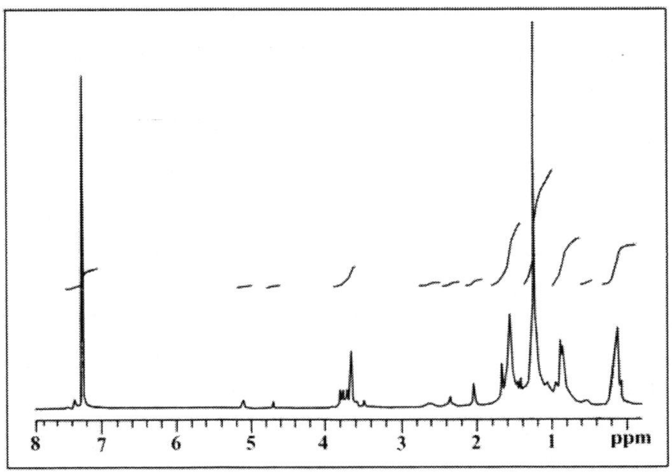

Figure 6. 1H NMR spectrum of oligomer VI.

In ^{1}H NMR spectrum of oligomer VI (Figure 6) one can observe a singlet signal characteristic for protons of methyl (Me-C≡, Markovnikov addition, scheme 1.1, structure VI^a) groups with chemical shift δ≈ 1.25 ppm, singlet signal of methyl protons (Me_3Si-) with chemical shift δ≈0.11 ppm; for methoxy groups (-COOMe) sharp singlet signal with chemical shift δ≈3.66 ppm. In ^{13}C NMR spectrum one can observe a resonance signal for ester carbonyl group in the weak field part, with chemical shift δ≈166.2 ppm; signal for methoxy groups (-COOMe) - δ≈41.4 ppm and resonance signal characteristic for quaternary carbon nucleus with chemical shift δ≈33.8 ppm; signal for methyl carbon (Me-CH) with chemical shift δ≈ 19.7 ppm (see Table 3 and 4).

Table 3. The chemical shifts of ^{1}H NMR spectrum of oligomer VI.

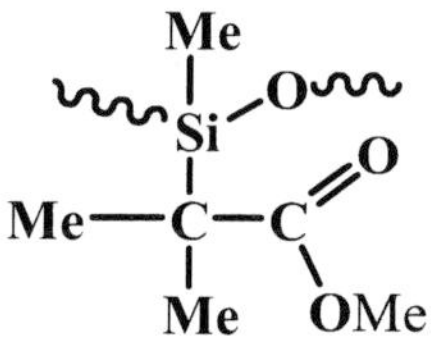

VI^a

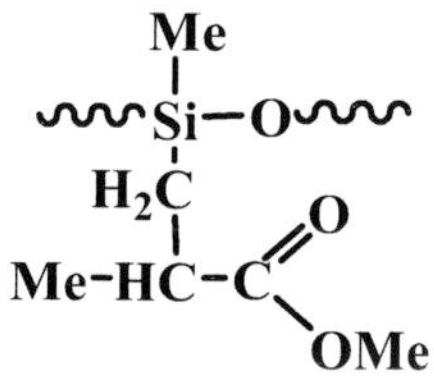

VI^b

Structure	Me-C	COOMe	Me_3Si	$SiCH_2$-	CH	Me-CH	Me-Si	%
VI^a	1.25s	3.66s	0.11s	-	-	-	0.07s	43
VI^b	-	3.72s	0.11s	0.84d J=6.62	2.6 sex	1.57d J=7.27	0.07s	57

Table 4. The chemical shifts of ^{13}C NMR spectrum of oligomer VI.

Structure	Me-C	-C≡	OMe	C=O	$SiCH_2$-	CH	Me-CH	$SiMe_3$	≡SiMe
VI^a	19.7	33.8	41.4	166.2	-	-	-	14.2	1.8
VI^b	-	-	41.2	166.2	18.2	30.2	19.7	14.2	1.8

For oligomers obtained by Anti-Markovnikov rule (scheme 1.2 structure VI^b) ^{1}H NMR spectrum contains a doublet signal for ≡Si-CH_2- groups with chemical shift δ≈0.84 ppm (J=6.62 Hz); sextet signal for protons of methine (≡CH) groups with chemical shift δ≈2.6 ppm; doublet signal for methyl protons (Me-CH) with chemical shift δ≈1.57 ppm and for methoxy groups resonance absorption at δ≈3.72 ppm.

In ^{13}C NMR spectrum of oligomer VI (structure VI^b, Figure 7) one can see the signal for ester carbonyl groups, which is analogically with structure VI^a resounds at δ≈166.2 ppm, additionally one can observe the resonance signal for methine (≡CH) groups with chemical shift δ≈30.2 ppm and for methyl groups a signal with chemical shift δ≈19.7 ppm. A signal with chemical shift δ≈18.2 ppm corresponds to methyllene groups in the fragment ≡Si-CH_2- (see table 4).

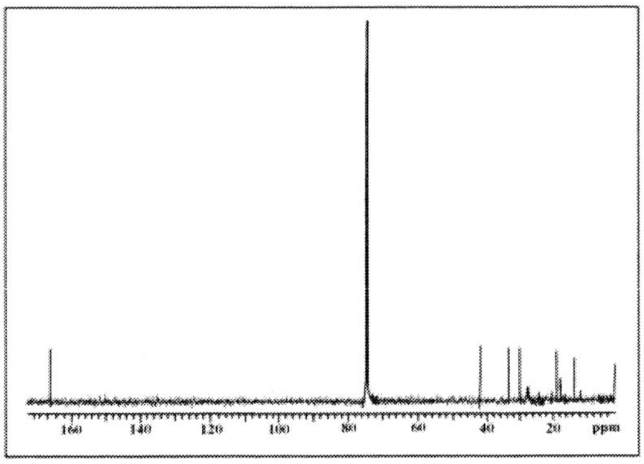

Figure 7. ^{13}C NMR spectrum of oligomer VI.

For oligomer VII analogically of oligomer VI, ^{1}H and ^{13}C NMR spectral investigation has been carried out. The results of NMR investigations are presented in tables 5 and 6.

Table 5. The chemical shifts of ^{1}H NMR spectrum of oligomer VII.

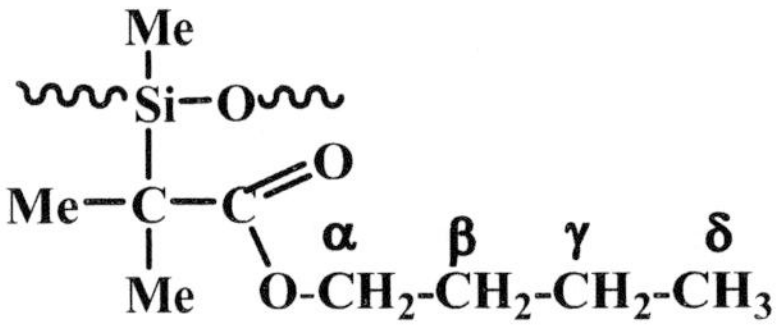

VII^{a}

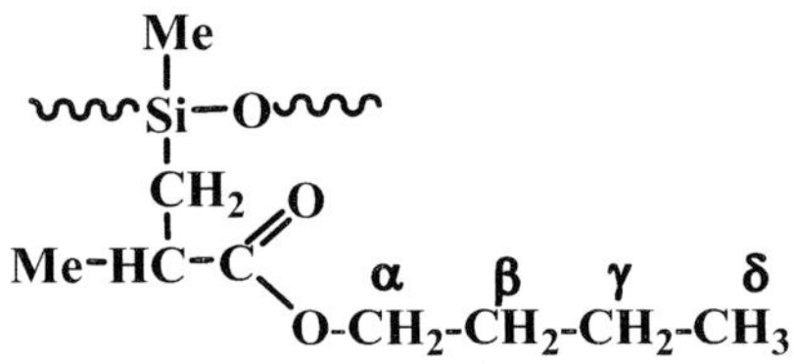

VII^{b}

#	Me	CH_2 (α)	CH_2 (β)	CH_2 (γ)	CH_3 (δ)	$SiCH_2$	CH	Me-CH	SiMe	$SiMe_3$	%
VII^{a}	0.88s	3.97m J=6.45	1.53 J=6.66	1.29q J=6.66 J=7.23	0.85m J=7.22	-	-	-	0.01	0.03	9
VII^{b}	-	3.99m J=6.64	1,53 J=6.66	1.29	0.85	0.66d J=7.62	2.51 m	1.13d	0.01	0.03	91

Table 6. The chemical shifts of ^{13}C NMR spectrum of oligomer VII.

#	Me	CH_2 (α)	CH_2 (β)	CH_2 (γ)	CH_3 (δ)	$SiCH_2$	CH	Me-CH	COO	SiMe	$SiMe_3$	-C≡
VII^{a}	19.0	64.6	19.4	21.9	14.1	-	-	-	177.1	1.9	2.03	34.8
VII^{b}	-	64.8	19.4	21.9	14.1	20.1	30.8	19.3	177.4	1.9	2.03	-

For full characterization of hydride addition of methylhydridesiloxane to methyl methacrylate by quantum-chemical half empiric AM1 method [18], for all initial, intermediate and final products, in model reaction of hydrosilylation of methyldimethoxysilane to methyl methacrylate, heats of formations (ΔH_f) and heat of reaction ($\Delta\Delta H$) proceeding with various concurrent scheme 2 were calculated.

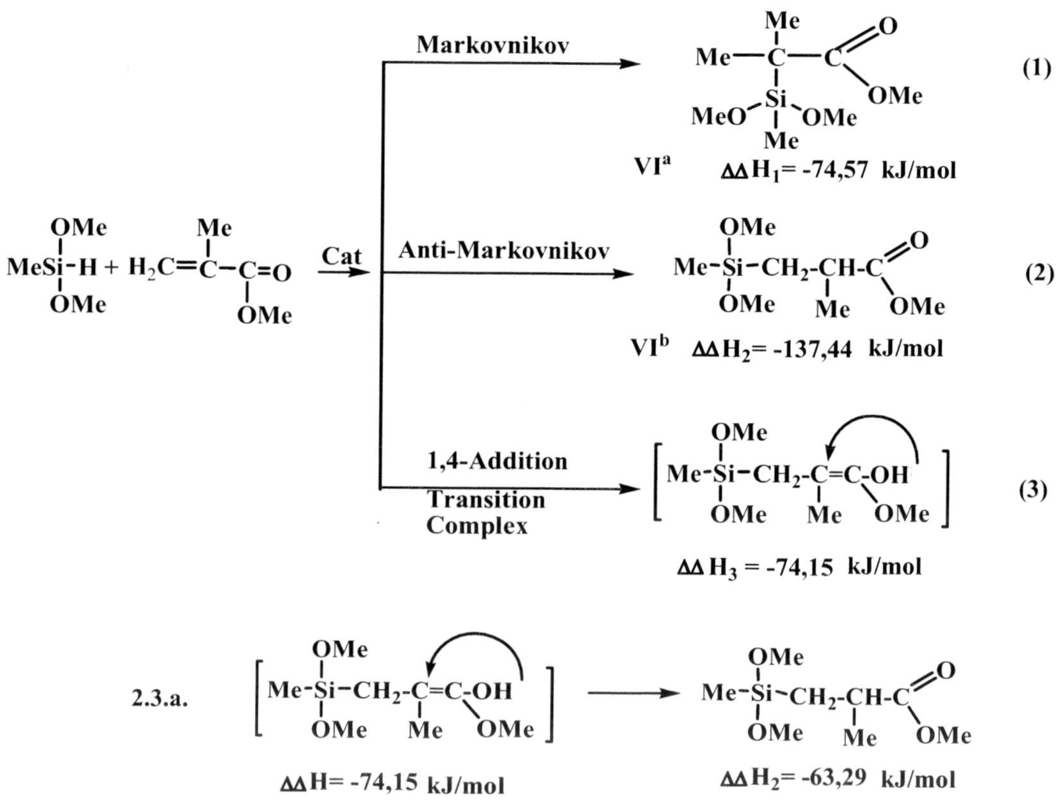

Scheme 2. The possible scheme of hydride addition of methylhydridesiloxane to methyl methacrylate.

As is seen from scheme 2, starting from that heat of reaction of hydrosilylation of methyldimethoxysilane to methyl methacrylate, thermodynamically the course of reaction according to the anti-Markovnikov rule (1,2-addition) is more profitable. It must be denoted that the product obtained by regrouping of the intermediate product (1,4-addition, scheme 2.3) is similar to the Eltekov rule (scheme 2.3a), identical to structure of the product obtained by anti-Markovnikov rule (scheme 2.2) and they have the same heat of reaction.

In Figure 8 dependence of reverse concentration of the reactant products on the time is presented.

Figure 8 shows dependence of reverse concentration on the time during hydride addition of methyl-hydridesiloxane to methacrylate. One can see that at the initial stages the hydrosilylation reaction is of second order. The hydrosilylation rate constants at various temperatures were calculated. For methacrylate: $k_{30^0C} \approx 0{,}3838$, $k_{40^0C} \approx 0{,}7813$ and $k_{50^0C} \approx 1{,}6003$ l/mol·s ($\gamma \approx 2$); for methyl methacrylate: $k_{50^0C} \approx$ 1,9314, $k_{55^0C} \approx 3{,}3286$ and $k_{60^0C} \approx 4{,}1006$ l/mol·s; for butyl methacrylate: $k_{60^0C} \approx 0{,}9754$, $k_{70^0C} \approx 1{,}5072$ and $k_{80^0C} \approx$ 3,3286 l/mol·s ($\gamma \approx 2$).

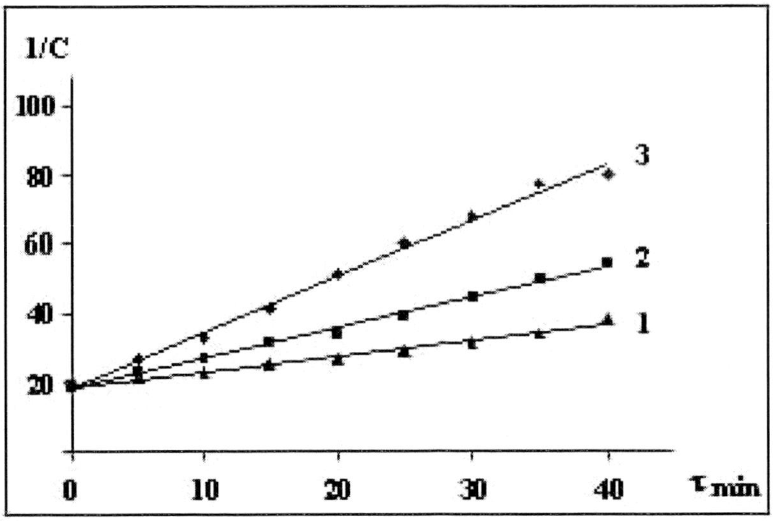

Figure 8. Dependence of the reverse concentration on the time, upon hydride addition of methylhydridesiloxane to methacrylate. Where curve 1 is at 30^0C, curve 2 is at40^0C and curve 3 is at 50^0C.

Figure 9 shows the dependence of hydrosilylation rate constants logarithm of the on the inverse temperature upon of hydride addition of methylhydridesiloxane to methyl acrylate. From these data the activation energy of hydrosilylation reaction was calculated. For methyl acrylate, it is equal to: $E_{act}\approx$ 55,7 kJ/mol; for methyl methacrylate - $E_{act}\approx$58,0 kJ/mol and for butyl methacrylate - $E_{act}\approx$62,1 kJ/mol. As is seen from above-mentioned values, the activation energies a little distinguished from each other.

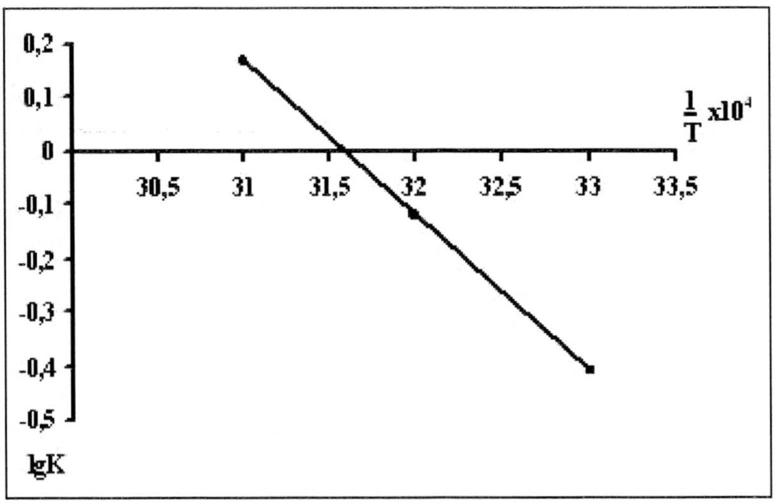

Figure 9. Dependence of hydrosilylation reaction rate constants logarithm of the on the in-verse temperature during hydride addition of methylhydridesiloxane to methyl acrylate.

The synthesized oligomers II-IV were studied by chromatographic gel permeation method. Figure 9-11 shows the molecular weight distribution of oligomers II, IV and V. From these Figures, it is evident that the oligomers have bi- and trimodal molecular weight distribution. The molecular masses of oligomers were determined.

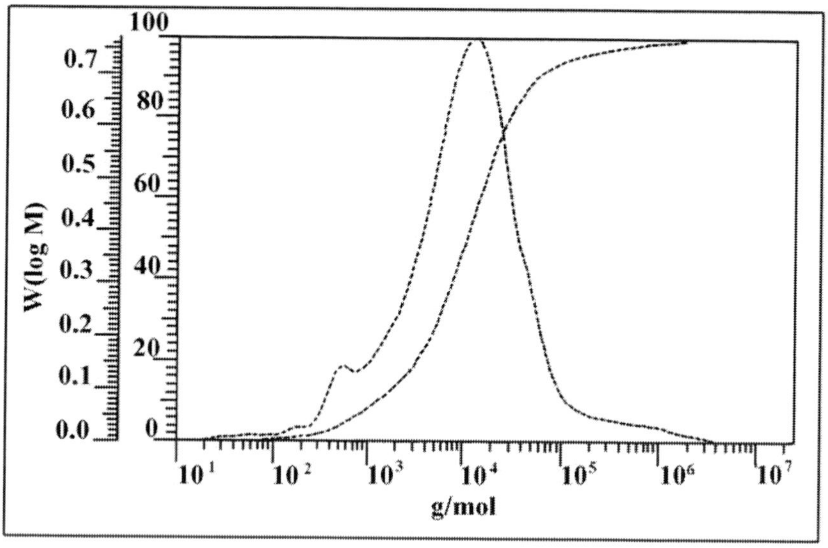

Figure 9. Gel permeation chromatographic curve of oligomer II.

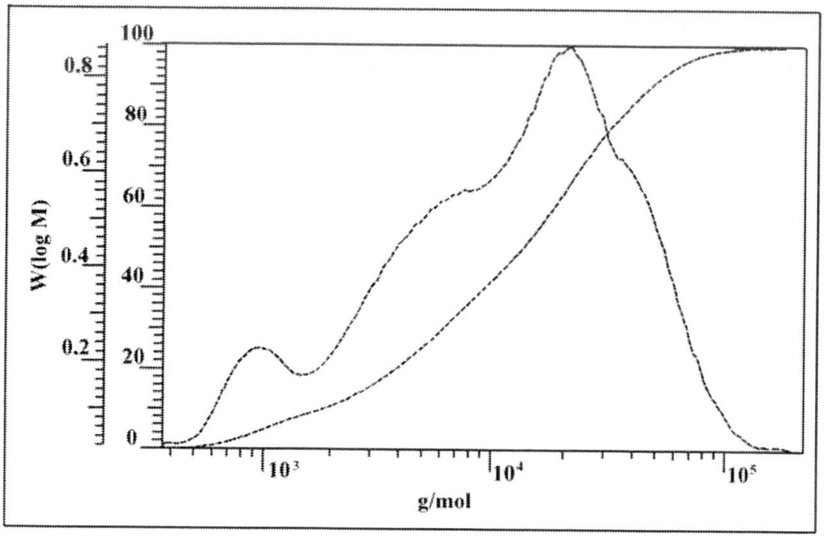

Figure 10. Gel permeation chromatographic curve of oligomer IV.

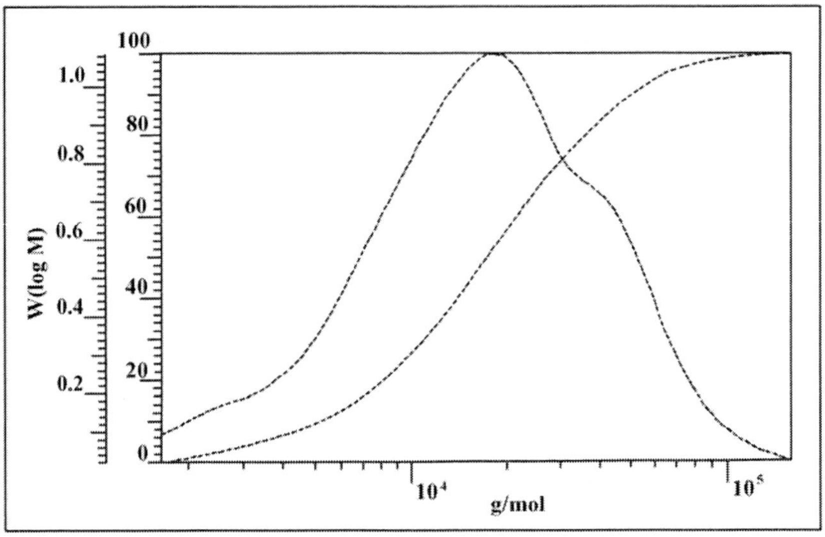

Figure 11. Gel permeation chromatographic curve of oligomer V.

The average molecular mass of oligomers II is~ $\bar{M}_\omega \approx 56.66x10^3$, $\bar{M}_n \approx 1.877x10^3$, $M_z \approx 2.6661x10^3$ and polydispersity D≈30. As it is seen from the values of average weight molecular masses it exceeds 11.4 times the theoretical calculated molecular mass (in case of full hydrosilylation). Average molecular masses for oligomers IV: $\bar{M}_\omega \approx 4.587x10^4$, $\bar{M}_n \approx 1.125x10^4$, $\bar{M}_z \approx$ $4.193x10^4$, D≈4.08; for oligomers V: $\bar{M}_\omega \approx 4.901x10^4$, $\bar{M}_n \approx 2.00x10^4$, $\bar{M}_z \approx 4.219x10^4$, D≈2.09 were determined analogically.. Increased molecular masses proved that the hydrosilylation reaction proceeds partially intermolecular reaction on the carbonyl group, which is proved with literature data [1]. It must be denoted that in the work [1] was not used for NMR investigation, which will uniquely be proved the structure of obtained products.

So, on the basis of NMR spectra and GPC data it is possible to prove that hydrosilylation mainly proceeds by 1,2-addition and partially proceeds intermolecular hydride addition of residual ≡Si-H gro-ups on the carbonyl group. Though, in NMR spectrum characteristic resonance signals for =CH-O-Si≡ bonds are not observed, which may be explained with low technical possibility of used spectrometer.

By wide-angle X-ray investigation in Figure 12, it was shown, that synthesized oligomers are one-phase amorphous systems. With an increase of the length of substituted side-groups in oligomers, the value of interchain distances rises from d_l=8.23 Å (oligomer VI) up to d_l=27.1 Å (oligomer V). The values of interchain distances of the above-mentioned oligomers are near to the value of interchain distances analogically structure organic polymers [19].

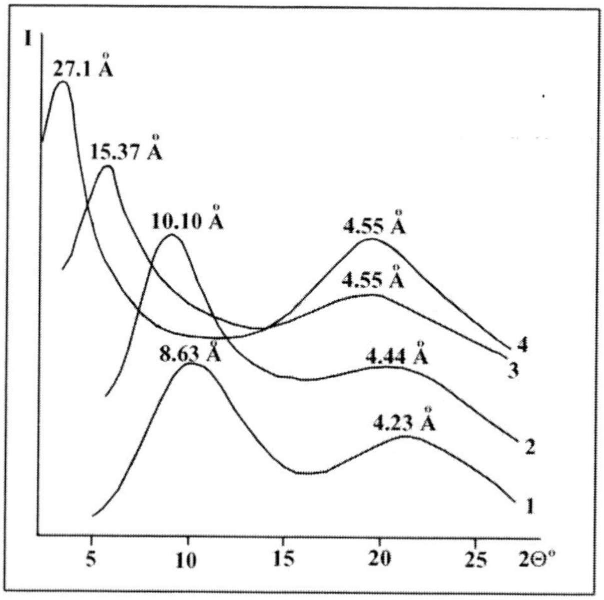

Figure 12. Wide angle X-ray diffraction patterns of oligomers, where curve 1 1 corresponds to oligomer I, curve 2 - II, curve 3 –IV and curve 4 – V.

By differential scanning calorimetric investigation it was shown, that in oligomers II and IV (Figure 13) one can observe one temperature transition in the range of -82÷-93^{0}C, which corresponds to the glass transition temperature of oligomers. On the calorimetric curve of oligomer V (Figure 14) one can see temperature transition at +24.9^{0}C.

Therefore, with an increase of the length of substituted side-groups the glass transition temperature rises. As is seen from DSC curves of oligomers, they do not reveal liquid-crystalline properties, in contrast to methylsiloxane oligomers containing various length oxyalkyl side groups [20].

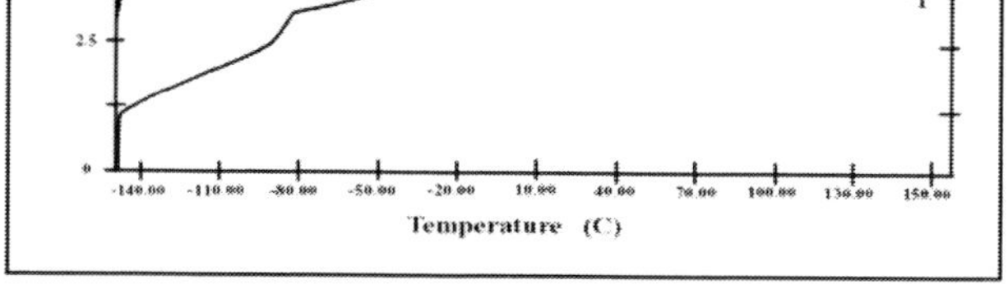

Figure 13. Differential scanning calorimetric curves of oligomers, where curve 1 corresponds to oligomer IV .and curve 2 – II.

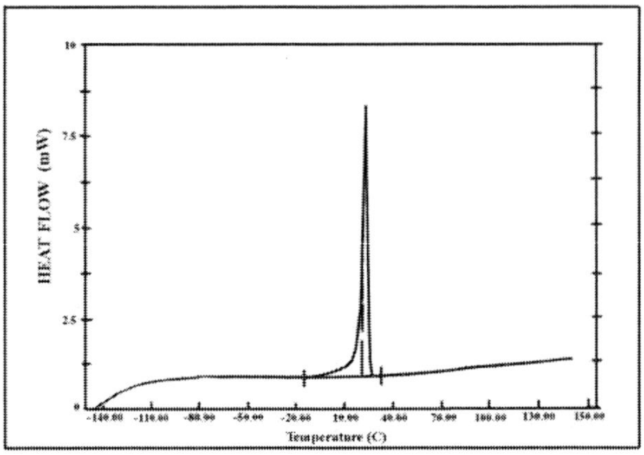

Figure 14. Differential scanning calorimetric curve of oligomer V.

So, for the first time the reaction of hydride addition of methylhydridesiloxane to acrylates and methacrylate has been studied and synthesized comb-type methylsiloxane oligomers with various length ester side groups through our research. They are interesting products, because by its hydrolysis it is possible to obtain water-soluble siliconorganic acids and their corresponding salts.

References

[1] A.D. Petrov, S.I. Sadikh-Zade and E.I. Filatova. *Zhurn. Obshch. Khim.,* 1959, 29, 2936 (in Russian).

[2] S.I. Sadikh-Zade and A.D. Petrov. *Chemistry and practical application siliconorganic compounds,* part 1, Leningrad, 1958, 1, 212 (in Russian).

[3] S.I. Sadikh-Zade and A.D. Petrov. Zhurn., *Obshch. Khim.*, 1959, 29(10), 3194 (in Russian).

[4] L.M. Sommer, F.P. Makray, D.W. Stevard, and P.G. Gampbeh. *J. Amer. Chem. Soc.*, 1957, 79, 2764.

[5] J.W. Curry and C.W. Harrison. *J. Org. Chem.,* 1958, 23, 627.

[6] J.I. Spier, J.A. Webster and G.H. Bernes. *J. Amer. Chem., Soc.*, 1957, 79, 974.

[7] B.P. 882101 (1961). (Union Carbide Corpor.). "Chlorosilanes and derivatives from Acrylates". Pike R.A., Black W.Th.

[8] K. Chengyou, K. Xiangzheng and L. Deshan. *Polymer Preprints,* 2000, 41(2), 1243.

[9] I.M. Gverdciteli, T.P. Doksopulo and E.I. Chikovani. *Zhurnal Obshch. Khim.*, 1975, 47(7), 352 (in Russian).

[10] S. Nitzsche. *Angew. Chem.,* 1960, 72(3), 112.

[11] H. Finkelmann, Y.J. Kock and G. Rehage Macromol. *Chem., Rapid Commun,* 1981, 2, 317.

[12] H. Rinsdorf and A. Schneller. Macromol. *Chem., Rapid Commun.*, 1982, 2, 557.

[13] N. Kawatsuki, S. Sakashita, K. Takatani, T. Yamomoto and O. Sangen. Macromol. *Chem., Phys.*, 1996, 197, 1919.

[14] Yu.K. Godovsky, I.I. Mamaeva, N.N. Makarova, V.S. Papkov and N.N. Kuzmin. Macromol. *Chem., Rapid Commun.,* 1985, 6, 797 .

[15] Yu.K. Godovsky, N.N. Makarova, V.S. Papkov, N.N. Kuzmin. *Vysokomol., Soed.,* 1985, 27B, 164 (in Russian).

[16] A. Blumshtein."Liquid Crystalline Order in Polymers", Edited by Academic Press, New York 1978, translated into Russian, *Mir,* Moscow 1981, 352.

[17] M. Hesse, H. Meier and B. Zeeh. "Spektroskopische Methoden in der Organischen Chemie". Georg Thieme Verlag, Stuttgart 1979, 480.

[18] M.I.S. Dewar, E.G. Zeobish, E.F. Healy and J.S. Stewart. *J. Am. Chem. Soc.*, 1985, 107, 3902.

[19] N.A. Plate and V.P. Shibaev V.P. "Comb-type polymers and liquid crystals"; Moscow, *"Khimia"*, 1980, 17 (in Russian).

[20] Gurgenidze G.N. Candidate dissertation, *"Synthesis and investigation of comb-type methylsiloxane oligomers"*, Tbilisi State University, Tbilisi Georgia 2003.

In: Modification Reactions of Oligomethylhydridesiloxanes ISBN: 1-60021-365-0
Editors: O.Mukbaniani, et al. pp. 77-94

Chapter 5

HYDROSILYLATION AND DEHYDROCONDENSATION REACTIONS OF METHYLHYDRIDESILOXANE TO ACRYLIC AND METHACRYLIC ACIDS

***O. Mukbaniani*[1], *G. Zaikov*[2], *N. Pirckheliani*[1], *and T. Tatrishvili*[1]**
[1]I. Javakhishvili Tbilisi State University, I. Chavchavadze Ave., 1, 0128 Tbilisi, Georgia
[2]Institute of Biochemical Physics Russian Academy of Sciences 4 Kosygin Street, Moscow 119991 (Russia)

ABSTRACT

The reaction of hydrosilylation of α,ω–bis(trimethylsiloxy)methylhydridesiloxane (n≈53) with acrylic acid at various ratio of initial compounds, in the presence of Platinum hydrochloric acid has been investigated. It was established that in the presence of Platinum hydrochloric acid competitive dehydrocondensation reactions take place with the formation of oligomers with various chain links. In high stages of reaction, three-dimensional systems were obtained. During dehydrocondensation the reaction order, reaction rate constants and activation energy were found.

The structures of synthesized oligomers were determined by IR and NMR spectra data. Gel chromatographic permeation, differential calorimetric scanning and X-ray analysis of synthesized oligo-mers were carried out.

Key words: dehydrocondensation, hydrosilylation, methylhydridesiloxane.

INTRODUCTION

In the literature there is much information about hydrosilylation of acrylic acid [1-4]. The reaction of hydride addition of triethylsilane to acrylic acid in the presence of Spier catalyst has been investigated by A.D. Petrov and it was shown that the reaction proceed in “1-4” position, with formation of an intermediate product, by rearrangement of this product trimethylsilylated ether of propionic acid with 56% yield was obtained. The reaction proceeds according to the following scheme [1]:

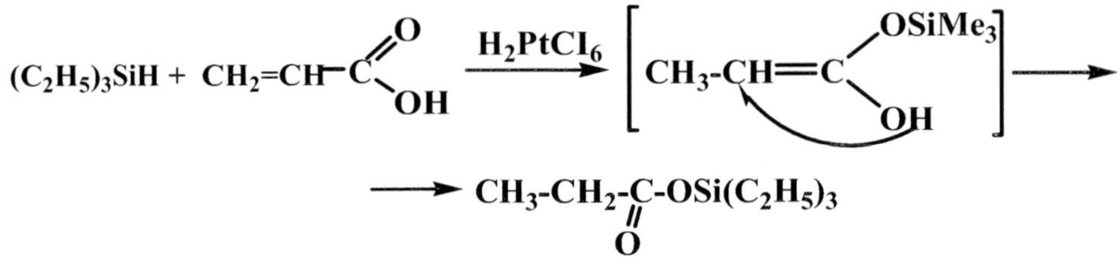

It was shown By the same author that the hydrosilylation acrylic methyl ether with triethylsilane proceeds in "1-4" position with formation of unsaturated silylated methyl ether [2].

The reaction of hydride addition of α,ω–dihydridedimethylsiloxanes with acrylic and methacrylic acid has been investigated by K.A. Andrianov and his co-workers, in the presence of Platinum hydrochloric acid and it was shown that hydrosilylation proceeds on an unsaturated C=C bond with Anti-Markovnikov rule receiving corresponding dicarboxylic acids. The reaction proceeds according to the following scheme [3,4]:

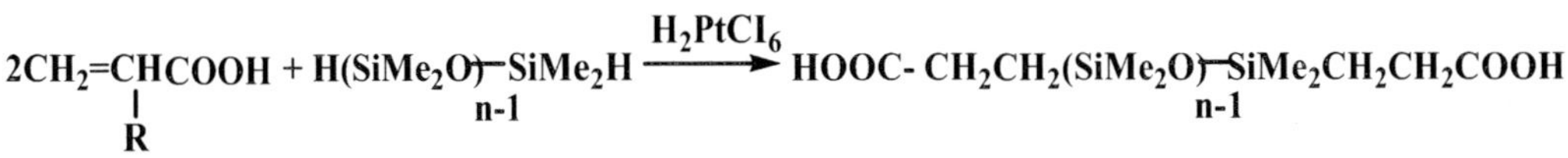

Where: R=H, Me; n=2÷11.

The structure of synthesized silicon containing dicarbon acids was determined by means of finding of molecular masses and IR spectra data.

The reaction of hydrosilylation of acrylic and methacrylic acid with 1,4-bis(dimethylsilyl)benzene in the presence of hydroquinone and Spier catalyst or platinum on the carbon was studied and it was shown that the reaction proceeds mainly by way of dehydrocondensation way, with formation of corresponding acyloxysilanes according to the following scheme [5]:

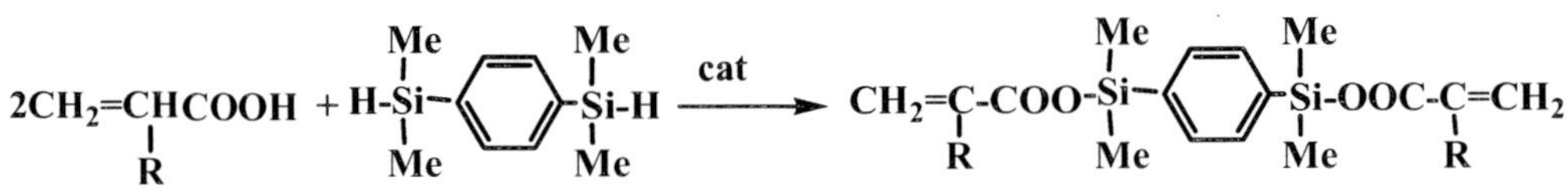

Where: R=H, Me.

In hydrosilylation reactions the hydroquinone was used for preventing the reactions of the polymerization process.

The reaction of hydrosilylation of methylhydridesiloxanes to unsaturated vinyl derivatives (alkenes, acrylic acid, acrylic ethers) in the presence of Spier catalyst has been investigated by the authors [6] and methylsiloxane oligomers with alkyl, carboxyl and ester side-groups has been obtained, accor-ding to the following scheme:

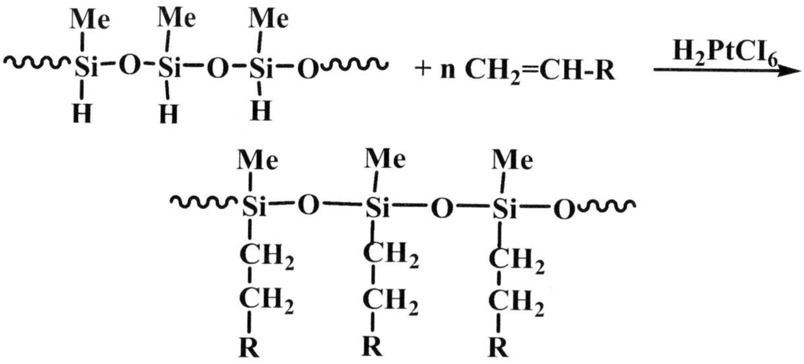

Where: R=C_nH_{2n+1}, -COOH, -COOAlk.

It is also known, that the reaction of catalytic dehydrocondensation between ≡Si-H containing compounds and hydroxyl containing organic compounds proceeds with isolation of hydrogen [7-9]. In the reactions of catalytic dehydrocondensation of hydrideorganosilanes with hydroxyl containing organic compounds are used such catalysts as colloid nickel, anhydrous zinc chloride, Platinum hydrochloric acid [10,11]. It was shown [9] that in the reaction of dehydrocondensation of dihydricdeorganocyclosiloxanes with dihydroxyorganocyclosiloxanes, platinum hydrochloric acid works as a caustic potassium and the activation energy has the same values. On the other hand, it is known that Platinum hydrochloric acid is catalyst of hydrosilylation [12]. The reaction capacity of ≡Si-H bond, determined by catalyst nature and amount, reaction temperature, reactant concentration, solvent character, etc [7, 8]. By the authors [9] it was shown that during dehydrocondensation of hydride-containing organosiloxanes with hydroxyl-containing organosilanes and siloxanes in the presence of powder-like caustic potassium as a nucleophilic catalyst the competing reaction of dialkylsiloxane elimination does not take place which is explained by a high rate of the reaction of the dehydrocondensation process.

Experimental Part

The starting materials for the synthesis of comb-type methylsiloxanes oligomers used were α,ω– bis(trimethylsiloxy)methylhydridesiloxane, acrylic and methacrylic acids.

The initial α,ω–bis(trimethylsiloxy)methylhydridesiloxane with the degree of polymerization n≈35, acrylic and methacrylic acids was received from Fluka. Before reaction, the acrylic and methacry-lic acids were distilled. The organic solvents used in the hydrosilylation reaction were cleaned by drying and distillation.

IR spectra all of samples have been taken on a UR-20 instrument from KBr pellets, while the ^{1}H NMR spectra was takenon a "Perkin-Elmer" at operating frequency of 250 MHz. All spectra were obtained with the use of $CDCl_3$ as solvent and internal standard. A "Perkin–Elmer" DSC-2 differential scanning calorimeter was used to determine TGA and the thermal transitions (T_g) were read at the maximum of the endothermic or exothermic peaks. Heating and cooling scanning rates were 10^{0}C/min.

Gel-permeation chromatography investigation was carried out with the use of Waters Model 6000A chromatograph with an R 401 differential refractometer detector. The column set comprised 10^3 and 10^4 Å Ultrastyragel columns. Sample concentration was approximately 3% by weight in toluene and typical injection volume for the siloxane was 5 μL. standardization of the GPC was accomplished by the use of styrene or polydimethylsiloxane standards with the known molecular weight.

Wide-angle X-ray diffractograms were taken on a "DRON-2" instrument. A-CuK_α was measured without a filter, the angular velocity of the motor $\omega \approx 2$ deg/min.

Dehydrocondensation Reaction of α,ω–Bis(Trimethylsiloxy)Methylhydridesiloxane with Acrylic Acid

Catalytic dehydrocondensation reaction was carried out in a two-necked flask equipped with a tube for catalyst inlet, reflux condenser connected with a gasometer. Between the gasometer and reflux condenser a cold trap and wash bottle were installed.

The reaction products were placed into the flask and thermostated in an oil bath until constant temperature was set up. Then the catalyst 0.1 M solution of Platinum hydrochloric acid in tetra-hydrofuran ($5 \div 9 \times 10^{-5}$ g per 1.0 g of starting substance) was introduced. After that, hydrogen started to release; hydrogen was collected in the gasometer. After the reaction has been finished 5 ml anhydrous toluene was added, filtered and the solvent was eliminated.

Hydride Addition of α,ω–Bis(Trimethylsiloxy)Methylhydridesiloxane to Acrylic Acid.

The hydrosilylation reaction was carried out in a three-necked flask equipped with stirrer, reflux condenser and calcium chloride drying tube. The initial reagents were placed in the flask and thermostated in an oil bath until constant temperature was achieved. 0.1 M solution of platinum hydrochloric acid in tetrahydrofuran ($5 \div 9 \times 10^{-5}$ g per 1.0 g of starting substance) was used as a catalyst. The reaction was carried out in an argon atmosphere in anhydrous toluene solution, at temperature range 60^0C. Then 5 ml anhydrous toluene to the reaction mixture was added, filtered connected to the vacuum and the toluene and un-reacted acrylic acid was removed at $30 \div 40^0$C.

The hydrosilylation reaction in the presence of platinum on the carbon was carried out in the same manner.

Results and Discussion

As it is seen from the literary review, the reactions of hydride addition ≡Si-H bond containing compounds to acrylic and methacrylic acid proceeds on different directions and disagrees to each other. The presented paper deals with synthesis and studies of the properties of comb-type organosiloxanes containing carboxyl groups in the side chain.

For the purpose of synthesis of organosiloxanes oligomers with carboxyl groups in the side chain the reaction of hydride addition of α,ω-bis(trimethylsiloxy)methylhydridesiloxanes ($n \approx 35$) to ac-rylic and methacrylic acids at 1:35 ratio of initial compounds, in the presence of catalysts Platinum hydrochloric acid or Pt/C (5%), at various temperatures in dry toluene solution (50-60%), so the melt condition was investigated.

Preliminary heating of the initial compounds for 4 h in the presence of 0,1 M of platinum hydrochloric acid solution in tetrahydrofuran (~5÷9x10^{-5} g per gram of reaction substances) in the temperature range 50-70^{0}C showed that polymerisation of acrylic or methacrylic acids, breaking of the siloxane backbone, elimination of methane or hydrogen in methylhydridesiloxanes or other changes do not take place. Besides, there are no changes in the NMR spectra of initial compounds. As it is known from the literature [13] Platinum hydrochloric acid, except the reactions of hydrosilylation, catalysed the reaction of dehydrocondensation. Because of that, for complete characterization of the process during the reaction the conversion of hydrogen was observed.

The reaction of hydrosilylation was investigated without solvent in the presence of catalysts at 70^{0}C temperature and it was shown, that at low stages of hydrogen conversion ~10÷20% the reaction proceeds with the formation of three-dimensional systems. In this case, the depth of the hydrosilylation reaction is about ~60%. For preventing the cross-linking processes, the reactions were conducted in an ambience of dry toluene solution.

During the reaction in solution, it was established, that at ~35% hydrogen conversions (Figure 1) the cross-linking processes take place. By determination of active ≡Si-H groups from the soluble part of reaction products it was established, that the depth of hydrosilylation is about ~50%. In the presence of Platinum hydrochloric acid in dilute solution of toluene, besides hydrosilylation, concurrent runs of the reaction of dehydrocondensation ~30% (Figure 1, curve 1) occur.

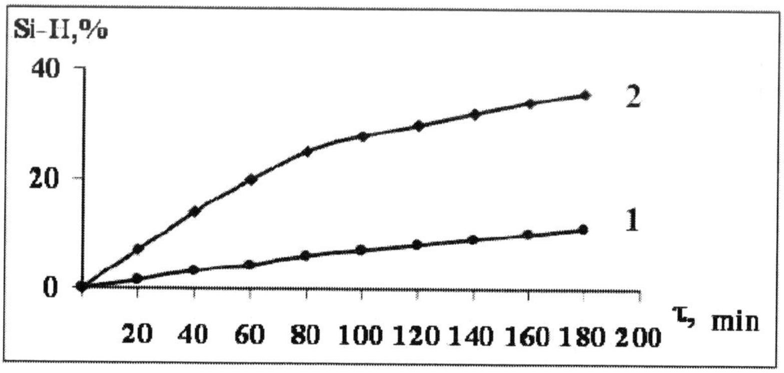

Figure 1. Dependence of changes of active ≡Si-H% groups on the time, upon of dehy-drocondensation reaction of α,ω–bis(trimethylsiloxy)methylhydridesiloxane to acrylicacid at 60^{0}C temperature. Where curve 1 corresponds to catalyst Pt/C, curve 2 corres-ponds to - H_2PtCl_6.

For the purpose of synthesis of oligomers, well soluble in organic solvents the same reaction was studied in the presence of platinum on the carbon (5%) in dry toluene solution, but it was shown that really the reaction of dehydrocondensation proceeds with less activity 9-10%, Figure 1 (curve 2), but the three dimensional systems are obtained. Hydrosilylation in this condition proceeds about ~54%. Therefore, in both cases the reaction proceeds by two concurrent ways, with formation of various- linked oligomers, according to the following scheme:

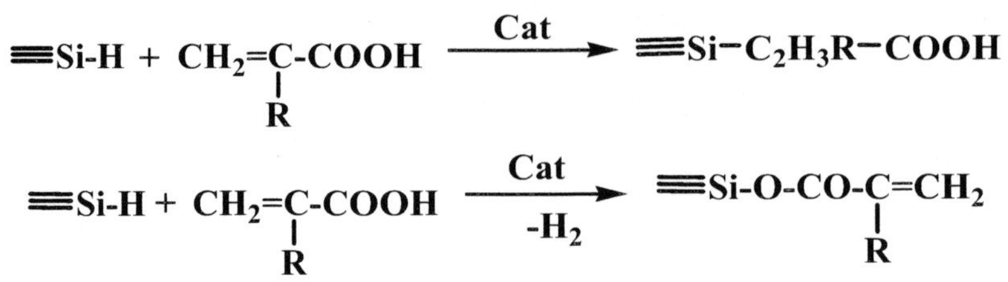

With an increase of dilution of reaction solution the hydrosilylation and dehydrocondensation reaction rates decrease, but the soluble part of reaction product increases. Generally the reaction proceeds according to the following scheme:

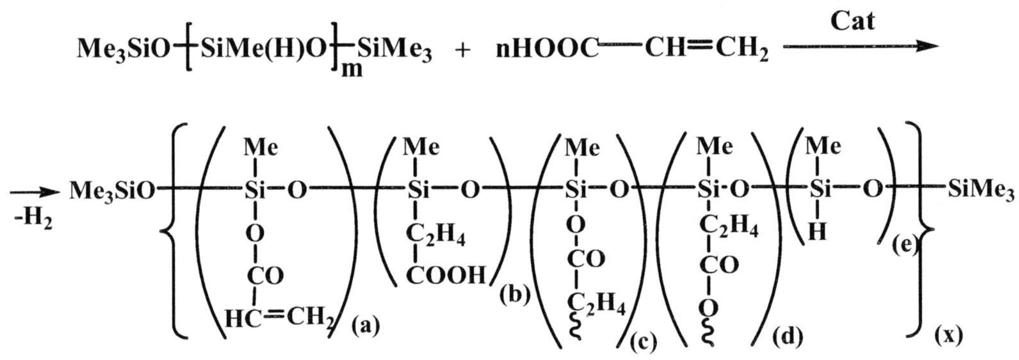

Where: m≈35; [(a)+(b)+(c)+(d)+(e)]x(x)=m. Cat – H_2PtCl_6, I (ratio 1:35); II (ratio 1:17,5) Pt/C – I^1 (ratio 1:35).

From cross-linked polymers the soluble part was isolated with the yield ~50-66%. They are colourless transparent viscous products soluble in ordinary organic aromatic type solvents. Some physical-chemical properties and the yields of synthesized branched oligomers are presented in table 1.

Table 1. Some physical and chemical properties and the yields of branched oligomers.

Oligo-mer, №	Yield, %	Ratio	η_{sp}^*	H_2 Conv., %	Depth of hydrosilylation, %	d_1, Å	$\bar{M}_\omega x10^{-3}$	T_g, ^0C
I	57	1:35	0.08	30	50	8.85	7.2	-67
II	49	1:17.5	0.06	23	49	-	-	-
I^1	66	1:35	0.08	10	54	8.23	-	-70
III	13	-	0.07	-	-	8.92	60.8	-23

*1% solution in Toluene at 25^{0}C.

It must be denoted that for a time the synthesized oligomers undergo a self-catalysed structuring process. From oligomer I after 24 hours was isolated soluble part III with the yield ~13%.

For oligomer III, ^{1}H and ^{13}C NMR investigation carried out. As it is seen from Figure 2 one can observe a triplet signal, with chemical shift δ≈1.25 ppm (J=6.0 Hz) characteristic for hydrogen protons in ≡Si–CH_2- group. One can see also a triplet signal for hydrogen protons in -CH_2-CO- group with chemical shift δ≈2.5 ppm and singlet signals with chemical shift δ≈0.1 and δ≈0.3 ppm for methyl protons in -$SiMe_3$ and ≡SiMe groups accordingly. In spectra one can see a complicated multiplet characteristic for vinyl groups with chemical shift δ≈5.80÷6.1 ppm, which indicate that the reaction proceeds partially by way of dehydrocondensation. There is a low intensity singlet signal with δ≈4.2 ppm characteristic for un-reacted ≡Si-H bond. One can see also, a characteristic signal for residual acids protons with chemical shift δ≈10,2 ppm and multiplet signals of phenyl protons in toluene with chemical shift δ≈7.2÷7.6 ppm, which indicate the traces of toluene in synthesized oligomers.

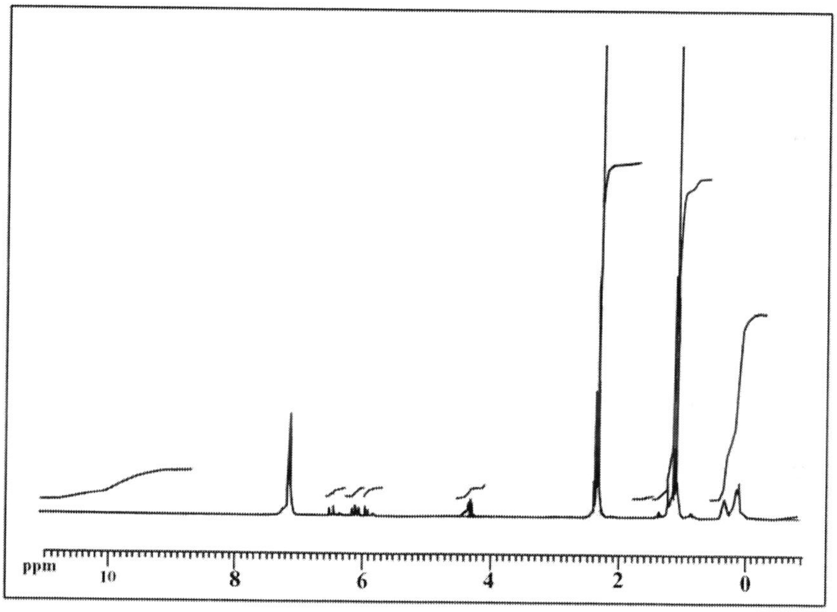

Figure 2. ^{1}H NMR spectrum of oligomer III (solvent and standard $CDCl_3$).

In ^{13}C spectra of oligomer III (figure III) the signal for carbon in residual acid car-bonyl group with chemical shift δ≈179.3 ppm is well observed; signal with chemical shift δ≈7.5 ppm in ≡SiMe group, signal with chemical shift δ≈6.9 ppm in -$SiMe_3$; signal with chemical shift δ≈25.9 ppm characteristic for carbon in ≡Si–CH_2- group; signal with chemical shift δ≈27.5 ppm characteristic for carbon in methylene group in -CH_2-COOH fragment. Concerning the reaction in the other direction, by way of dehydrocondensation, it is proven by presence in ^{13}C spectrum the signal with chemical shift δ≈171.7 ppm characteristic for carbon in free from proton carboxyl group ≡Si–O-CO-. Obtaining of cross-linked systems may be explained by intermolecular hydrosilylation and dehydrocondensation reactions:

$$-C_2H_4-COOH + H-Si\equiv \xrightarrow[-H_2]{Cat} \equiv Si-C_2H_4-CO-O-Si\equiv$$

$$\equiv Si-O-CO-CH=CH_2 + H-Si\equiv \xrightarrow{Cat} \equiv Si-O-CO-C_2H_4-Si\equiv$$

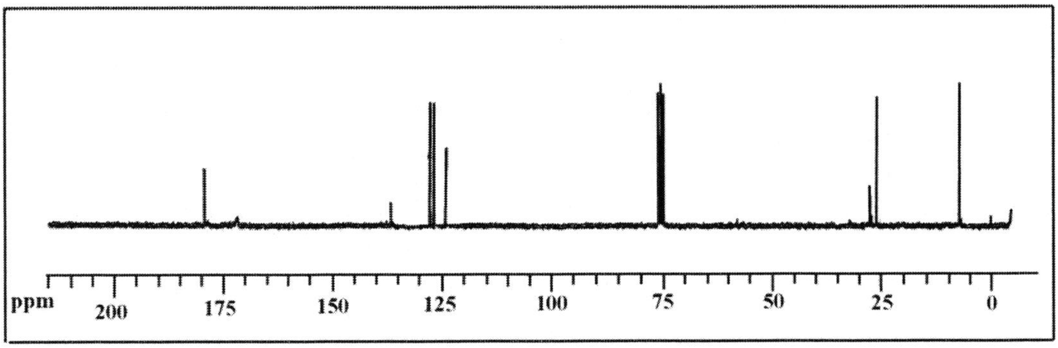

Figure 3. ^{13}C NMR spectrum of oligomer III (solvent and standard $CDCl_3$).

In the IR spectra of oligomers one can see the absorption bands characteristic for ≡Si-O-Si≡ and ≡Si-O-C≡ bonds at 1020 and 1080 cm^{-1}. In the IR spectra one can see the absorption bands for ≡SiMe, for un-reacted ≡Si-H bonds and intensive signal for carbonyl ethereal groups, signals for double bonds at 1275, 2165, 1720 and 1640 cm^{-1} accordingly.

It must be denoted, that in IR spectra there is no absorption band characteristic for associated and non associated hydroxyl groups. From which it is evident, that hydroxyl group takes part in intermolecular dehydrocondensation as it is stated above, which led to the formation of three dimensional or cross-linked systems and in intramolecular cyclization according to the following scheme:

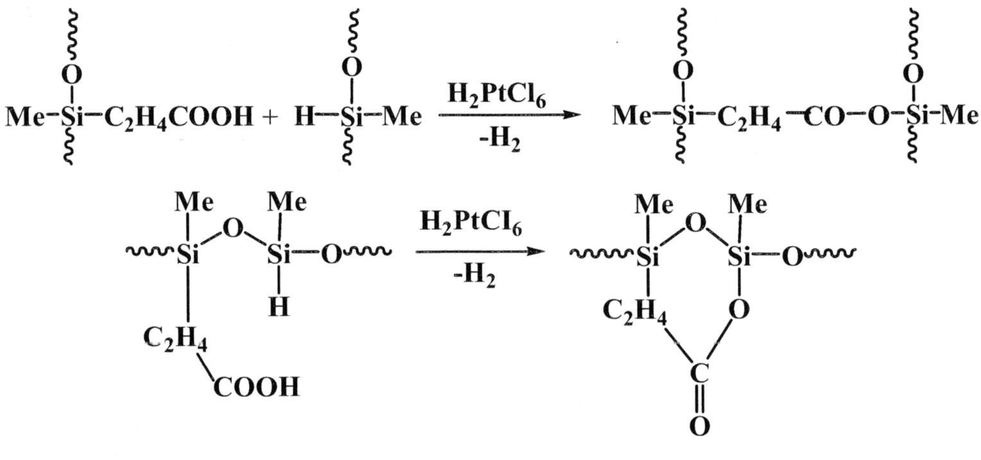

Where: $-C_2H_4- = -(CH_2)_2-$.

On the other side if the reaction proceeds by way of dehydrocondensation, in this case in synthesized oligomers resonance signals for unsaturated vinyl fragments, with chemical shift in the range 5,5÷6,2 ppm, must be observed. But in the spectra of Figure 2 and 3 there are low intensity resonance signals, which show that the fragment **≡Si-O-CO-CH=CH$_2$** may be entered in intra- and inter-molecular hydride addition by the scheme:

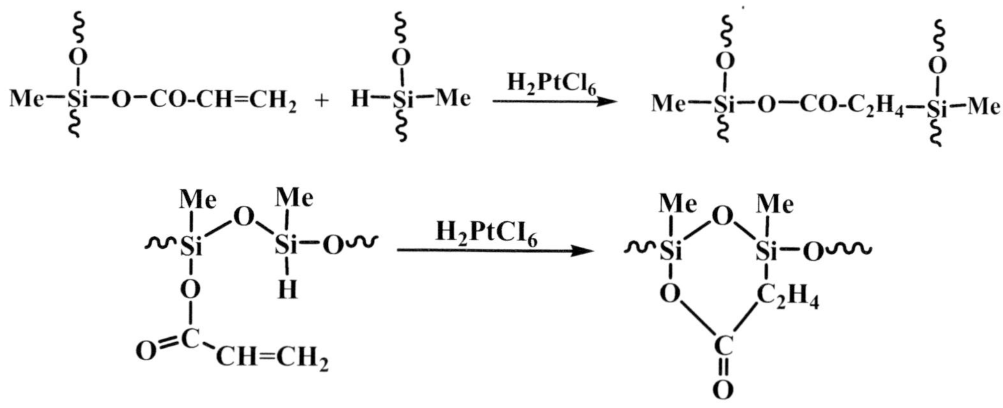

Where: $-C_2H_4- = -(CH_2)_2-$.

For oligomer, I and III wide angle X-ray investigation has been carried out. It was shown that oligomers are represented as one phase amorphous products, where the value of interchain distances changes in the range $d_I \approx 8.23$-8.92 Å.

The synthesized oligomers were studied by GPC method. As it is seen from Figure 4 the GPC curves of the soluble part of oligomers I and III sharply distinguished from each other. Molecular weight distribution of oligomer I has a bimodal character (Figure 4, curve 1) and oligomer III has a polymodal character (Figure 4, curve 2), from which it's evident that during hydrosilylation reaction dehydrocondensation reactions take place, that is in contradiction with earlier published data [3,4]. Average weight molecular masses of oligomer I is equal to $\overline{M}_\omega \approx 7.2x10^3$ and polydispersity D≈9.4. For irregular branched oligomer III average weight of molecular masses of oligomer III is equal to $\overline{M}_\omega \approx 60.8x10^3$ and polydispersity D≈35,3.

Analogical hydrosilylation reaction of α,ω–bis(trimethylsiloxy)methylhydridesiloxane to metacrylic acid in the presence of catalyst – Platinum hydrochloric acid H_2PtCl_6 (0,1 M solution in tetrahydrofuran) at 1:35 ratio of initial compounds, at various temperature, both in a melt and in a dry toluene solution (50-60%) has been investigated.

It was shown that at 400C in the case of 2% conversion of hydrogen the cross-linking processes takes place. From soluble part of cross-linked products by determination of amount of active ≡Si-H group it was established, that the depth of hydrosilylation is about ~84%. With an increase of temperature from 500C and to 600C the part of the reaction of dehydrocondensation increases from ~3.8% up to ~15.0% (see Figure 5).

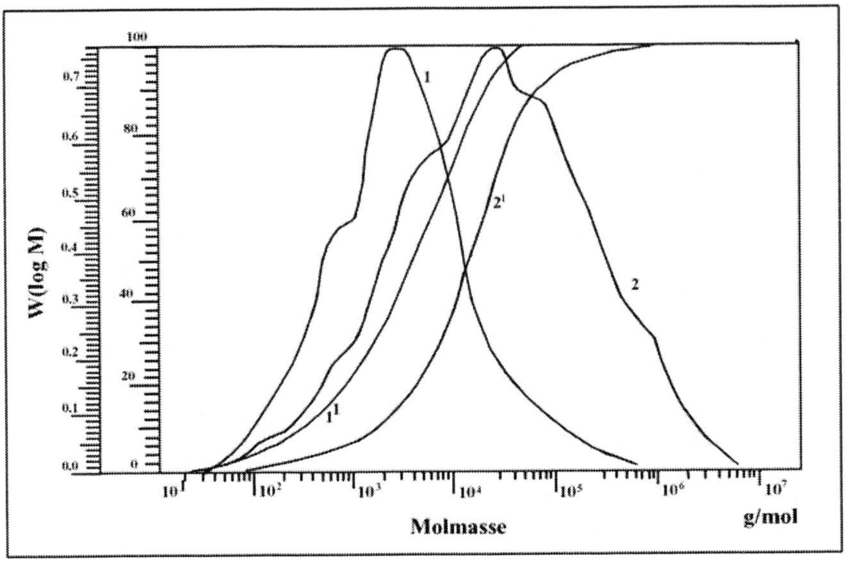

Figure 4. Gel permeation chromatographic curves of oligomers. Where curve 1 corresponds to oligomer I and curve 2 corresponds to oligomer III.

The part of hydrosilylation reaction at 60^0C is equal to ~82%. So, during the reaction of α,ω–bis(trimethylsiloxy)methylhydridesiloxane to methacrylic acid in the presence of catalyst - H_2PtCI_6, aside from hydrosilylation reaction (both directions – anti-Markovnikov rule and Markovnikov rule), catalytic dehydrocondensation takes place.

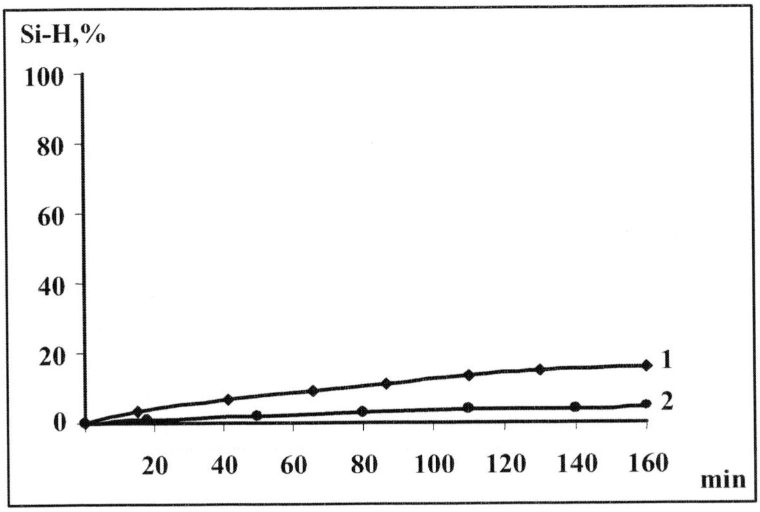

Figure 5. Dependence of changes of concentration active ≡Si-H% groups on the time, upon of dehydrocondensation reaction of α,ω–bis(trimethylsiloxy)methylhydridesiloxa-ne to methacrylic acid at 60^0C temperature. Where curve 1 corresponds to 50^0C, curve 2 60^0C (in the presence of H_2PtCI_6).

During the dilution the depth of dehydrocondensation and hydrosilylation reactions decreases and the yield of the soluble parts rise. The reaction of methylhydridesiloxane with methacrylic acid proceeds with the formation of various-linked products, according to the following scheme:

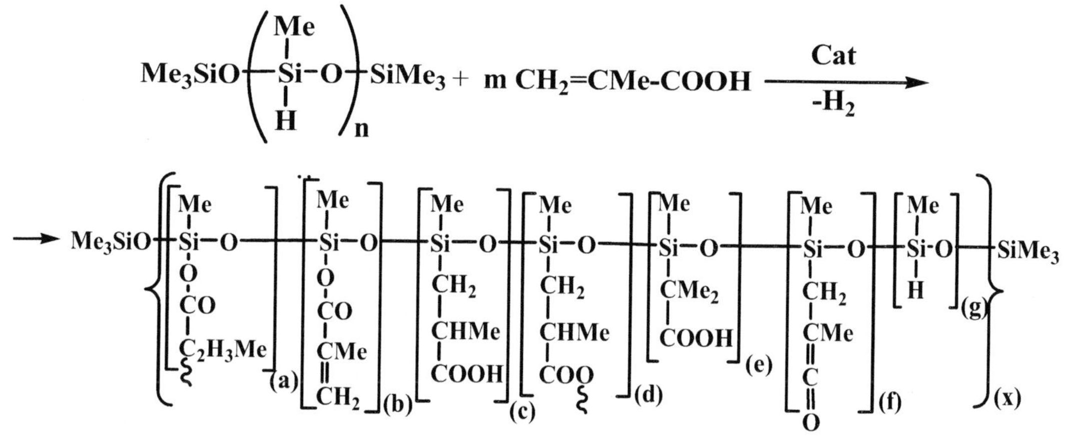

Where: [(a)+(b)+(c)+(d)+(e)+(f)+(g)](x)=n; 40°C - IV^1; 50°C - IV^2 and 60°C - IV^3.

From cross-linked polymers, obtained branched oligomers are viscous products well soluble in aromatic solvents, with $\eta_{sp}\approx 0.04$-0.06. Some physical-chemical properties and the yield of oligomers IV^1-IV^3 are presented in table 2.

Table 2. Some physical and chemical properties and the yield of oligomers IV^1-IV^3.

#	Yield, %	Ratio	$\eta^*_{sp.}$	Reaction t-re T°C	Convers-ion of H_2, %	Hydrosilylation, %	d_1,Å
IV^1	68	1:35	0.04	40	2.0	84	-
IV^2	66	1:35	0.05	50	3.8	88	-
IV^3	64	1:35	0.06	60	15	82	9.07

*In 1% solution of toluene at 25°C.

Synthesized oligomers were studied by ^1H and ^{13}C NMR spectral data. Based on spectral data of reaction products proposed, that in general hydride addition of methylhydridesiloxane to acrylic and methacrylic acid may proceed by several ways, see scheme 1:

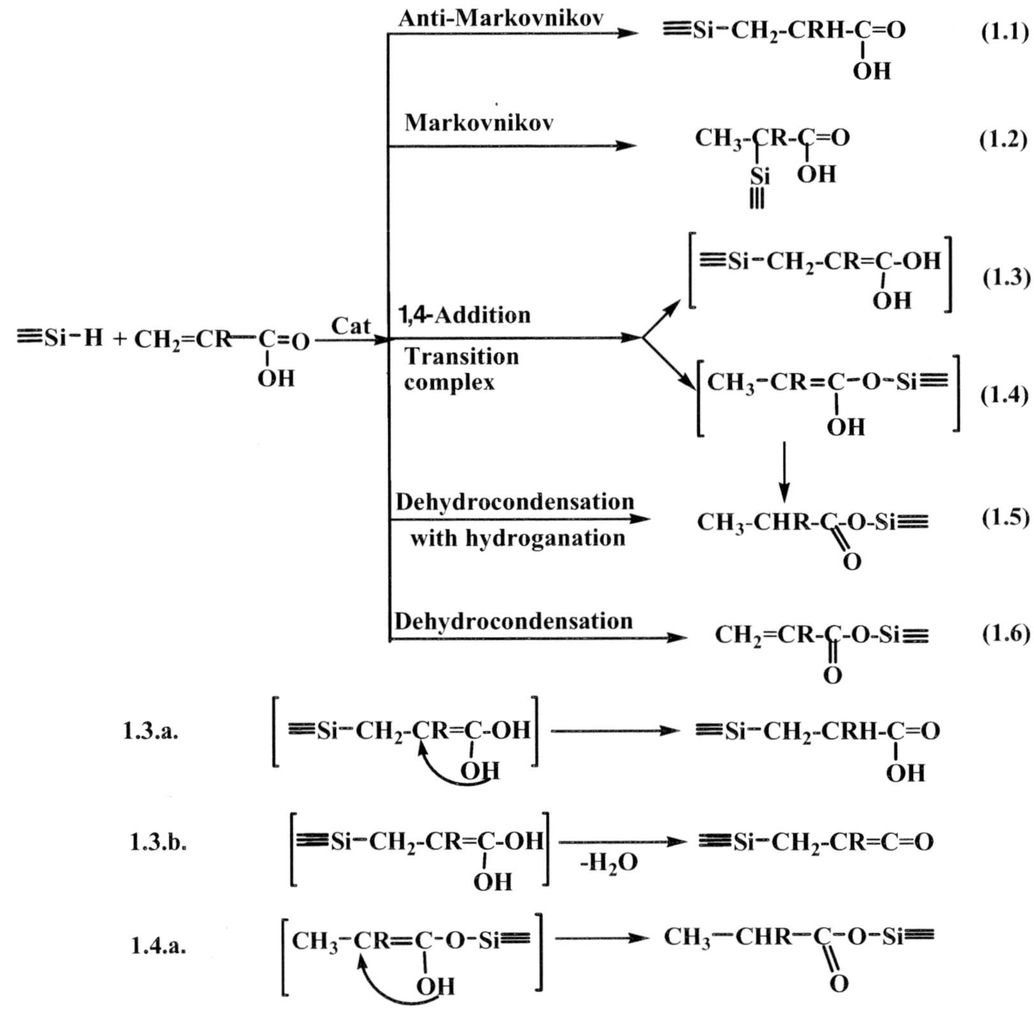

Scheme 1. The possible scheme of hydrosilylation of ethylhydridesiloxane to acrylic and methacrylic acids.

In ^{1}H NMR spectra of oligomer IV[3] one can observe singlet signal with chemical shift $\delta\approx1.21$ ppm for methyl protons in =CH-Me group, broadened singlet signal for protons of methyl group in C-Me group, signal with chemical shift $\delta\approx1.22$ ppm in α-addition product (Markovnikov rule, scheme 1.2), signal with chemical shift $\delta\approx0.8$ ppm corresponds to protons of methylene group, β-addition product in $\equiv Si\text{-}CH_2\text{-}$ fragment, which creation may be examined as a structure obtained by regrouping of an intermediate product (transition complex - 1,4-addition according to the Eltekov rule (scheme 1.3a).

The structure of the product obtained by the above-mentioned scheme 1.3a is similar to the structure of the product obtained by the anti-Markovnikov rule (1,2-addition). Chemical shifts and multiplicity of remaining protons are: for methine group (≡CH) sestet (in case R=Me) with $\delta\approx2.6$ ppm (J=6.83 Hz), for methyl group singlet at $\delta\approx1.21$ ppm, (J=7.07 Hz) and for protons in carboxyl group a broadened signal with chemical shift at $\delta\approx10\text{-}11$ ppm. In case of another course of reaction, after which elimination of the molecule of water under the scheme 1.3 takes place, with formationb of the ketene group, their chemical shifts are following: for protons in methylene ≡Si-CH_2- group singlet signal with chemical shift δ≈0.78 ppm, for methyl protons in –CMe=C=O fragment singlet signal with chemical shift δ≈1.96 ppm. It must be denote, that the authors [1] suggested the direction of the reaction with obtaining of a transition complex according to the scheme 1.4 earlier. The obtained transition complex then undergoes regrouping according to the scheme 1.4a with obtaining a simultaneous dehydrocondensation-hydrogenation product similar to the scheme 1.5. However, processing the reaction to above-mentioned scheme is improbable, because released hydrogen during dehydrocondensation reaction, is eliminated away from the reaction mixture.

Obtaining three various hydrosilylation products with methacrylic acid is proven by ^{13}C NMR spectra data. Each adduct characterized with the following chemical shifts: for hydrosilylation product by Markovnikov rule (scheme 1.2) resonance signal for methyl protons at δ≈18.11 ppm, signal for quaternary carbon nucleus at δ≈34.11 ppm and signal for carbon nucleus in carboxyl group at δ≈172.74 ppm. In the hydrosilylation product by anti-Markovnikov rule (1,2-addition, scheme 1.1), one can observe the signal for carbon nucleus with chemical shift at δ≈18.98 ppm in ≡Si-CH_2- fragment, signal at δ≈29.58 ppm for carbon nucleus in =CH- group and signal for methyl group at δ≈22.92 ppm in Me-CH- fragment.

One can analogically observe the value of signals for compound containing ketene groups. Signals with chemical shift δ≈18.98 ppm characteristic for carbon nucleus of methylene group (≡Si-CH_2-), signal at δ≈30.2 ppm for carbon in methyl group [-C(Me)=C=O],, signal for tertiary carbon nucleus at δ≈ 135.6 ppm [-C(Me)=C=O] and signal at δ≈183.43 ppm for carbon nucleus in the group =C=O.

As it is seen from ^{1}H and ^{13}C NMR spectra data hydrosilylation reaction of methylhydridesiloxanes to acrylic and methacrylic acid distinguished from each other and from literature data [1]. With acrylic acid hydrosilylation mainly proceeds by dehydrocondensation and hydrosilylation (anti-Markovnikov rule (scheme 1.1 and 1.3a). In contrast to this, hydrosilylation to methacrylic acid proceeds almost in all directions presented on the scheme 1.

For full characterization of hydride addition of methylhydridesiloxane to acrylic and methacrylic acid by quantum-chemical half empiric AM1 method [14], for all initial, intermediate and final products, in the model reaction of hydrosilylation of methyldimethoxysilane to acrylic and methacrylic acid, the heats of formations (ΔH_f) and change of enthalpy of model reactions (ΔΔH) were calculated proceeding with various concurrent scheme 2:

As it is seen from the scheme 2, starting from calorific effects of all possible directions of hydrosilylation reaction of methyldimethoxysilane to acrylic acid thermodynamically it is more profitable course of the reaction according to the anti-Markovnikov rule (1,2-addition). It must be denoted that the product obtained by regrouping of the intermediate product (1,4-addition, scheme 2.3) similar to Eltekov rule (scheme 2.3b), identical to the product obtained by anti-Markovnikov rule (scheme 2.1) and they have the same value of calorific effects [ΔΔH1=ΔΔH3+ΔΔH3a= -82.51 kJ/mol+(-73.35 kJ/mol)= -135.85 kJ/mol]. Obtaining of ketene by regrouping of the intermediate product according to the scheme 2.3a is most probable.

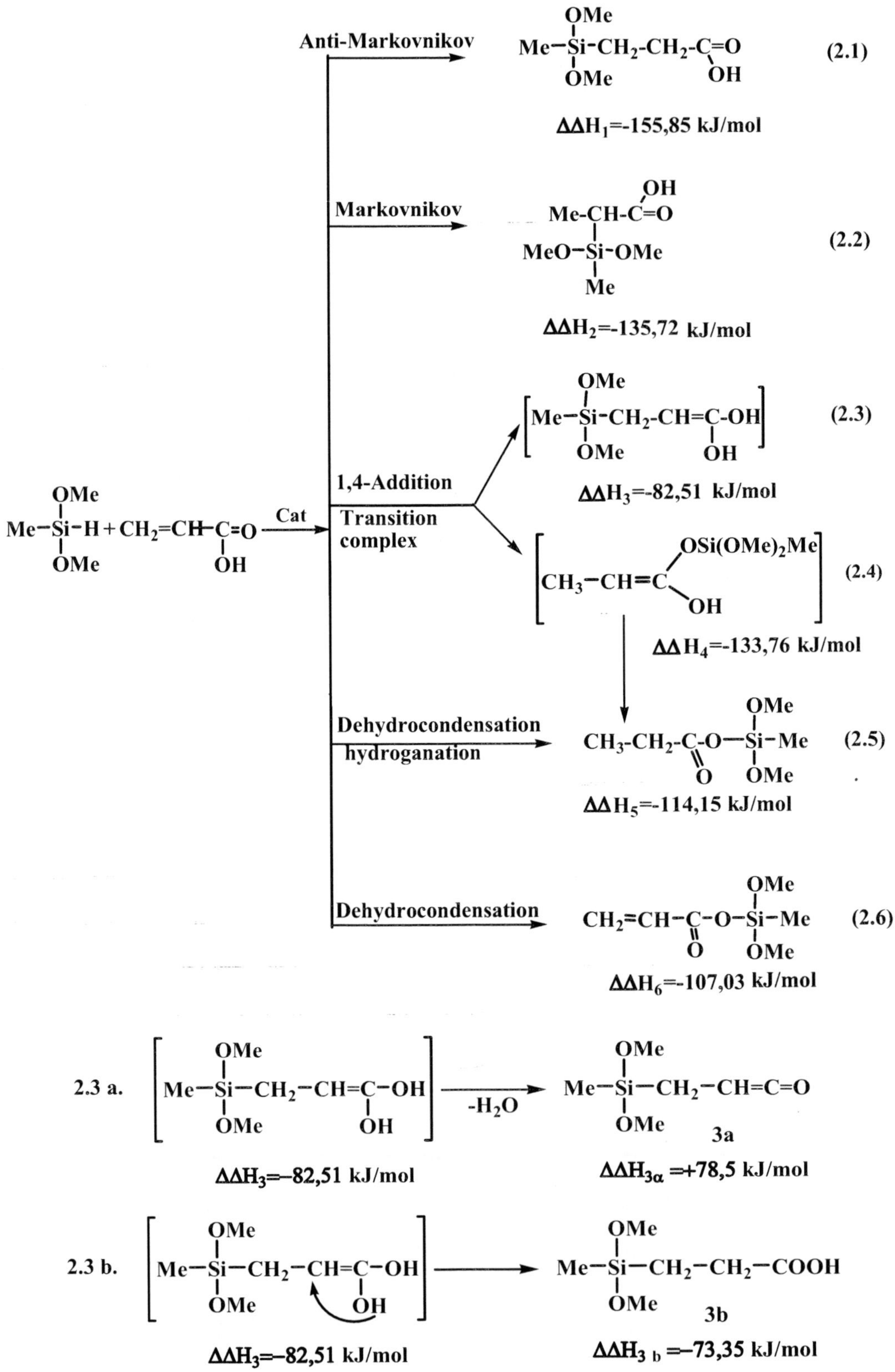
Anti-Markovnikov
(2.1)
ΔΔH1=-155,85 kJ/mol
Markovnikov
(2.2)
ΔΔH2=-135,72 kJ/mol
(2.3)
ΔΔH3=-82,51 kJ/mol
Cat
1,4-Addition
Transition complex
(2.4)
ΔΔH4=-133,76 kJ/mol
Dehydrocondensation hydroganation
(2.5)
ΔΔH5=-114,15 kJ/mol
Dehydrocondensation
(2.6)
ΔΔH6=-107,03 kJ/mol
2.3 a.
-H2O
3a
ΔΔH3=–82,51 kJ/mol
ΔΔH3α =+78,5 kJ/mol
2.3 b.
3b
ΔΔH3=–82,51 kJ/mol
ΔΔH3 b =–73,35 kJ/mol

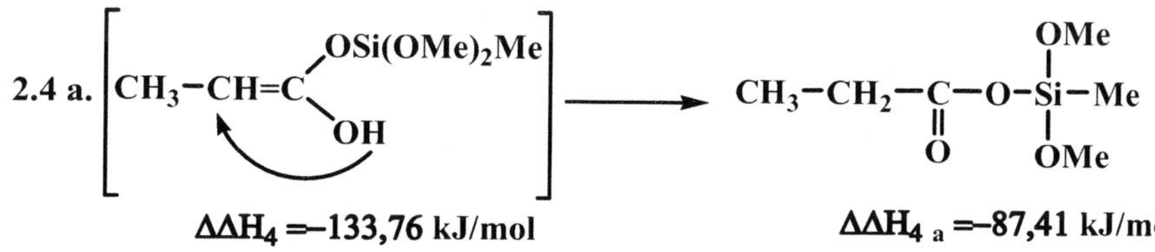

Scheme 2. Change of enthalpy of model reactions of hydrosilylation of methyldimethoxysilane to acrylic acid .

As it is seen from ^{1}H and ^{13}C NMR spectral data of oligomer III, in the reaction of hydrosilylation of methylhydridesiloxanes to acrylic acids obtaining of ketenes is not observed.

The structure of the product obtained in dehydrocondensation-hydrogenation reactions (scheme 2.5) is identical to the product obtained by regrouping of the intermediate product (scheme 2.4) according to the Eltekov rule. A possible scheme of the reaction in an enlarged aspect may be represented as:

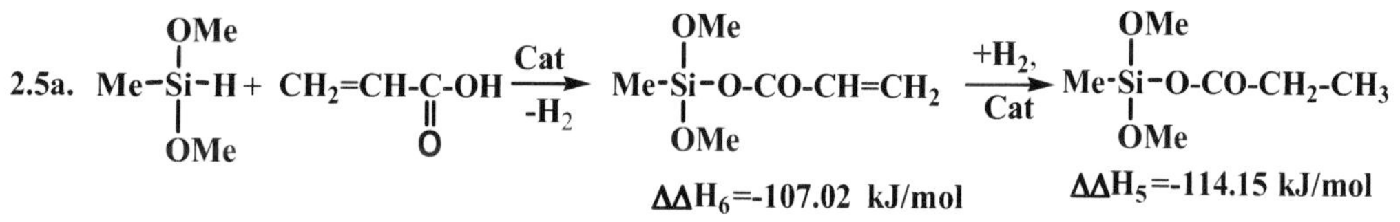

The structure of the product obtained in this way is identical to the structure of the product obtained by regrouping of intermediate product (scheme 2.4) according to the scheme 2.5a and they have the same values of calorific effects ($\Delta\Delta H=\Delta\Delta H_4+\Delta\Delta H_{4a}=\Delta\Delta H_6+\Delta\Delta H_5=-221.17$ kJ/mol). The course of the reaction according to the schemes 2.4 and 2.4a compared with another course is energetically profitable, the dehydrocondensation-hydrogenation course of the reaction according to the scheme 2.5a is improbable, because during the dehydrocondensation reaction released hydrogen is isolated from the reaction mixture, that is fixed from reaction mixture during the process, but strong denial of course of the reaction by the scheme 2.5a is impossible. Besides, it is possible that sensitivity of the instrument and conditions of received spectra was insufficient for identification of all possible products obtained with a low concentration.

In scheme 3 one can see a change of enthalpy of all reactions proceeding with various concurrent ways during the model hydrosilylation reaction of methyldimethoxysilane to methacrylic acid:

On the assumption of heat of model reaction of methyldimethoxysilane with methacrylic acid (scheme 3), course of hydrosilylation according to anti-Markovnikov rule (1,2-addition) thermodynamically is more profitable. The structure of the product, which is obtained by, regrouping of transition complex (1,4-addition, scheme 3.3) according to the Eltekov rule, scheme 3.3a, is similar to the structure of the product obtained by anti-Markovnikov rule. Their heat of reaction $\{\Delta\Delta H_1= \Delta\Delta H_3+\Delta\Delta H_{3b}=[-83.72$ kJ/mol+(-56.94 kJ/mol)]=-140.66 kJ/mol$\}$ are same too.

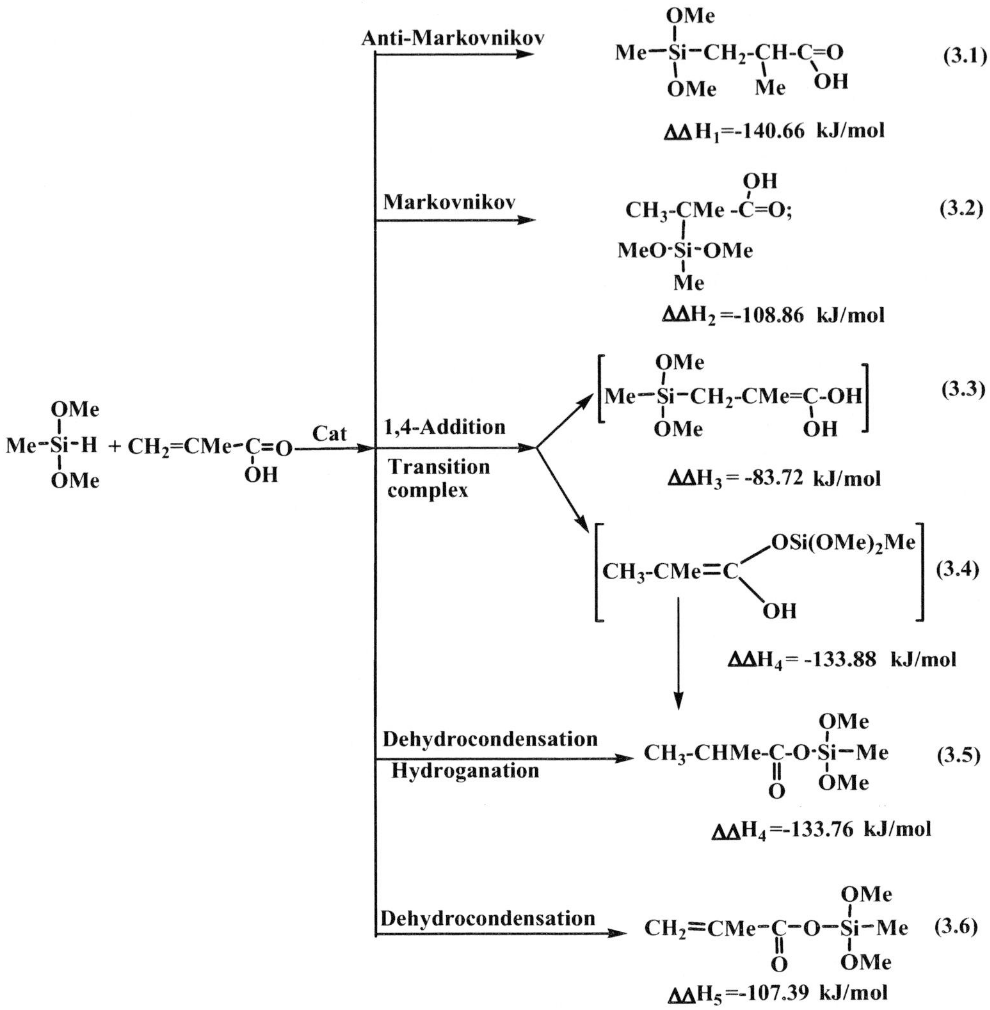
Anti-Markovnikov
(3.1)
ΔΔH1=-140.66 kJ/mol
Markovnikov
(3.2)
ΔΔH2=-108.86 kJ/mol
Cat
1,4-Addition
Transition complex
(3.3)
ΔΔH3= -83.72 kJ/mol
(3.4)
ΔΔH4= -133.88 kJ/mol
Dehydrocondensation
Hydroganation
(3.5)
ΔΔH4=-133.76 kJ/mol
Dehydrocondensation
(3.6)
ΔΔH5=-107.39 kJ/mol

3.3a.

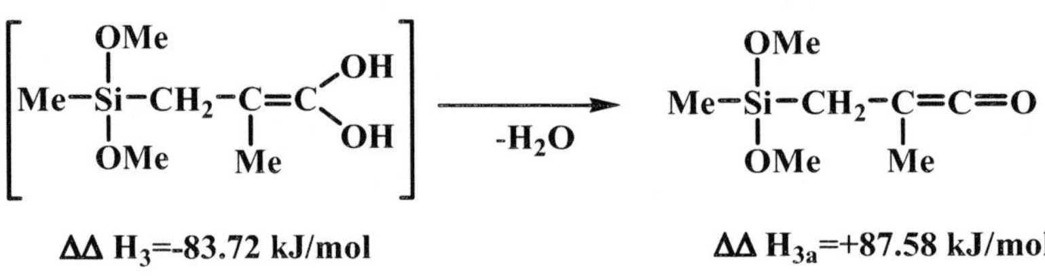
-H2O
ΔΔ H3=-83.72 kJ/mol
ΔΔ H3a=+87.58 kJ/mol

3.3.b.

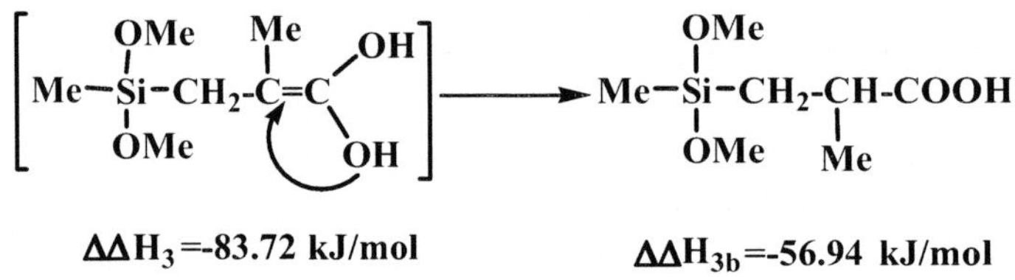
ΔΔH3=-83.72 kJ/mol
ΔΔH3b=-56.94 kJ/mol

3.4.a. 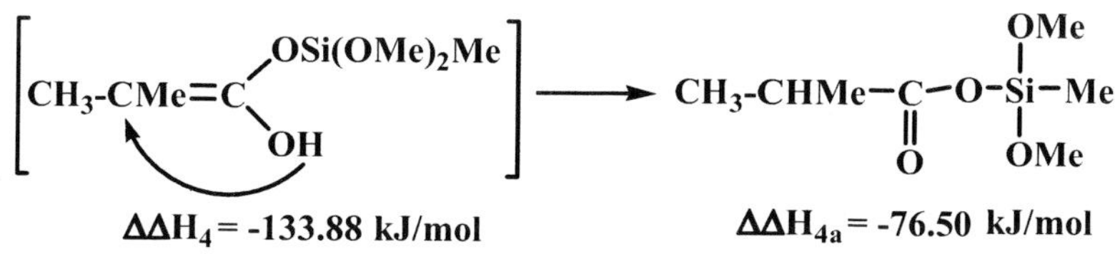

Scheme 3. Change of enthalpy of model reactions of hydrosilylation of methyldimethoxysilane to methacrylic acid .

Formation of a ketene group from the transition complex according to the scheme 3.3a is most probable because its heat of reaction is equal to $\Delta\Delta H_{3a}$= +87.58 kJ/mol. In spite of this, unlike acrylic acid, in methacrylic acid obtaining of ketene group is proved by ^{1}H and ^{13}C NMR spectra data.

With respect to the dehydrocondensation-dehydrogenation reaction of methyldimethoxysilane with methacrylic acid, a more comprehensive aspect is represented as:

3.5a. 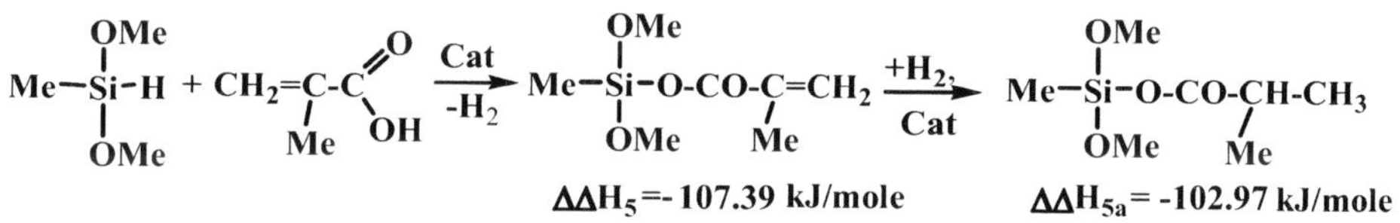

Analogically the course of the reaction according to the above-mentioned scheme 3.5 is improbable, though its exclusion is not possible.

By wide-angle X-ray analyses of oligomer IV[3] it was shown, that oligomer is one phase amorphous system with a value of interchain distances $d_l \approx 9.07$ Å.

So, for the first time it was studied that hydride addition reactions of methylhydridesiloxane to acrylic and methacrylic acid in the presence of catalysts Platinum hydrochloric acid and Pt/C and differed from existing literature data and aside from hydrosilylation reactions (1,2- and 1,4-addition), dehydrocondensation reactions take place with the formation of various-linked thermo reactive oligomers.

REFERENCES

[1] A.D. Petrov, S.I. Sadikh-Zade, E.I. Filatova. *Zhurn. Obshch., Khimii,* 1959, 29, 2936 (in Russian).

[2] S.I. Sadikh-Zade, A.D. Petrov. *Zhurn., Obshch., Khimii, Chemistry and practical application siliconorganic compounds, part 1*, Leningrad, 1958, 212 (in Russian).

[3] K.A. Andrianov, I.I. Abkhazava, L.M. Khananashvili, G.D Bagratishvili. *Zhurn., Obshch. Khimii,* 1973, 43(7), 1505 (in Russian).

[4] K.A. Andrianov, I.I. Abkhazava, L.M. Khananashvili, G.D. Bagratishvili, T.P. Doksopulo, V.A. Chagelishvili. *Zhurn., Obshch. Khimii,* 1973, 41(1), 606 (in Russian).

[5] I.M. Gverdtsiteli, T.P. Doksopulo, E.I. Chikovani. Zhurn., Obshch. Khimii, 1975, 47(7), 352.
[6] S. Nitzsche. Angew. Chem., 1960, 72(3), 112 (in Russian).
[7] B.N. Dolgov. Chemistry and Practical Application of Organosilicon Compounds Central Library of Technical Information, Leningrad, 1958, 1, 18 (in Russian).
[8] A.I. Nogaideli, R.Sh. Tkeshelashvili, L.I. Nakaidze, O.V. Mukbaniani. Procedings of Tbilisi State University, 1976, 167, 69 (in Russian).
[9] Mukbaniani, L.M. Khananashvili, I.G. Esartia, S.D. Khaduri. *Intern. Journ. of Polymeric Materials,* 1994, 24(1), 131 (in Russian).
[10] B.N. Dolgov, I.P. Kharitonov, M.G. Vorokvov. *Zhurn., Obshch., Khimii,* 1954, 24, 861 (in Russian).
[11] J. Baines and C.J. Eaborn. *Chem. Soc.,* 1955, 4023.
[12] Mukbaniani, U. Scherf, G.N. Gurgenidze, M.G. Karchkhadze, S.M. Meladze, L.M. Khananashvili. *International Journal of Polymeric Materials*, 2001, 48(3), 267.
[13] Mukbaniani, G.N. Gurgenidze, S.M. Meladze and L.M. Khananashvili. *Russian Polymer News,* 2001, 6(4), 18.
[14] M.I.S. Dewar, E.G. Zeobish, E.F. Healy, J.S. Stewart. *J. Am. Chem. Soc.*, 1985, 107, 3902.

In: Modification Reactions of Oligomethylhydridesiloxanes ISBN: 1-60021-365-0
Editors: O.Mukbaniani, et al. pp. 95-107

Chapter 6

DEHYDROCONDENSATION AND HYDROSILYLATION REACTIONS OF METHYLHYDRIDESILOXANE TO ALLYL ALCOHOL

O. Mukbaniani[1]***, G. Gurgenidze***[1] ***and S. Meladze***[2]
[1]I.Javakhishvili Tbilisi State University, I. Chavchavadze Ave.,1, 0128 Tbilisi, Georgia
[2]The Georgian Institute of Consumers' Association, V. Pshavela Ave., 44, 0177 Tbilisi, Georgia

ABSTRACT

The reaction of dehydrocondensation of α,ω–bis(trimethylsiloxy) methylhydridesiloxane (n≈53) with allylalcohol at various ratio of initial compounds, in the presence of anhydrous powder like caustic potassium and platinum hydrochloric acid has been investigated. It was established that in the presence of caustic potassium, a dehydrocondensation reaction takes place and methylsiloxane oligomers with axiall groups in the side chain have been obtained. In the presence of platinum hydrochloric acid competitive dehydrocondensation and hydrosilylation reactions take place with the formation of oligomers with various chain links. In high stages of reaction three-dimensional systems were obtained. During dehydrocondensation the reaction order, reaction rate constants and activation energy were found.

The structure of synthesized oligomers was determined by IR and NMR spectra data. Gel permeation chromatography, differential scanning calorimetry and X-ray analysis of synthesized oligomers were carried out.

Key words: dehydrocondensation, hydrosilylation, and methylhydridesiloxane.

INTRODUCTION

It is known that compounds with ≡Si-H bonds react with water to cause limination of hydrogen in the presence of alkali catalysts [1].

The reaction of catalytic dehydrocondensation between ≡Si-H containing compounds and hydroxyl-containing organic compounds proceeds similarly with the isolation of hydrogen [2.3]. In the reactions of catalytic dehydrocondensation of hydrideorganosilanes with hydroxyl containing organic compounds, catalysts are used such as colloid nickel, anhydrous zinc chloride, platinum hydrochloric acid [3.4]. It was shown [5], that in the reaction of dehydrocondensation of dihydrideorganocyclosiloxanes with dihydroxyorganocyclosiloxanes, platinum hydrochloric acid works as a caustic potassium and the activation energy has the same values. However, on the other hand it is known that platinum hydrochloric acid is catalyst of hydrosilylation [6]. The reaction capacity of ≡Si-H bond is determined by the catalyst nature and amount, reaction temperature, reactant concentration, solvent character, etc [7.8]. By the authors [9] it was shown that during dehydrocondensation of hydride-containing organosiloxanes with hydroxyl-containing organosilanes and siloxanes in the presence of powder-like caustic potassium as a nucleophilic catalyst in the competing reaction of dialkylsiloxane, elimination does not take place, which is explained by a high rate of the reaction of the dehydrocondensation process.

EXPERIMENTAL PART

The starting materials for the synthesis of comb-type methylsiloxane with axiall and propanol groups in the side chain were α,ω–bis (trimethylsiloxy) methylhydridesiloxane and allyl alcohol.

The initial α,ω–bis(trimethylsiloxy)methylhydridesiloxane with the degree of polymerization $n \approx 53$ was used as received from Nuncritz GMBS. The organic solvents were cleaned by drying and distillation.

IR spectra all of samples have been taken on an UR-20 instrument from KBr pellets, while the ^{1}H NMR spectra was taken on a "Perkin-Elmer" at an operating frequency of 250 MHz. All spectra were obtained with the use of $CDCl_3$ as solvent and internal standard. A Perkin–Elmer DSC-2 differential scanning calorimeter was used to determine TGA and the thermal transitions (T_g) were read at the maximum of the endothermic or exothermic peaks. Heating and cooling scanning rates were 10^0C/min.

Gel permeation chromatography investigation was carried out with the use of a Waters Model 6000A chromatograph with an R 401 differential refractometer detector. The column set comprised 10^3 and 10^4 Å Ultrastyragel columns. Sample concentration was approximately 3% by weight in toluene and typical injection volume for the siloxane was 5 μL. standardization of the GPC was accomplished by the use of styrene or polydimethylsiloxane standards with the known molecular weight.

Diffractograms were taken on a "DRON-2" instrument; (Burevestnik, Saint-Petersburg, Russia) instrument using A-CuK_α radiation. The values were measured without a filter, and the angular velocity of the motor was ~2 deg/min.

Dehydrocondensation Reaction of α,ω–Bis(Trimethylsiloxy)Methylhydridesiloxane with Allyl Alcohol

A catalytic dehydrocondensation reaction was carried out in a two-necked flask equipped with a tube for catalyst inlet, reflux condenser connected with a gasometer. Between the gasometer and reflux condenser a cold trap and wash bottle were installed.

The reaction products were placed into the flask and thermostated in an oil bath until constant temperature was set up. Then the catalyst was introduced. After that, hydrogen started releasing. Hydrogen was collected in the gasometer. After the reaction finished the solvent was eliminated (if caustic potassium was used as a catalyst, the reaction products were washed to neutral medium). The hydrosilylation reaction was carried out by the above-mentioned reaction using platinum hydrochloric acid as a catalyst.

Hydride Addition of α,ω–Bis (Trimethylsiloxy)Methylhydridesiloxane to Allyl Alcohol

The hydrosilylation reaction was carried out in a three-necked flask equipped with stirrer, reflux condenser and calcium chloride drying tube. The initial reagents were placed in the flask and thermostated in an oil bath until constant temperature was achieved. 0.1 M solution of platinum hydrochloric acid in tetrahydrofuran ($5\div9x10^{-5}$ g per 1.0 g of starting substance) was used as a catalyst. The reaction was carried out in an argon atmosphere in an anhydrous toluene solution, at temperature range $30\div40^{0}$C. Then the reaction mixture was connected to the vacuum and the toluene and un-reacted allyl alcohol were removed at $30\div40^{0}$C up to constant weight.

Results and Discussion

The main task of our work was the synthesis of comb-type silicon organic oligomers both with unsaturated groups and with hydroxyl groups in the side chain. As initial products α,ω–bis (trimethylsiloxy)methylhydridesiloxane and allyl alcohol were used.

For the purpose of synthesis comb-type silicon organic oligomers with oxyallyl-fragments in the side chain the reaction of dehydrocondensation at 1:53 and 1:26.5 mole ratio of initial components in the presence of 0.1% anhydrous powder-like caustic potassium has been carried out. Because the initial compounds are represented as inter insoluble liquids, the reaction of dehydrocondensation was carried out in anhydrous toluene (C=0.0856 mole/l), in the temperature range $20\div40^{0}$C.

Preliminary heating of the initial compounds for 4 h in the presence of KOH (0.1% of total mass of the initial components), or 0,1 M of platinum hydrochloric acid solution in tetrahydrofuran ($\sim5\div9x10^{-5}$ g per gram of reaction substances), in the temperature range 40-50^{0}C showed that polymerization of allyl alcohol, or other changes in the break of the siloxane backbone, or elimination of methane in methylhydridesiloxane do not take place. Besides, there are no changes in the NMR spectra of initial compounds.

The reaction of catalytic dehydrocondensation was investigated at different temperatures (20, 30 and 40^{0}C) during which the conversion of hydrogen was observed and measured.

Figure 1 showed the dependence of the ≡Si-H % concentration decreases on the time. As it is seen from Figure 1, in the reaction of catalytic dehydrocondensation, with the temperature rise the hydrogen conversion increases from 90% (20^{0}C) up to 97% (40^{0}C). At 1:53 ratio of initial compounds, the reaction mainly proceeds during the first thirty minutes. As it is seen from Figure 2 during dehydrocondensation in 1:26.5, mole ratio of initial compounds the rate of dehydrocondensation increases up to 100%.

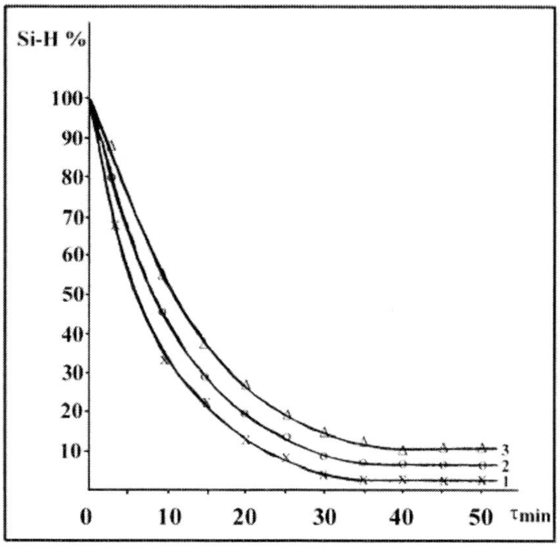

Figure 1. Dependence of changes of active ≡Si-H% groups on time upon hydride addition of α,ω–bis(trimethylsiloxy)methylhydridesiloxane to allyl alcohol, where curve 1 is at 40^{0}C, curve 2 is at 30^{0}C and curve 3 is at 20^{0}C (ratio 1:53).

The dehydrocondensation reaction proceeds according to the following scheme 1. Where: [(a)+(b)](k)=m≈53; n=53: I (20^{0}C), II (30^{0}C), III (40^{0}C). n=26.5: IV (20^{0}C), V (30^{0}C), VI(40^{0}C).

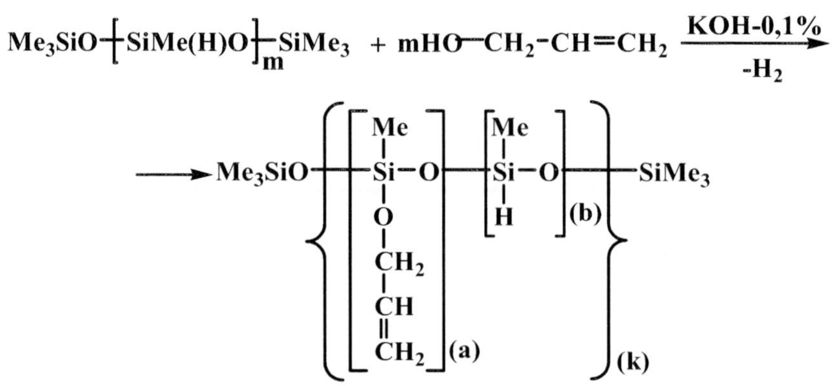

Scheme 1. Hydrosilylation reaction of α,ω-bis(trimethylsiloxy)methylhydridesiloxane to allyl alcohol.

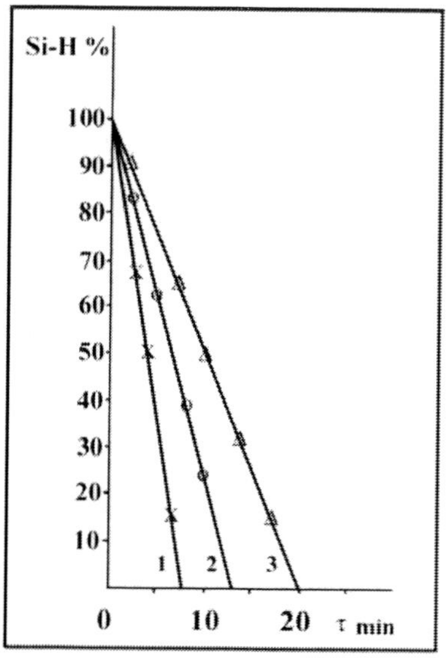

Figure 2. Dependence of changes of active ≡Si-H% groups on time upon hydride addition of α,ω–bis(trimethylsiloxy)methylhydridesiloxane to allyl alcohol, where curve 1 is at 40^0C, curve 2 is at 30^0C and curve 3 is at 20^0C (ratio 1:26.5).

As a result transparent colourless oils soluble in ordinary organic solvents with η_{sp}=0.09÷0.13 were received . Some physical-chemical properties of synthesised oligomers are presented in table 1. It was established that in the beginning stages the reaction of dehydrocondensation is the reaction of a second order. Figures 3 and 4 present a direct dependence of the reverse concentration of the reactants versus time. From which the reaction rate constants at various temperatures and at various ratios of initial compounds were calculated. The reaction rate constants at 1:53 ratio is $k_{40^0C}\approx 2{,}4751$; $k_{30^0C}\approx 1{,}7322$; $k_{20^0C}\approx 1{,}1917$ l/mol·s and temperatural coeficient of this reaction is about $\gamma\approx 1{,}45$ and the reaction rate constants at 1:26.5 ratio is $k_{40^0C}\approx 1{,}4825$; $k_{30^0C}\approx 1{,}0355$; $k_{20^0C}\approx 0{,}7265$ l/mol·s and the temperature coefficient of this reaction is $\gamma\approx 1{,}42$.

Table 1. Some physical-chemical properties of ethylhydridesiloxanes with organic branching side groups.

№	η_{xv}*	d_1, Å	Tg, ^{0}C	$\overline{M}_\omega$**	5% mass losses, T,^{0}C	Residual mass,%	Yield, %
I	0.04	-	-116	6030	-	-	90
II	0.04	-	-116	-	-	-	94
III	0.05	9.30	-110	6500	250	21	98
IV	0.05	9.23	-110	-	240	23	32
V	0.03	-	-	-	-	-	100
VI	0.04	9.12	-123	-	-	-	100
VII	0.04	-	-	4800	230	17	100

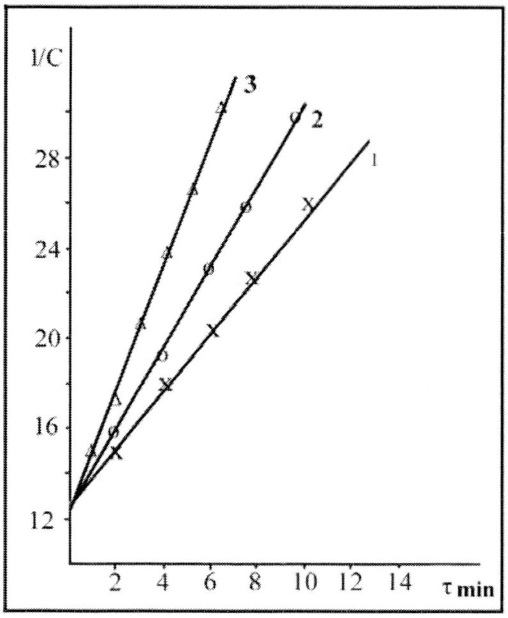

Figure 3. Dependence of reverse concentration on the time, where curve 1 is at 20^0C,curve 2 at 30^0C and curve 3 is at 40^0C (ratio 1:53).

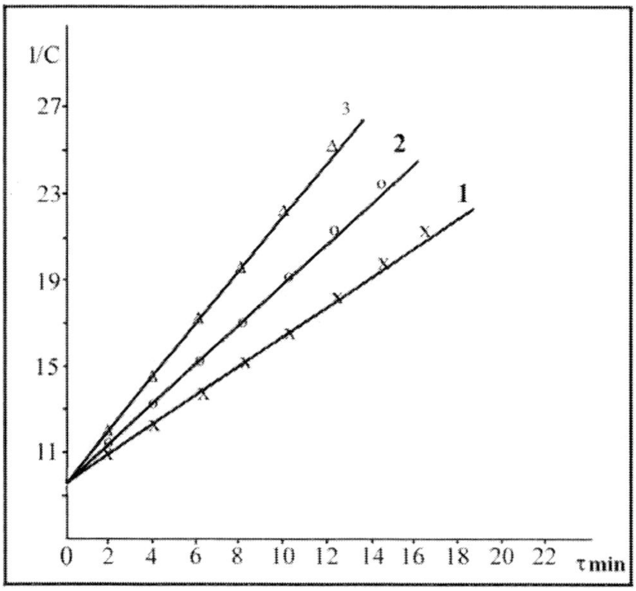

Figure 4. Dependence of reverse concentration on the time, where curve 1 is at 20^0C,curve 2 at 30^0C and curve 3 is at 40^0C (ratio 1:26.5).

Figure 5 and 6 show the dependence of the reaction rate constants logarithm on the reverse temperature, from which the activation energies of the reaction of dehydrocondensation were calculated. For the 1:53 ratio the activation energy is equal to $E \approx 29.2$ kJ/mol. This value of activation energy is near to the activation energy of the reaction of dehydrocondensation of hydroxy(alkyl)thioethanes to methylhydridesiloxane, which is equal to $E \approx 31.6$ kJ/mol [10]. For the 1:26.5 ratio the activation energy is equal to $E \approx 28,5$ kJ/mol.

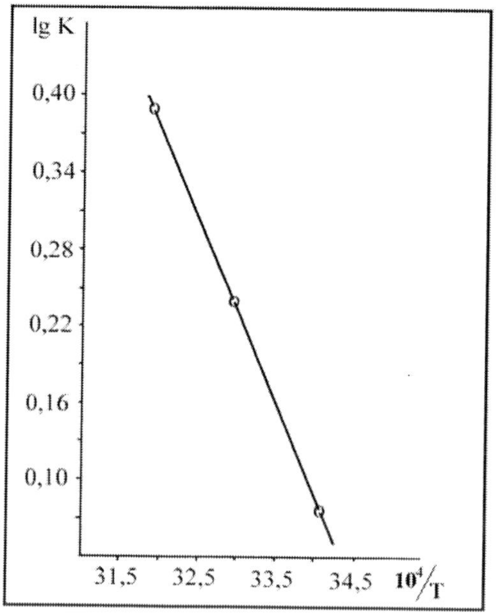

Figure 5. Dependence of reverse temperature on the rate constants of upon hydride addition of α,ω–bis(trimethylsiloxy)methylhydridesiloxane to allyl alcohol (ratio 1:53).

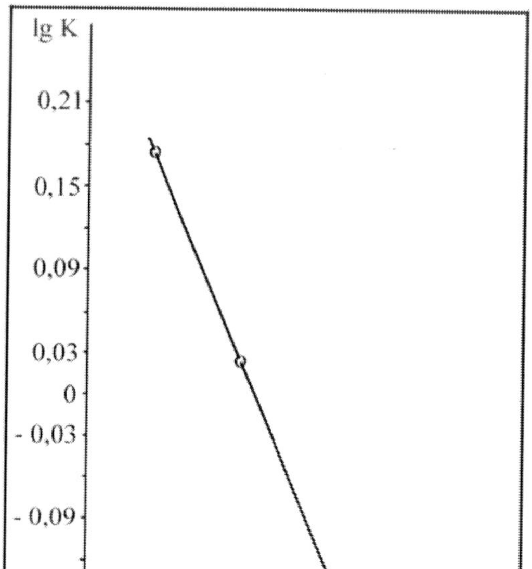

Figure 6. Dependence of reverse temperature on the rate constants of upon hydride addition of α,ω–bis(trimethylsiloxy)methylhydridesiloxane to allyl alcohol (ratio 1:26.5).

In the IR spectra of synthesized oligomers one can observe absorbtion band characteristic of asymmetric valence oscilation of ≡Si-O-Si≡ and ≡Si-O-C bonds in region 1020 and 1080 cm^{-1}. In the region 1275 and 2160 cm^{-1} , one can observe absorption band characteristic for ≡Si-Me and unreacted ≡Si-H bonds accordingly. In ^1H NMR spectra of oligomer III (Figure 7) one can observe singlet signals of methyl protons with chemical shift δ≈0,1 ppm, methylene protons in CH_2O group with chemical shift δ≈4.12 ppm, complicated multiplet for methylene protons in CH_2= group with chemical shift in the range δ≈4.9÷5.2 ppm, complicated multiplet for -CH= group with chemical shift δ≈5.7÷5.9 ppm, with centre δ≈5.8 ppm.

Figure 8 showed ^{13}C NMR spectra of oligomer III. In the spectra one can observe the signal for a carbon atom in the CH_2= fragment with chemical shift δ≈134.94 ppm, for the carbon in the CH= fragment δ≈113,06 ppm and in the CH_2O fragment δ≈61.79 ppm.

For the purpose of synthesis of methylsiloxane oligomers with hydroxy groups in the side chain the reaction of hydrosilylation of α,ω–bis(trimethylsiloxy)methylhydridesiloxane to allyl alcohol has been carried out in the ratio of initial components 1:53, in the presence of 0,1 M solution of platinum hydrochloric acid in tetrahydrofurane in the temperature range 30-40^{0}C.

Hydrosilylation reaction was investigated both in melt condition and in solution. During the reaction in melt condition the reaction proceeds very energetically and for a minute, gelation takes place for a little while. As it's known from the literature [11] platinum hydrochloric acid is a catalyst of hydrosilylation as well as the catalyst of dehydrocondensation.

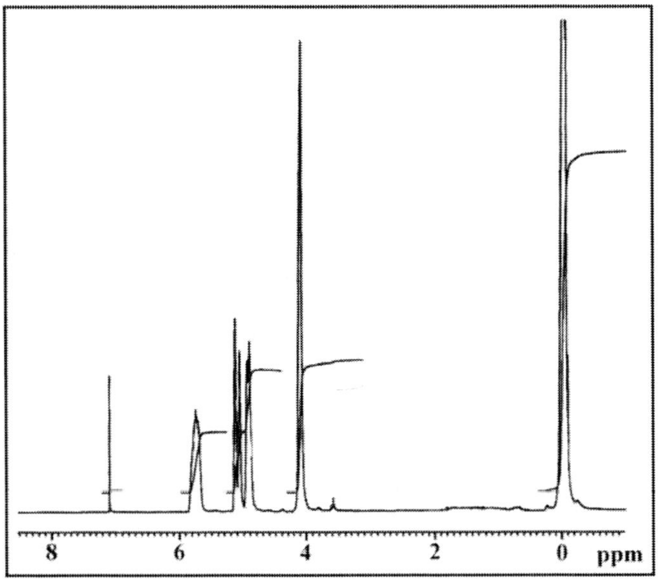

Figure 7. ^1H NMR spectrum of oligomer III.

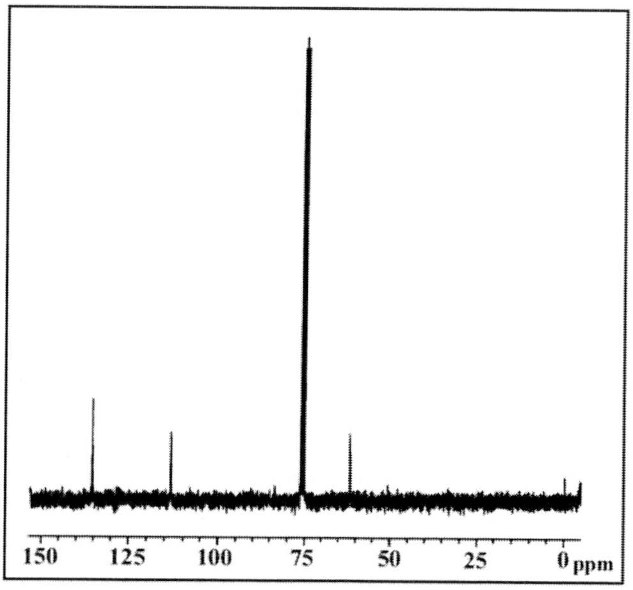

Figure 8. ^{13}C NMR spectrum of oligomer III.

$$\equiv Si-H+CH_2{=}CH-CH_2OH \xrightarrow{H_2PtCl_6} \equiv Si-CH_2-CH_2-CH_2OH$$

$$\equiv Si-H+HO-CH_2-CH{=}CH_2 \xrightarrow[-H_2]{H_2PtCl_6} \equiv Si-O-CH_2-CH{=}CH_2$$

Scheme 2. competitive reactions of ≡Si-H bonds with allyl alcohol in the presence of platinum hydrochloric acid.

So, during this reaction competitive reactions may proceed according to the following scheme 2.

During investigation, it was established that in the presence of platinum hydrochloric acid along the hydrosilylation reaction, the dehydrocondensation reaction takes place. To minimize gelation processes and to receive soluble oligomers the above-mentioned reactions were carried out in a dilute 40% solution of anhydrous toluene. In spite of that, the reaction was carried out in a dilute toluene solution at high stages of the reaction, a sewing reaction takes place. It must be denoted that during the hydrosilylation reaction, of releasing of hydrogen was observed. It was determined that during this reaction hydrosilylation proceeds about ~20%, and catalytic dehydrocondensation proceeds about ~50%, after that the sewing reactions began.

Therefore, the above-mentioned hydrosilylation-dehydrocondensation reactions at the first stages proceeds with the formation various linked oligomers, according to the following scheme 3. Where: [(a)+(b)+(c)](l)=m≈53 – VII.

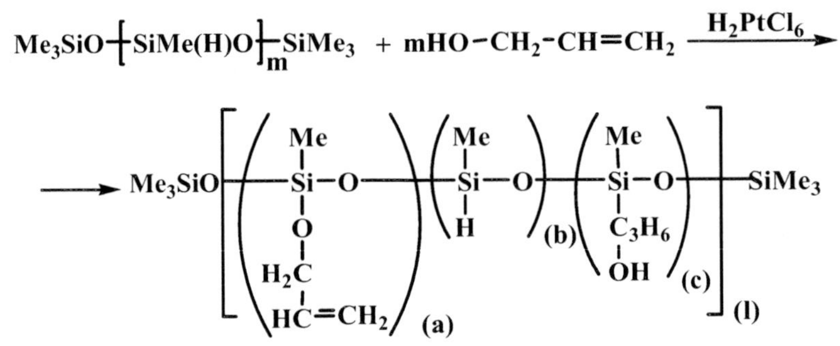

Scheme 3. Hydrosilylation-dehydrocondensation reactions of α,ω-bis(trimethylsiloxy)-methylhydridesiloxane with allyl alcohol, in the presence of platinum hydrochloric acid.

After this, the cross-linking processes take place. The formation of three-dimensional systems may proceed with competitive reactions according to the scheme 4.

$$\equiv Si{-}C_3H_6{-}OH + H{-}Si\equiv \xrightarrow[-H_2]{H_2PtCl_6} \equiv Si{-}C_3H_6{-}O{-}Si\equiv$$

$$\equiv Si{-}O{-}CH_2{-}CH{=}CH_2 + H{-}Si\equiv \xrightarrow{H_2PtCl_6} \equiv Si{-}O{-}C_3H_6{-}Si\equiv$$

Scheme 4. Competitive intermolecular macromolecular hydrosilylation-dehydrocon-densation reactions.

From the sewing reaction, the three-dimensional systems soluble part was taken, for which ^1H NMR investtigation was carried out. From Figure 9 it was seen that the hydride addition proceeds the Farmer rule as well as the Markovnikov rule with the ratio 3:1. In ^1H NMR spectra in the soluble part of oligomer VII, one can observe the singlet signal characteristic for methyl protons at silicon atoms with chemical shift δ≈0.1 ppm for methyl protons, in CH_3-CH= fragment; and for CH protons with chemical shifts: δ≈0.8 and δ≈1.7 ppm, these fragments may be obtained during Markovnikov addition; for CH_2 protons in -CH_2O- and ≡Si-CH_2- fragment (Markovnikov rule), with chemical shifts δ≈1.7 and δ≈3.7 ppm accordingly. During the reaction of the Farmer rule in the NMR spectra one can observe singlet signals with chemical shifts δ≈1.2; δ≈1.5 and δ≈3.8 ppm characteristic for protons in ≡Si-CH_2-, -CH_2- and –CH_2O- protons accordingly. In the spectra there are no observed resonance signals for hydroxyl groups, from which it's evident that in synthesized oligomers intra- and inter-molecular reactions of dehydrocondensation, take place according to the following scheme 5,6.

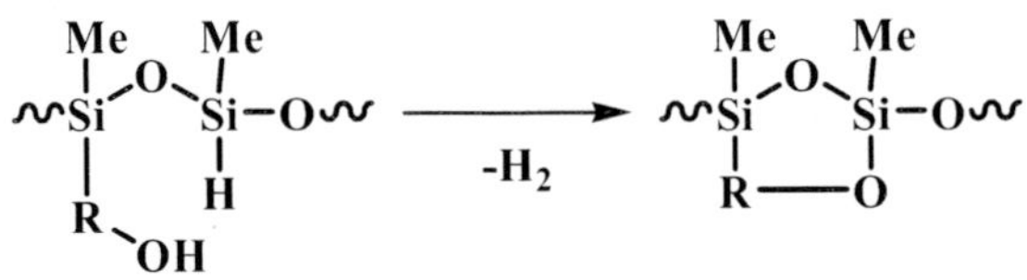

Scheme 5. Intra-molecular dehydrocondensation-cyclization.

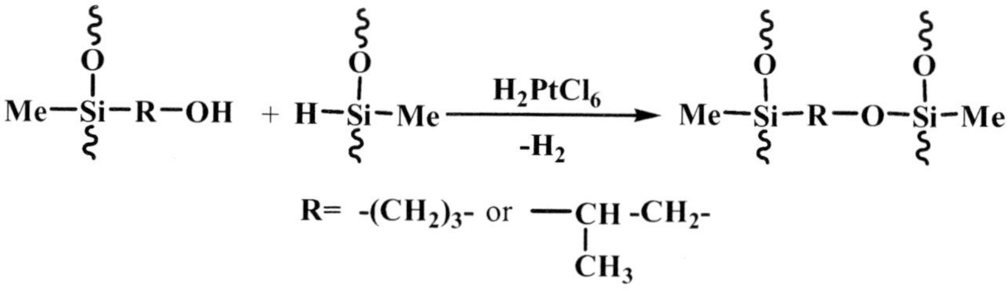

Scheme 6. Inter-molecular dehydrocondensation reaction.

Another side of the reaction proceeds by way of dehydrocondensation, in this case in synthesized oligomers resonance signals for unsaturated allyl fragments in the range 5.0÷6.0 ppm must be observed. But in the spectra Figure 9 there are no such resonance signals, which show that the fragment **≡Si-O-CH$_2$-CH=CH$_2$** may be entered in intra- and inter-molecular hydride addition by the scheme 7,8.

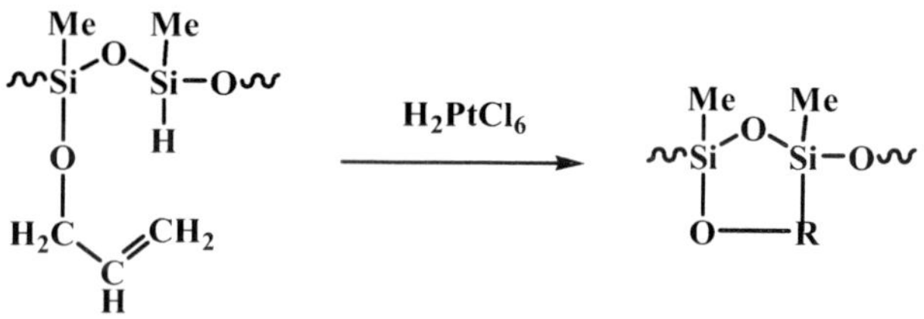

Scheme 7. Intra-molecular hydride addition.

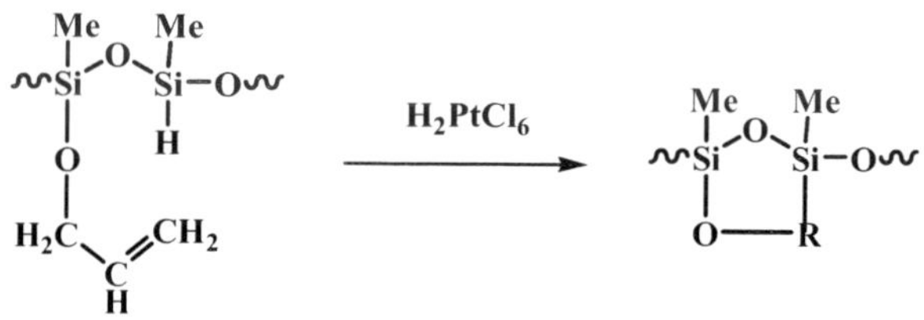

Scheme 8. Inter-molecular hydride addition, where: R= -(CH$_2$)$_3$- , or -CH(CH$_3$)CH$_2$-.

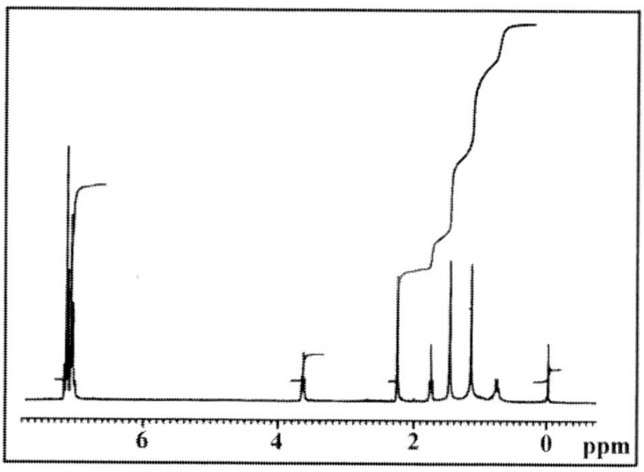

Figure 9. ^{1}H NMR spectrum of oligomer VII.

Gel permeation chromatographic analysis for oligomers was carried out. On Figure 10 molecular weight and molecular weight distribution of oligomer III are represented. Average molecular mass of oligomers changes in range $\bar{M}_{\omega}\approx 4.8\text{x}10^{-3}\div 6.5\text{x}10^{-3}$ and polydispersity D=2.5÷4.7. By differential scanning calorimetric investigation it was shown, that oligomers I, III and VI are characterized with one glass transition temperature (T_g). The value of this T_g is in the temperature range -123÷-110^{0}C (see table 1). The oligomers are characterized with small thermal-oxidative stability.

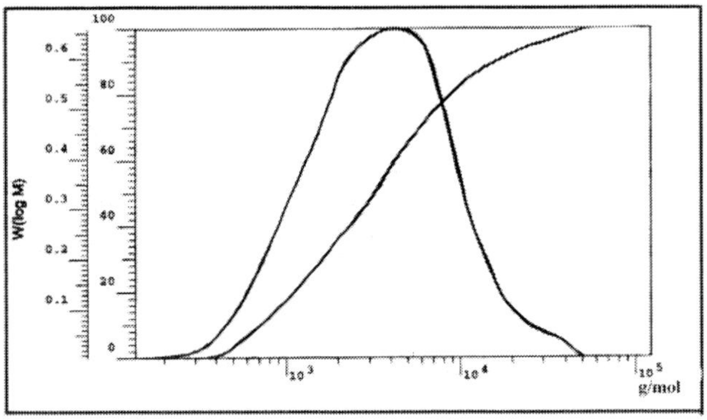

Figure 10. Gel permeation chromatographic curves of oligomer III.

One can observe 5% mass losses in the temperature range 230-250^{0}C. The main destruction process proceeds in the temperature range 270-540^{0}C. After 550^{0}C mass losses do not occur.

Wide-angle X-ray roentgenographic analysis of synthesized oligomers was carried out. As it's seen from Figure 11 the oligomers with oxyallyl groups in the side chain are one phase amorphous systems and maximum values of interchain distances of oligomers changes in the range $d_1\approx 9.30\div 9.40$ Å.

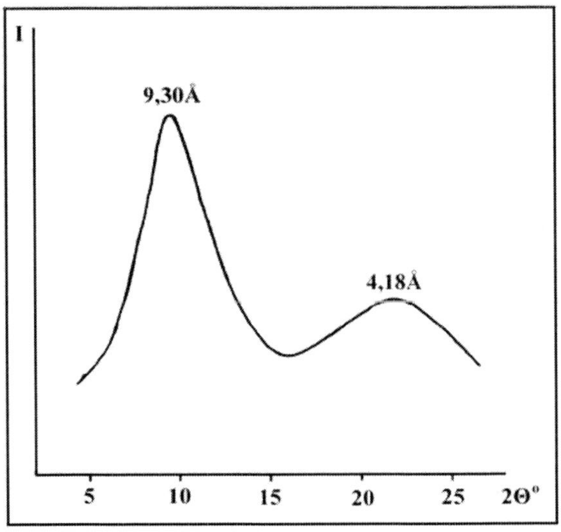

Figure 11. X-ray pattern of oligomer III.

So, by catalytic dehydrocondensation of α,ω–bis(trimethylsiloxy)methylhydridesiloxane with allyl alcohol methylhydridesiloxanes with oxyallyl side chain groups were synthesized by us for the first time. By hydride addition of α,ω–bis(trimethylsiloxy)methylhydridesiloxane with allyl alcohol in the presence of platinum hydrochloric acid in a low stage of conversion of active ≡Si-H groups (~77%) soluble methylhydridesiloxane with oxyallyl and propanol groups in the side chain were obtained. The synthesized comb-type methylhydridesiloxane with oxyallyl and propanol side groups are interesting products for modification of carbo chain elastomers.

REFERENCES

[1] P.P. Price. *J. Am. Chem. Soc.*, 1947, 69, 2600.

[2] B.N. Dolgov. Chemistry and Practical Application of Organosilicon Compounds; Chemical Library of Technical Information, Leningrad, 1958, 1, 18 (in Russian).

[3] K.A. Andrianov, B.G. Zavin. *Zhurn. Neorg. Khim.*, 1960, 12, 2216 (in Russian).

[4] S.N. Borisov, I.G. Sviridova, V.S. Orlova. Zhurn. Obshch. Khim., 1966, 36, 687 (in Russian). *Chem. Abstr.*, 1966, 65, 8946b.

[5] O.V. Mukbaniani, L.M. Khananashvili, N.A. Koiava, G.V. Porchkhidze, Yu.I. Tolchinski. Inten. *J.Polym. Mater.*, 1992, 17,113.

[6] L.M. Khananashvili, O. Mukbaniani, I.A. Inaridze, G.V. Porchkhidze, Kh.E. Koberidze. *Eur. Polym. J.*, 1998, 34, 581.

[7] B.N. Dolgov, I.P. Kharitonov, M.G.Voronkov. Zhurn. Obshch. Khim., 1954, 24, 861 (in Russian). *Chem. Abstr.*, 1955, 49, 8094i.

[8] J. Beines, C. Eaborn. *J. Chem. Soc.* 1955, 4023.

[9] Lee, C.L. *J. Organomet. Chem.*, 1966, 6, 603.

[10] E. Chachua, O. Mukbaniani, D. Gadzhiev-Shengelia, M. Karchkhadze, L. Khananashvili, *Bull. of the Georgian Academy of Sci.*, 1998, 3, 58 (in Russian).

In: Modification Reactions of Oligomethylhydridesiloxanes ISBN: 1-60021-365-0
Editors: O.Mukbaniani, et al. pp. 109-118

Chapter 7

FORMATION OF NEW THERMOREACTIVE POLYSILOXANES

O. Mukbaniani, T. Tatrishvili, G. Titvinidze and N. Mukbaniani
Iv. Javakhishvili Tbilisi State University, I. Chavchavadze Ave. 1,
0128 Tbilisi, Georgia

ABSTRACT

Hydrosilylation of α,ω–bis(trimethylsiloxy)methylhydridesiloxane to propargyl alcohol in the presence of the catalyst, platinum hydrochloric acid, was performed at several temperatures. Comb-type methylsiloxane oligomers with substituted reactive groups in the side chain have been obtained. At the high stages of reaction gelation takes place. Except for hydrosilylation, dehydrocondensation reaction also proceeds with a low yield. The synthesized oligomers were characterized by ^{1}H NMR spectroscopy. Calculations by quantum-mechanical semi empirical AM1 method for modelling reaction between methyldimethoxysilane [$Me(MeO)_2SiH$] and propargyl alcohol were performed for full characterization of hydride addition of polymethylhydrosiloxane to propargyl alcohol. For all initial, intermediate and final products enthalpies of formation as a function of the distance between ≡C-Si≡ bonds are calculated. Comb-type oligomers were characterized by gel-permeation chromatography, differential scanning calorimetry, thermogravimetric and wide-angle X-ray diffractometry analyses. Cross-linking of synthesized oligomers using γ-aminopropyltriethoxysilane was investigated and studied using DSC and TGA.

Key words: polymethylhydridesiloxane, propargyl alcohol, hydrosilylation, comb-type oligomers, cross-linking.

1. INTRODUCTION

Polysiloxanes are characterized by a number of interesting properties, e.g. excellent heat resistance, low toxicity, biocompatibility, high oxygen permeability, poor wettability, extremely low surface tension, low melting point and glass transition temperature, resistance to radiation, special rheological properties, outstanding electrical isolating properties and others [1, 2]. Polydimethylsiloxanes in which some methyl groups are replaced by reactive organofunctional groups are widely used. They have attracted much research and found useful applications, because it is a simple way to add new properties to the exceptional properties of silicones. Functional groups may be used to obtain new organic-silicone architectures. If the functional groups are distributed along the main chain, graft copolymers are obtained [3]. Macromolecular grafting is a synthetic approach that involves the functionalization of a preformed polymer backbone containing reactive groups with an appropriate reagent. There is an increasing interest in catalytic grafting of polyhydrosiloxane to an alternative route for the synthesis of organic–inorganic hybrid polymers, owing to the complexities involved in direct generation of well-defined linear hybrid polymers [4]. Besides the process of equilibration, the hydrosilylation process is the most widely used method of preparing organofunctional polymethylsiloxanes starting from poly(methylhydro)siloxanes [5, 6]. Hydrosilylation of alkynes results in versatile materials for natural products [7] and polymers [8]. Selectivity is also an important consideration in the hydrosilylation of alkynes because in addition to regioisomers and stereoisomers as legitimate products there is also the probability of obtaining diaducts [9]. Hydrosilylation of alkynes offers a greater synthetic challenge of the triple bond and the potential for a broader product distribution [10]. The majority of functionalized polysiloxanes were designed to undergo thermally induced cross-linking, and cure by hydrolytic processes or photochemically induced polymerization [1].

The properties of organosilicon polymers depend on the structure of macromolecular chains and on the nature of organic groups surrounding the silicon atom [11].

2. Experimental Part

2.1. Materials and Techniques

The starting materials for the synthesis of thermoreactive comb-type methylorganosiloxane oligomers with functional groups in the side chain were α,ω–bis(trimethylsiloxy)methylhydrosiloxane and propargyl alcohol.

The initial α,ω–bis(trimethylsiloxy)methylhydridesiloxane had the degree of polymerization $n \approx 35$, propargyl alcohol and catalyst H_2PtCl_6 were obtained from Aldrich and used as received. The organic solvents were cleaned by drying and distillation.

FTIR spectra were obtained on Nicolet Nexus 470 with MCTB detector. ^{1}H spectra were recorded on a 400 MHz Bruker NMR spectrometer, using $CDCl_3$ as solvent and internal standard. Thermal gravimetric analysis was performed on a Perkin-Elmer TGA-7, the thermogravimetric analyzer at ~10 ^{0}C/min. Differential scanning calorimetric analysis was performed on a Perkin-Elmer DSC-7. Heating and cooling scanning rates were ~5 ^{0}C/min.

Gel-permeation chromatography (GPC) investigations were performed using a Waters Model 6000A chromatograph with an R 401 differential refractometer detector. The column set comprised 10^3 and 10^4 Å Ultrastyragel columns. Sample concentration was approximately 3% by weight in toluene; a typical injection volume for the siloxane was 5 μL. standardization of the GPC was accomplished by the use of styrene or polydimethylsiloxane standards with the known molecular masses.

Wide-angle X-ray diffractograms were taken on a DRON-2 (Burevestnik, Saint-Petersburg, Russia) instrument using A-CuK_{α} radiation. The values were measured without a filter, and the angular velocity of the motor was ~2 deg/min.

2.2. Hydrosilylation of α,ω–Bis(Trimethylsiloxy)Methylhydrosiloxane to Propargyl Alcohol

The hydrosilylation reaction was carried out in a two-necked flask equipped with a magnetic stirrer, tube for catalyst inlet and reflux condenser, which was connected to the equipment for measuring the volume of the eliminated hydrogen.

The reaction products 0.8677g (0.00038 mmole) α,ω–bis(trimethylsiloxy) methylhydrosiloxane and 0.7519 g (0.0134 mmole) propargyl alcohol and 4.2 ml dry toluene were placed into the flask and then into an oil bath until constant temperature was reached. Then the catalyst 0.1 M solution of platinum hydrochloric acid in tetrahydrofuran ($5 \div 9 \cdot 10^{-5}$ g per 1.0 g of starting substance) was introduced. During the reaction volume of eliminated hydrogen was measured. After the reaction completion the solvent was eliminated on a vacuum pump. During the reaction partial gelation occurs ; after evaporation of the solvent 3 ml toluene has been added and the soluble part was extracted, the yield of soluble polymer is 60% (0.88 g).

2.3. Cross-Linking Using Γ-Aminopropyltriethoxysilane

The soluble part of the hydrosilylation product was dissolved in toluene at room temperature and γ-aminopropyltriethoxysilane was added (5% of oligosiloxane by weight). The solution, after a 10 minute stirring, was poured into a pan and placed in an oven for curing at 120 ^{0}C for 12 h.

3. Progress of the Reaction

There is extensive literature on hydrosilylation reactions of monomeric organohydrosilanes to propargyl alcohol. Products may contain three isomeric organosilanes obtained according to anti-Markovnikov rule (β-(E)- and β-(Z)-isomers) and Markovnikov rule (α-isomer). In general, we can represent this by the following schemes [12, 13]:

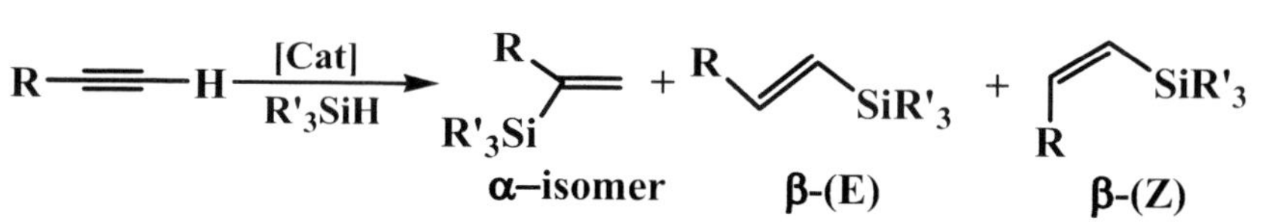

(1)

For synthesis of comb-type, thermo reactive siliconorganic oligomers with functional groups in the side chain via hydrosilylation reaction of α,ω–bis (trimethylsiloxy) methylhydridesiloxane to propargyl alcohol we have used 0.1 M solution of H_2PtCl_6 in THF. 1:35 was the ratio of the initial compounds.

Preliminary heating of initial compounds in the temperature range of 50 ÷ 60^{0}C in the presence of catalyst showed that in these conditions there is no polymerization of propargyl alcohol, nor scission of siloxane backbones, nor elimination of methane from methylhydridesiloxane. There were no changes in the gel-permeation chromatography, also in NMR and FTIR spectra of initial compounds were found.

From the literature it is known, that platinum hydrochloric acid represents a catalyst of hydrosilylation reactions, as well as of dehydrocondensation reactions [14]. Because of this, we have investigated hydrosilylation reactions in detail. As we mentioned above in the literature are known the hydrosilylation reactions of propargyl alcohol with hydrosilanes. According to the literature data during hydrosilylation does not take place during dehydrocondensation reactions [12]. In spite of this we established, that the next concurrent reactions proceed:

$$\equiv Si{-}H + CH{\equiv}C{-}CH_2OH \longrightarrow \equiv Si{-}CH{=}CH{-}CH_2OH$$
$$\equiv Si{-}H + HO{\cdot}CH_2{-}C{\equiv}CH \xrightarrow[-H_2]{} \equiv Si{-}O{-}CH_2{-}C{\equiv}CH \qquad (2)$$

In general we can represent hydrosilylation of α,ω-bis (trimethylsiloxy)methylhydride-siloxane with propargyl alcohol by the following scheme:

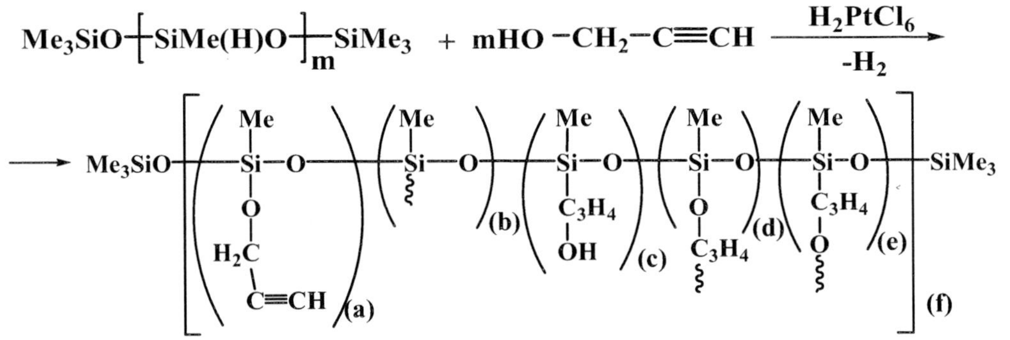

(3)

Where: [(a)+(b)+(c)+(d)+(e)] · (f) = m ≈ 35; I[1] – 50 ^{0}C, I – 60 ^{0}C

The synthesized oligomers are orange liquids. They are well soluble in ordinary aromatic type organic solvents with the specific viscosity $\eta_{sp} \approx 0.06 \div 0.08$. Structure and composition of the oligomers were established by elementary analysis, determination of molecular masses, by ^{1}H NMR spectral data. Some physical –chemical properties of synthesized oligomers are presented in table 1. We observe that after a definite time at room temperature synthesized oligomers autocatalytically are cross-linked.

Table 1. Some physical-chemical properties of synthesized oligomers.

№	Reaction temperature / °C	yield %	η_{sp} *	$\overline{M}_{\omega} \cdot 10^{-3}$, $M_n \cdot 10^{-3}$, (D)	T_g,°C	Temperatures / °C for 5 % mass losses	d_1, Å	Elementary* composition, %		
								C	H	Si
I[1]	50	65	0.06	29.45 6.4 (4.6)	+ 76	270 °C	7.8	38.47 38.11	6.90 6.72	27.91 27.79
I	60	60	0.08	30.55 6.5 (4.7)	+ 77	270 °C	7.8	39.86 39.69	6.88 6.79	27.01 25.55

* in 1% toluene solution at 25 °C. N

** In numerator there are calculated values, in denominator – found values.

The reaction was carried out in a solution of toluene and without solvent. Without solvent, the reaction proceeds vigorously by gelation, because of this, a further reaction was carried out in the toluene solution. During the hydrosilylation reaction a branching process takes place with participation of residual ≡Si-H groups in intermolecular cross-linking reactions by the following general scheme:

$$\equiv Si-C_3H_4-OH + H-Si\equiv \xrightarrow[-H_2]{} \equiv Si-C_3H_4-O-Si\equiv$$
$$\equiv Si-O-CH_2-C\equiv CH + H-Si\equiv \longrightarrow \equiv Si-O-C_3H_4-Si\equiv \quad (4)$$

From the cross-linked system, the soluble part was extracted and investigated by NMR spectral analysis. In Table 2 we present NMR spectra data for oligomer I:

$\equiv Si-CH=CH-CH_2-OH$	$\equiv Si-C(=CH_2)-CH_2-OH$	$\equiv Si-O-CH_2-C\equiv CH$
anti-Markovnikov rule	**Markovnikov rule**	**Dehydrocondensation**

Table 2. ^{1}H NMR δ (ppm) of oligomer I.

Si-CH$_3$	Si(CH$_3$)$_3$	-C≡CH	-CH$_2$-O-Si≡	-CH$_2$-OH	-CH$_2$-OH	=C=CH$_2$	≡Si-CH=	Si-CH=CH-
0.1	0.2	2.5	3.9	4.1	4.7	5.5	5.8	6.3

From ^{1}H NMR data we infer that the hydrosilylation proceeds according to the anti-Markovnikov and Markovnikov rule. The ratio of the fragments obtained from the anti-Markovnikov and Markovnikov rule is equal to 1 : 1. Also spectra data show the proceeding of dehydrocondensation, which is in agreement with the experimental investigation. The signal for hydroxyl group with low intensity in the range δ≈4.7 ppm is caused by participation of the hydroxyl group in the branching process according to the intermolecular and intramolecular dehydrocondensation reactions according to the following scheme:

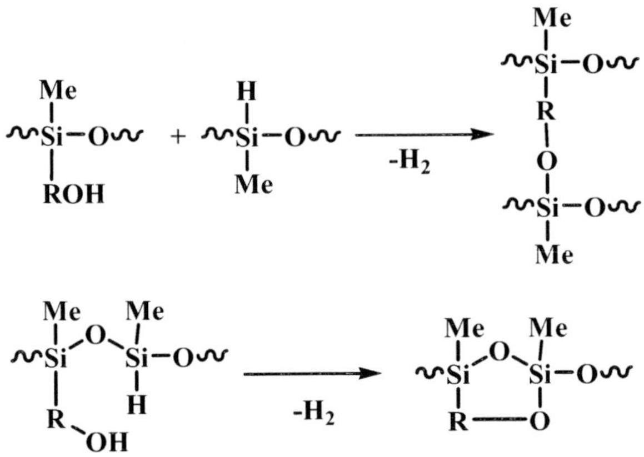

(5)

R = $-CH_2-CH=CH-$; $-CH(Me)=CH-$

We have carried out and investigated cross-linking of synthesized oligomer using γ-aminopropyltriethoxysilane according to the technique described in the experimental part.

4. Quantum-Chemical Calculations

Semi empirical quantum-mechanical methods are simplified versions of the Hartree-Fock approach using empirical (= derived from experimental data) corrections in order to improve performance. These methods are usually referred to through acronyms encoding some of the underlying theoretical assumptions. We apply one of the most frequently used AM1 methods. It is based on the neglect of differential diatomic overlap (NDDO) integral approximation. This approach belongs to the class of zero differential overlap (ZDO) methods, in which all two-electron integrals involving two-center charge distributions are neglected. A number of additional approximations are made to speed up calculations and a number of parameterized corrections are made in order to correct for the approximate quantum-mechanical model. For AM1 the parameterization is performed so that we obtain enthalpies of formation H^{form} instead of total enthalpies [15] as a function of the distance R_{C-Si}. The calculations also provide us with P_{x-y} values which represent bond orders.

We have performed calculations by quantum- chemical half empiric AM1 method for modelling reaction between methyldimethoxysilane [$Me(MeO)_2SiH$] and propargyl alcohol for full characterization of hydride addition of polymethylhydrosiloxane to propargyl alcohol using software Chem3D Ultra 9.0 from Cambridge Soft.

We consider the hydrosilylation of [$Me(MeO)_2SiH$] with propargyl alcohol in view of hydrosilylation according to the anti-Markovnikov rule and Markovnikov rule; we also consider the dehydrocondensation reaction. We calculate activation energies (U_{act}) and formation enthalpies (H^{form}). Calculation method is described in our previous article [16]. In table 3 are presented activation energies and formation enthalpies for the reaction according to hydrosilylation (anti-Markovnikov rule and Markovnikov rule) and dehydrocondensation.

Table 3. Formation enthalpies and activation energies modelling hydrosilylation-dehydrocondensation reaction with methyldimethoxysilane to propargyl alcohol.

Reaction Direction	formation enthalpies H^{form} kJ/mol	activation energy U_{act} kJ/mol
Markovnikov rule	-191.0	185.5
Anti-Markovnikov rule	-190.2	185.0
Dehydrocondensation	-76.0	224.0

We now compare formation enthalpies. Clearly reaction of methyldimethoxysilane to propargyl alcohol is more favourable according to hydrosilylation. Also the possibility of proceeding hydrosilylation according the anti-Markovnikov and Markovnikov rule is equal. This result is in agreement with experimental data.

5. Gel-Permeation Chromatography Results

The synthesized oligomers were studied by GPC. Figure 1 shows the molecular weight distribution of oligomer I. A trimodal curve is seen. We find the number of average molecular mass $\overline{M}_n \approx 6.5 \times 10^3$ and the weight average molecular mass $\overline{M}_\omega \approx 30.55 \times 10^3$. Thus, the polydispersity index is equal to 4.7.

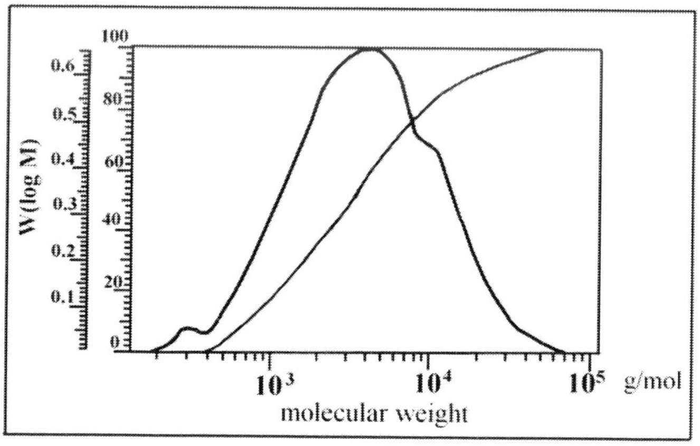

Figure 1. Gel-permeation chromatographic curves for oligomer I.

The average molecular weight of synthesized oligomer I exceed ~ 8 times the theoretical value calculated for the full hydrosilylation. This indicates that branching processes take place during the reaction. Branching process is caused by intermolecular reactions, which we discussed before (scheme 4). Also the difference in appearance and degree of crosslinking can be attributed to the effectiveness of the different Pt - species of H_2PtCl_6, which cannot be separated from oligomers after reaction. Pt is partially in the form of Pt (II). It reacts with the ≡Si-H group of the PMHS backbone with elimination of hydrogen and producing Pt metal.

This fact leads us to suggest that crosslinking also occurs according to the following scheme [16-18]:

$$2 \equiv Si\text{-}H + Pt\ (II) + H_2O \rightarrow \equiv Si\text{-}O\text{-}Si\equiv + Pt\ (0) + H_2\uparrow \quad (5)$$

In the FTIR spectrum we find a signal with low intensity in the region 1187 cm^{-1}, which is characteristic for silsesquioxane bonds ($SiO_{3/2}$) and tells us about side reactions due to crosslinking. Crosslinking in such a way proceeds with a low degree.

6. Calorimetric Results: DSC and TGA

We show in Figure 2 DSC results for oligomer I (curve 1) and for the cross-linked system (curve 2). In the case of oligomer I there is only one endothermic peak, which corresponds to the glass transition temperature $T_g \approx 77\ ^0C$. High glass transition temperature shows cross-linking in polymer by formation of a rigid system. After cross-linking, using γ-aminopropyl-triethoxysilane, glass transition temperature of the polymeric system increases up to 159 ^0C.

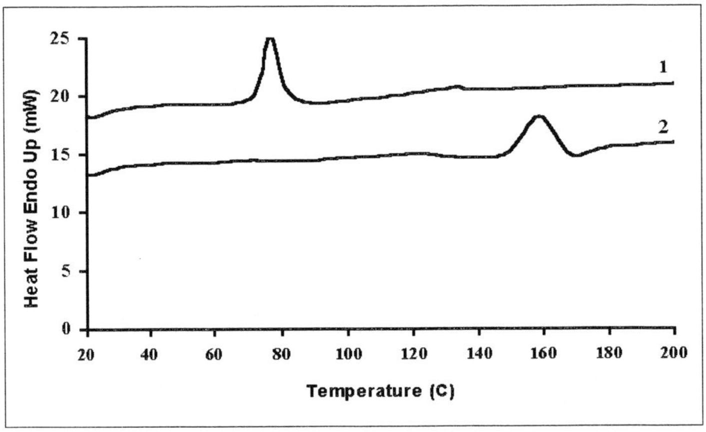

Figure 2. DSC curve of oligomer I (1) and for cross-linked system (2).

We have performed TGA in air and also in a nitrogen atmosphere. Return to table 1. We present that 5 % mass losses in air atmosphere are observed at ~ 270 ^0C. The main thermo-oxidative destruction process proceeds in the range 300-480 ^0C. In Figure 3 we show thermogravimetric curves for oligomer I and a cross-linked system in the nitrogen atmosphere.

TGA provides several kinds of information [19]. One of them is the temperature at which 5 % weight loss occurs. For synthesized oligomer I in a nitrogen atmosphere this takes place at ~320 ^0C. Cross-linking using γ-aminopropyltriethoxysilane increases thermal stability of the polymer (~480 ^0C).

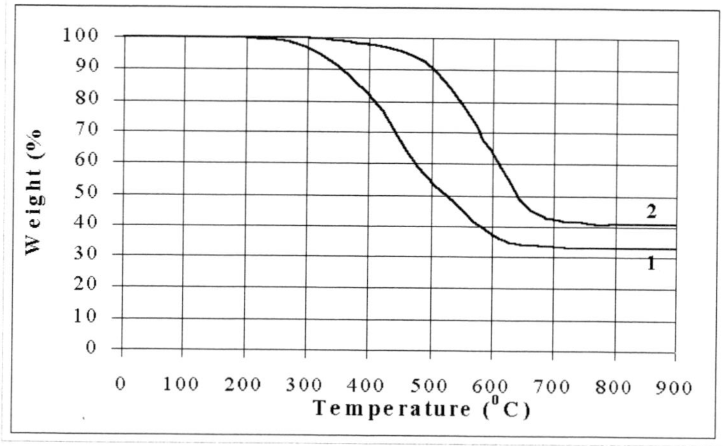

Figure 3. Thermogravimetric curves of oligomer I (1) and cross-linked polymer (2).

7. Possible Application

Functionalized siloxane monomers and oligomers, which undergo rapid crosslinking reactions, are technologically important materials. These siloxanes find specific uses in such diverse areas as elastomers, caulks and sealing agents, adhesives, and as release agents [20]. We are testing obtained oligomers as thermal crosslinking agents for several thermoplastic polymers and for several composites.

References

[1] H.R. Kricheldorf, (Ed.), *Silicon in Polymer Synthesis*, Springer, Berlin, 1996.

[2] C.G. Cofer, J. Economy. Chapter 2 in *Mechanical and Thermophysical Properties of Polymer Liquid Crystals*, Brostow, W. (Ed.), Chapman and Hall, London, 1998.

[3] K. Kazmierski, N. Hurduc, G. Sauvet, J. Chojnowski. *J. Polym Sci : Part A : Polym. Chem.*, 2004, 42, 1682.

[4] B.P.S. Chauhan, J.S.Rathore, N. Glloxhani. *Appl. Organometal., Chem.* 2005, 19, 542.

[5] M.P. McGrath, E.D. Sall, S. Tremontm. *J. Chem. Rev.* 1995, 95, 381.

[6] J.M. Buriak. *Chem. Rev.* 2002, 102, 1271.

[7] C.J. Flann, L.E. Overman. *J. Am. Chem. Soc.*, 1987, 109, 6115.

[8] J. Oku, M. Taksuchi, A. Saito, R. *Asami. Polym. J.*, 1992, 24, 1409.

[9] R.A. Benkeser, D.F. Ehler. *J. Organomet. Chem.*, 1974, 69, 193.

[10] M. Chauhan, B.J. Hauck, L.P. Keller, P. Boudjouk. *J. Organomet. Chem.*, 2002, 645, 1.

[11] O.V. Mukbaniani, and L.M. Khananashvili. Internat. *J. Polym. Mater.* 1994, 27, 31.

[12] B. Marciniec, in *Comprehensive Handbook on Hydrosilylation*, Pergamon Press, Oxford, 1992.

[13] B. Marciniec, J. Guli, H. Maciejewski, *in Encyclopedia of Catalysis*, Wiley, New York, 2002.

[14] I. Toulokhonova, B. Bjerke-Kroll, R. West. *J. Organomet. Chem* 2003, 686, 101.

[15] M.I.S. Dewar, E.G. Zeobish, E.F. Healy, J.S. Stewart. *J. Amer. Chem. Soc.*, 1985, 107, 3902.
[16] O Mukbaniani, G. Zaikov, U. Scherf, S. Meladze, T. Tatrishvili, Z. Pachulia, Kh. Koberidze, M. Labartkava. J. Balk. *Trib. Assoc.* 2005, 11, 1.
[17] G.W. Gray, D. Lacey, G. Nestor, M.S. White. *Makromol. Chem. Rapid Commun.*, 1986, 7, 71.
[18] J. Lipowitz, S.A. Bowman. *J. Org. Chem.*, 1973, 38, 162.
[19] K.P. Menard, Chapter 8 in *Performance of Plastics*, W. Brostow, (Ed.), Hanser, Munich-Cincinati, 2000.
[20] Y. Abe, T. Gunji. *Prog. Polym. Sci.*, 2004, 29, 149.

In: Modification Reactions of Oligomethylhydridesiloxanes ISBN: 1-60021-365-0
Editors: O.Mukbaniani, et al. pp. 119-137

Chapter 8

COMB-TYPE SILICONORGANIC COMPOUNDS WITH EPOXY GROUPS IN THE SIDE CHAIN

O. Mukbaniani[1], U. Scherf[2], G. Gurgenidze[1], S. Meladze[1], L. Khananashvili[†]

[1]I.Javakhishvili Tbilisi State University, 1, I.Chavchavadze ave., 0128, Tbilisi, Georgia.

[3]Max-Planck-Institute in Polymer Research, Ackermannweg 10, D-55128 Mainz, Germany

ABSTRACT

The reaction of hydride addition of α,ω–bis(trimethylsiloxy)methylhydridsiloxanes with allyl glycidyl ether in the presence of platinum hydrochloric acid and Pt/C at the 1:30 ratio of initial compounds has been studied. It was established that when the reaction proceeds without solvent, three-dimensional systems are obtained.

The optimal condition of hydride addition reaction, concentration of the solution, the reaction temperature has been determined. The reaction order, activation energy and hydride addition rate constants were found. In the case of platinum hydrochloric acid, hydride addition proceeds with inductive an period, and faster than in case of the catalyst platinum on the carbon Pt/C.

The reactions of primary and secondary amines with epoxy groups containing organosiloxanes were carried out and comb-type organosiloxanes with aminohydroxyl groups in the side chain have been obtained. After protonation of amino nitrogen groups in the side chain, water soluble organosiloxane oligomers were obtained.

The structure of all synthesized comb-type organosiloxanes were determined by IR, ^{1}H and ^{13}C NMR spectra data.

Key words: epoxy organosiloxane, hydride addition, platinum hydrochloric acid.

INTRODUCTION

Epoxy containing organosiloxane may be obtained by hydride addition of mono- and dihydride containing linear organosilanes and siloxanes to allyl glycidyl ether[1÷4] in the presence of platinium hydrochloric acid in anhydrous isopropanole.

During the hydrosilylation of allyl glycidyl ether with 1,4-bis(dimethylsilyl)benzene, or diphenylsilane in isothermal condition by calorimetric method at various concentrations of initial compounds and catalysts, the hydrosilylation rate constants and reaction order have been calculated [5].

The epoxy containing organosiloxanes, (with Si-C bonds) are obtained by oxidative epoxydation of α,ω-divinylorganosiloxanes the yield of synthesized diepoxyorganosiloxanes in this case is about ~85% in the tetrachlorocarbon area [6].

Epoxy organosiloxanes with various chain lengths may be obtained by anionic cooligomarization of organocyclosiloxanes with 1,3-bis(3-glycidoxypropyl)tetramethyl disiloxane [7].

The epoxy ring-opening reaction with organic amines [8] and organic acids [8,9] showed that it proceeds with the formation of β and α-adducts.

The epoxy ring-opening reactions with primary and secondary amines showed that in the case of primary amines the reaction proceeds completely and has a macromolecular character[10].

Another side increases the possibilities which give us the of modification reactions of methyl(ethyl)hydrogensiloxanes[11]

EXPERIMENTAL PART

The starting materials for the synthesis of comb-type organosiloxanes with epoxy groups in the side chain were α,ω-bis(trimethilsiloxy)methylhydridsiloxanes with various degrees of polymerization n≈30,53, and ally glycidyl ether.

The organic solvents were purified by drying and distillation. The purity of the starting compounds was controlled by gas-liquid chromatography "LKhM-8 MD" (Russian model), phase SKTF-100(10%), the NAW chromosorb, carrier gas He and the 2M column.

IR spectra all of samples have been taken on an UR-20 instrument, while the ^{1}H NMR spectra was taken on a "Perkin-Elmer" at operating frequency of 250 MHz. All spectra were obtained with the use of $CDCl_3$ or D_2O as the solvent and internal standard.

Gel permeation chromatography was carried out with the use of a Waters Model 6000A chromatograph with an R 401 differential refractometer detector. The column set comprised10^3 and 10^4 Å Ultrastyragel columns. Sample concentrations were approximately 3% by weight in toluene, and typical injection volume for the siloxane was 5 μL. Standardization of the GPC was accomplished by the use of styrene or PDMS standards with the known molecular weight.

Hydride Addition of Methylhydridesiloxane to Allyl Glycidyl Ether

The hydrosilylation reaction was carried out in a three-necked flask equipped with stirrer and reflux condenser, with a calcium chloride tube. The initial reagents were placed in no space to the flask and thermostated in an oil bath until constant temperature was achieved. 0.1M solution of platinum hydrochloric acid in tetrahydrofuran ($5 \div 10 \times 10^{-5}$ g per 1g of the starting substance) was used as a catalyst. The reaction was performed in the argon atmosphere in the presence of toluene at temperature region 85÷95°C. Then the reaction mixture was connected to the vacuum and the toluene with un-reacted allyl glycidyl ether was removed. The oligomers I-II were isolated after the solvent was removed in a vacuum at 70-80°C and they are clear transparent oils. The other epoxyorganosiloxane II was obtained by the same method.

The Addition of Butyl Amine to Epoxyorganosiloxane

The reaction was carried out in a three-necked flask equipped with a reflux condenser and chlorocalcium tube, dropping funnel and mechanical stirrer. The solution of 2.1370 g (0.3969 mmole) of epoxyorganosiloxane in 15 ml of anhydrous solution of toluene and 2.6076 g (0.03572 mole) butyl amine was refluxed under argon, until epoxide resonances disappeared at δ=2.7 and δ=3.1 ppm in the ^{1}H NMR spectrum. Then the solvent and excess amine was removed in a vacuum at 80^{0}C to give 2.9 g (96%) the product V.

The Addition of Diethylamine to Epoxyorganosiloxane

The reaction was carried out in a three-necked flask equipped with a reflux condenser and chlorocalcium tube, dropping funnel and mechanical stirrer. The solution of 2.5825 g (4.796 mmole) of epoxyorganosiloxane in 17 ml of toluene anhydrous solution and 3.1509 g (0.043 mole) diethyl amine were refluxed under argon, until the disappearance of epoxide resonances at δ=2.7 and δ=3.1 ppm in the ^{1}H NMR spectrum. Then the solvent and excess amine was removed in a vacuum at 80°C to give 3.5 g (97%) the product X.

THE SYNTHESIS OF OTHER HYDROXYAMINOSILOXANES WAS CARRIED OUT USING THE SAME METHOD

The Reaction of Hydrogen Bromide with Diethylaminohydroxyorganosiloxane

The reaction was carried out in a three-necked flask equipped with a reflux condenser and chlorocalcium tube, dropping funnel and mechanical stirrer. To the solution of 0.9828 g (0.1297 mmole) of diethyalaminohydroxyepoxysiloxane in 10 ml anhydrous toluene solution 0.3145 g (3.89 mmole) (48%) hydrogen bromide was added in 3 ml water at 0^{0}C temperature range. From water solution 1.2 g (93%) transparent vitreous product XIII was received.

The Reaction of Acetic Acid with Diethylaminohydroxyorganosiloxane

The reaction was carried out in a three-necked flask equipped with a reflux condenser and chlorocalcium tube, dropping funnel and mechanical stirrer. To the solution of 1.7662 g (0.233 mmole) of diethylaminohydroxyepoxysiloxane in 10 ml dry CH_2Cl_2 0.4199 g (6.993 mmole) acetic acid (99.8%) in 15 ml water at 0^0C temperature range was added, after that the solution clouded. After processing of the water solution 2.1 g (95.5%) transparent oil type the compound XV was received.

RESULTS AND DISCUSSION

In the literature there is some information about synthesis of polyfunctional organosiloxane containing glycidoxypropyl groups in the side dimethylsiloxane chain [12].

The present paper deals with synthesis and studies of the properties of comb-type organo-siloxanes containing epoxy groups in the side chain.

For the purpose of synthesis of organosiloxanes and with the epoxy group in the side chain, the reaction of hydride addition of α,ω-bis(trimethylsiloxy)methylhydridsiloxanes ($n \approx 30,53$) to allyl glycidyl ether at 1:30 or 1:53 ratio of initial compounds, in the presence of platinum hydrochloric acid or Pt/C as a catalyst has been investigated.

The reaction of hydrosilylation was investigated without solvent in the presence of catalysts at temperature range 80-90^{0}C. In both cases the hydrosilylation proceeds with an inductive period, after that the reaction proceeds autocatalytic with the formation of three-dimensional systems.

For the purpose of synthesis of soluble, epoxy group containing comb-type organosiloxanes, hydrosilylation reactions were investigated in dilute solutions of anhydrous toluene. The optimal concentration is about 0.1 mole/l. Lower than this concentration; sewing processes does not take place. The hydrosilylation was carried out at various temperatures 80÷90^{0}C. The course of the reaction was observed by decrease of amount of active ≡Si-H groups. As it is clear from Figure 1, with the rise of the temperature, the rate of hydride addition increases. At one and the same temperature the rate of hydrosilylation in the presence of platinum hydrochloric acid is tighter than in the case of platinum on carbon. As it is obvious from Figure 1 the reaction mainly proceeds vigorously during the first hours. After 2h the conversion slowly changes.

From Figure 1 we can see that conversion of active ≡Si-H groups is not complete and decreases from 98% ($t=90^0C$) to 92% ($t=80^0C$).

The synthesized oligomers with epoxy groups in the side chain, after removal of solvents are liquid, colorless (or light yellow) transparent products well soluble in ordinary organic solvents, with $\eta_{sp} \approx 0,04 \div 0,08$. Some physical-chemical properties of synthesized oligomers are presented in table 1.

Table 1. Some physical-chemical properties methylhydridesiloxanes with organic branching side groups.

№	Yield. %	η_{sp}*	d_1, Å	$\bar{M}_\omega$	Elementary composition, %				
					C	H	Si	N	Br/S
I	92	0.05	-	5384 / 5200	48.13 / 47.74	8.19 / 8.01	16.77 / 16.30	-	-
II	94	0.04	-	-	48.13 / 47.98	8.19 / 8.08	16.77 / 16.45	-	-
II'	92	0.04	-	-	48.13 / 47.72	8.19 / 7.93	16.77 / 16.56	-	-
III	97	0.05	-	-	48.13 / 48.01	8.19 / 8.21	16.77 / 16.14	-	-
IV	93	0.06	-	9384 / 9000	48.13 / 48.81	8.19 / 7.87	16.77 / 16.60	-	-
V	97	0.06	-	7576 / 7500	53.21 / 53.30	10.21 / 10.13	11.85 / 11.60	5.54 / 5.12	-
VI	97	0.07	-	8596 / 8600	59.46 / 59.13	8.29 / 7.87	10.44 / 10.13	4.88 / 4.41	-
VII	98	0.06	-	9016 / 8900	60.68 / 60.31	8.58 / 8.30	9.96 / 9.63	4.66 / 4.20	-
VIII	-	-	-	9134 / 8700	52.02 / 51.76	7.09 / 6.70	9.80 / 9.31	4.59 / 4.70	10.51 / 9.78
IX	97	-	-	8654 / 8300	54.91 / 54.28	7.48 / 6.97	10.35 / 9.86	4.85 / 4.41	-
X	98	0.07	6.94	7576 / 7600	53.21 / 52.78	10.21 / 9.89	11.85 / 11.56	5.54 / 5.08	-
XI	97	0.07	6.94	8596 / 8500	59.46 / 59.22	8.29 / 8.45	10.44 / 10.20	4.88 / 4.61	-
XII	98	0.06	6.90	9258 / 9100	59.10 / 58.76	10.96 / 10.57	9.70 / 9.34	4.53 / 4.23	-
XIII	98	0.08	-	11026 / 10900	46.36 / 46.12	6.69 / 6.45	8.12 / 8.33	3.80 / 3.54	21.76 / 20.89
XIV	98	0.08	7.08	10006 / 9900	40.29 / 40.02	7.97 / 7.55	8.95 / 8.76	4.19 / 3.80	23.98 / 23.40
XV	97	0.08	7.02	9436 / 9400	50.35 / 50.11	9.40 / 9.16	9.49 / 9.13	4.45 / 4.09	-

*In 1% toluene solution (XIV and XV in water) at 250C.

In numerator there are calculated values, in denominator experimental values.

The IR spectra of oligomers revealed that the absorption bands characteristic for ≡Si-H bonds in the range 2160-2170 cm^{-1} do not vanish completely. In the IR spectra one can see the absorption band at 850, 920, and 4545-4550 cm^{-1} characteristic for epoxy groups, the absorption bands at 1020-1100

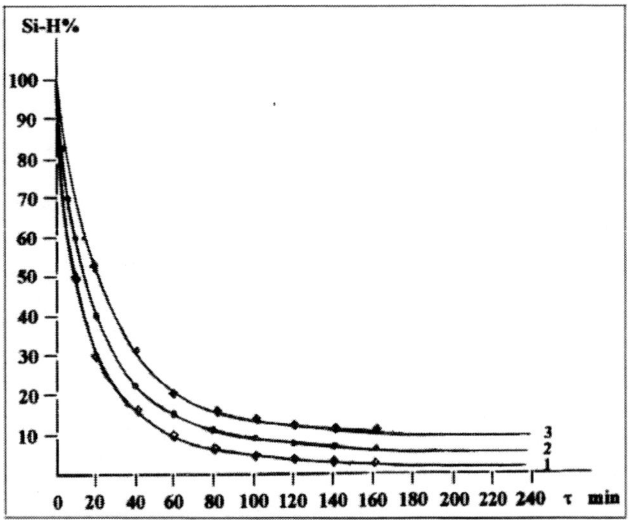

Figure 1. Dependence of changes of active Si-H% groups on time upon hydrid addition of allyl glycidyl ether to α,ω–bis(trimethylsiloxy)methylhydridsiloxane, where curve 1 is at 90^{0}C, curve 2 is at 85^{0}C and curve 3 is at 80^{0}C.

and 1150 cm^{-1} corresponds to asymmetric valence oscillation for the linear ≡Si-O-Si≡ and C-O-C simple ether bonds, as well as absorption bands in the range 2900-2950 cm^{-1}, typical for valence oscillations of the CH bonds in the CH_2 groups in the ethylene bridges and absorption bands at 1270 cm^{-1} characteristic for ≡Si-Me bonds.

The hydrosilylation reaction proceeds according to the following scheme 1.

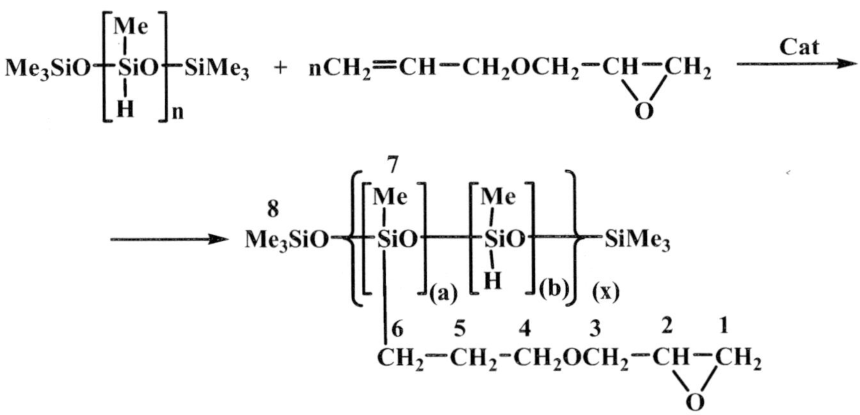

Scheme 1. Hydrosilylation reaction of methylhydridesiloxane to allyl glycidyl ether, in the presence of catalysts.

Where: [(a)+(b)](x)=n; n≈30,53. n=30, I (80°C); II (85°C); III (90°C) - [H_2PtCl_6]. n=53, IV (85°C) - H_2PtCl_6. n=30, II' (85°C) - Pt/C.

In the ^{1}H NMR spectra of oligomer III Figure 2 one can observe the singlet signals for methyl protons in the trimethylsiloxy group with chemical shift δ=0.05 ppm, for methyl group at δ=0.01 ppm, 0.47 ppm (s- C^6H_2), 1.50 ppm (s- C^5H_2), 3.20; 3.30; 3.45; 3.50 ppm (m- C^3H_2, C^4H_2) ; 3.0 ppm (s- C^2H) and 2.41 and 2.65 ppm (d-C^1H_2).

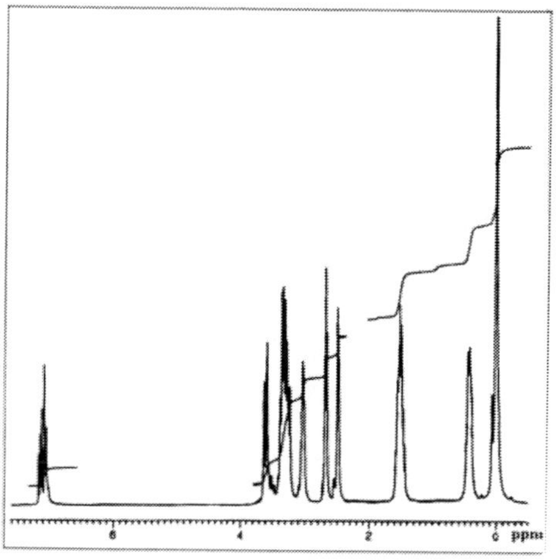

Figure 2. ^{1}H NMR spectra of oligomer III.

In the ^{13}C NMR spectrum of oligomer III Figure 3 one can observe the signals 1.85 ppm (C^7); 13.9 ppm (C^6); 23.56 ppm (C^5); 71.86 ppm (C^4); 74.43 ppm (C^3); 51.29 ppm (C^2) and 44.70 ppm (C^1).

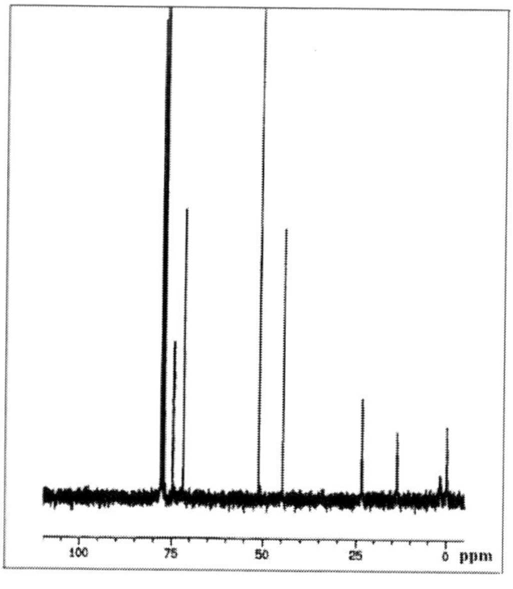

Figure 3. ^{13}C NMR spectra of oligomer III.

Figure 4 shows the time dependence of the reverse concentration, from which it is evident, that at the beginning stages the hydrosilylation reaction is of a second order. This result is in opposition with the data literature [5], where the authors proved that depending on the origin concentration of reacting compounds the hydrosilylation reaction may be at first either at a low value of concentration, or zero-at higher value of concentration.

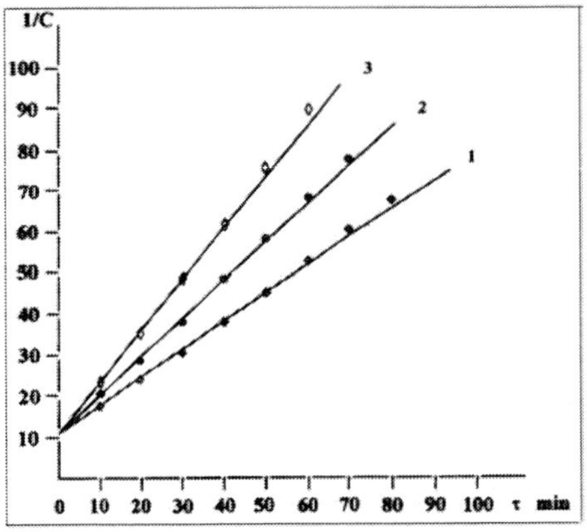

Figure 4. Dependence of reverse concentration on the time, where curve 1 is at 80^{0}C, curve 2 at 85^{0}C and curve 3 is at 90^{0}C.

From Figure 4 the hydrosilylation rate constants at various temperatures were calculated: $k_{85°C} \approx 0.6878$, $k_{90°C} \approx 0.9325$ and $k_{95°C} \approx 1.2571$ l/mol·s. It was found that for each increase in temperature of 10°C the reaction rate constants increase approximately 1.35 times.

Figure 5 shows the dependence of logarithm of hydrosilylation rate constants on the reverse temperature, from which the activation energy was calculated $E_{activ} \approx 30.6$ kJ/mol.

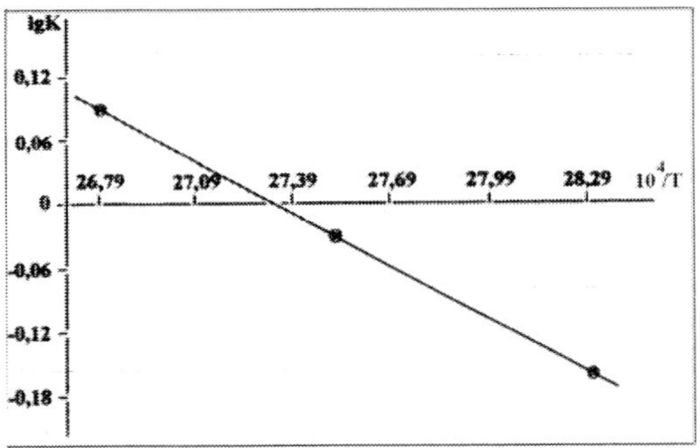

Figure 5. Dependence of reverse temperature on the rate constants of upon hydride addition of allyl glycidyl ether to α,ω–bis(trimethylsiloxy)methylhydridesiloxane.

Comb-type organosiloxanes with epoxy groups in the side chain reacted with primary amines. During the epoxy ring-opening reaction in melting condition the reaction proceeds with the formation of three dimensional systems, this fact may be explained after the ring-opening reaction with the present of reactionable secondary (–NH-) groups, which enter the ring-opening intra- and intermolecular condensation. In order to receive soluble comb-type organosiloxanes contained along a hydrophobic methylsiloxane chain, hydrophilic hydroxyl and amino groups the reaction of epoxyorganosiloxanes with primary amines were carried out in 10% solution of anhydrous toluene. The reaction products were heated up to the boiling point of toluene for 12 hours. The completeness of the reaction was confirmed by ^{1}H NMR spectra by disappearance of the resonance signals characteristic for epoxy groups. As it's seen from table 1 aminohydroxysiloxanes were obtained in good yields. The reaction proceeds according to the following scheme 2.

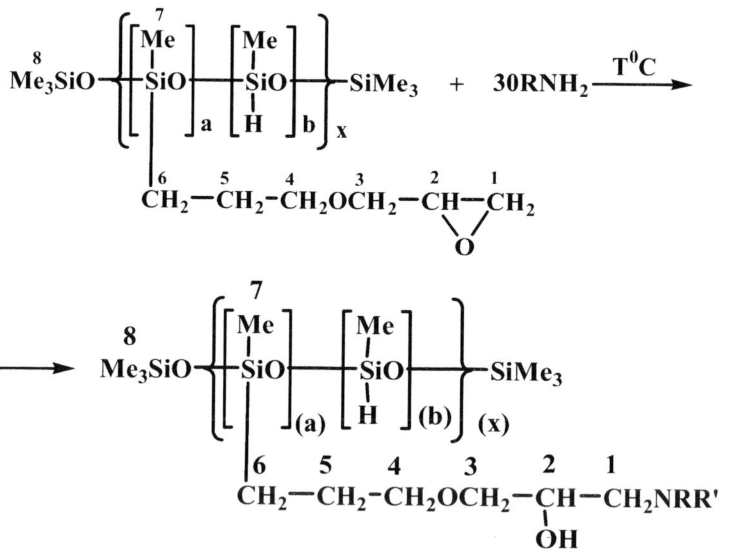

Scheme 2. Reactions of epoxyorganosiloxanes with primary amines.

Where: [(a)+(b)](x)≈30. R=C_4H_9 (V), $C_6H_5CH_2$ (VI), 2,3-$Me_2C_6H_3$ (VII), 1,4-HOC_6H_4 (VIII), o-HSC_6H_4(IX).

Synthesized secondary aminohydroxysiloxanes (V, VI, VII and IX) are colorless or slight yellow compounds soluble in ordinary organic solvents. Unfortunately all experiments of receiving the soluble product VIII were without success. It's possible that during the reaction of p-aminophenol with epoxymethyhydridelsiloxanes the secondary intra- and intermolecular reactions take place. The composition and structure of synthesized compounds were determined by elementary analysis, by finding the molecular masses, by IR and NMR spectra data.

In the ^1H NMR spectra of all secondary aminohydroxymethylhydridesiloxanes one can observe the singlet signals with chemical shifts 0.1-0.15 ppm for methyl protons at silicon (7 and 8 position), δ=0.5 ppm (C^6H_2); δ=1.5 ppm (C^5H_2); δ=1.9÷2.1 ppm (NH); δ=2.9÷3.0 ppm (C^4H_2); δ=3.2 ppm (C^3H_2); δ=3.3 ppm(C^2H); δ=3.6 ppm (C^1H_2); δ=3.7-3.8 ppm (OH) (see Figure 6 for compound VI).

In the ^1H NMR spectra of compound IX one can see a broadened singlet signal which may correspond to labile protons linked to electronegative nitrogen and sulfur groups. By integral intensity this signal corresponds approximately to two proton absorption intensities, but this does not exclude the existence of -SH groups, because there is a possible proton exchange reaction which may be seen from broadened singlet signals. That is why in compound IX there is possible existence of $-NH_2$ or =NH and -SH groups.

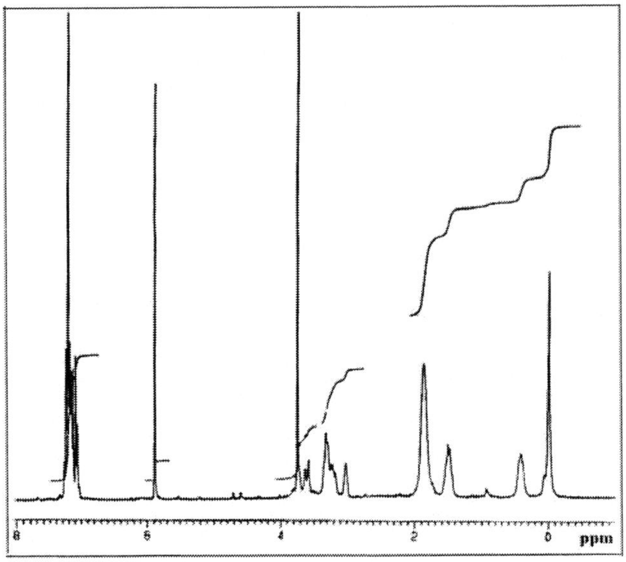

Figure 6. ^{1}H NMR spectrum of oligomer VI.

Full identification of aminohydroxymethylsiloxanes was received for example from benzylaminohydroxylsiloxane VI by using COSY NMR spectra. On Figure 7 one can see the resonance signals on the surface. Where the resonance signals at $\delta \approx 0.1$ corresponds to methyl groups at 7 and 8 positions. The resonance signals with chemical shift at $\delta \approx 0.5$; 1.5 and 2.2 correspond to methylene groups in 7,8; 6 and 5 positions and the resonance signal with chemical shift in the range $\delta \approx 7.0 \div 7.4$ ppm corresponds to phenyl protons.

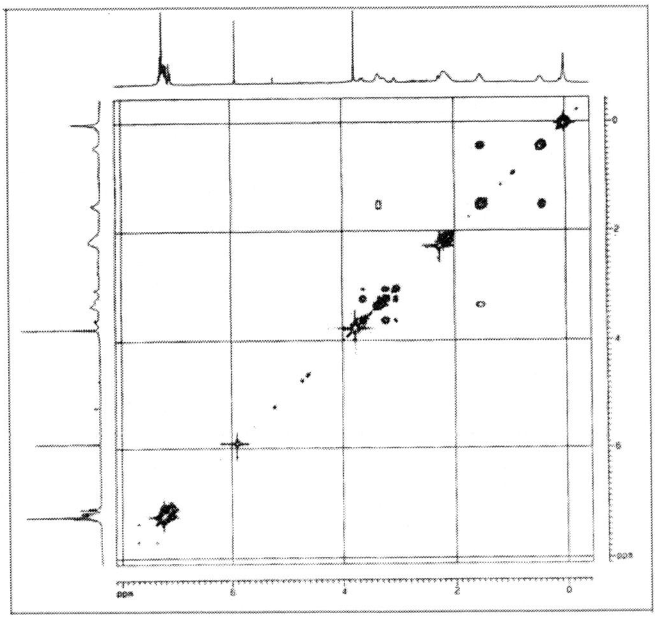

Figure 7. COSY NMR spectrum of oligomer VI.

In the IR spectra of all aminohydroxymethylsiloxanes one can see the absence of absorption bands characteristic for epoxy bonds at 910 cm^{-1} and absorption bands characteristic for –NH- and -OH bonds at 3400-3600 cm^{-1} (those absorption bands are put on each other). Hence it follows that all epoxy groups take place in the ring-opening reaction. There the absorption bands for trimethylsiloxy are observed, ≡Si-Me, ≡Si-O-Si≡ and COC bonds at 840, 1275, 1000-1130 cm^{-1} accordingly.

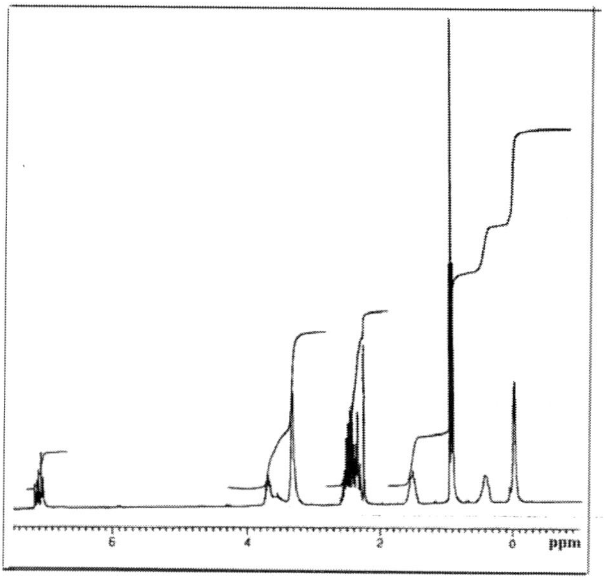

Figure 8. ^{1}H NMR spectrum of oligomer X.

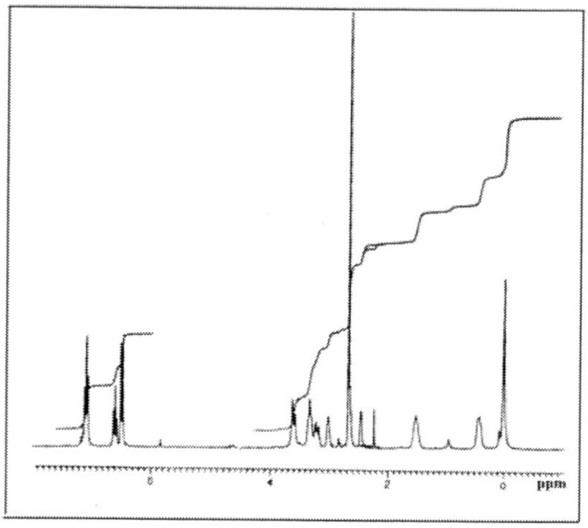

Figure 9. ^{1}H NMR spectrum of oligomer XI.

Comb-type organosiloxanes with epoxy groups in the side chain reacted with secondary amines. The reaction was carried out in 10% solution of anhydrous toluene. The reaction products were heated up to the boiling point of toluene for 12 hours. The completion of the reaction was confirmed by ^{1}H NMR spectra by disappearance of the resonance signals characteristic for epoxy groups. As it's seen from table 1 the yields of comb-type aminohydroxysiloxanes are higher because this reaction proceeds without any complication, according to the following scheme 3.

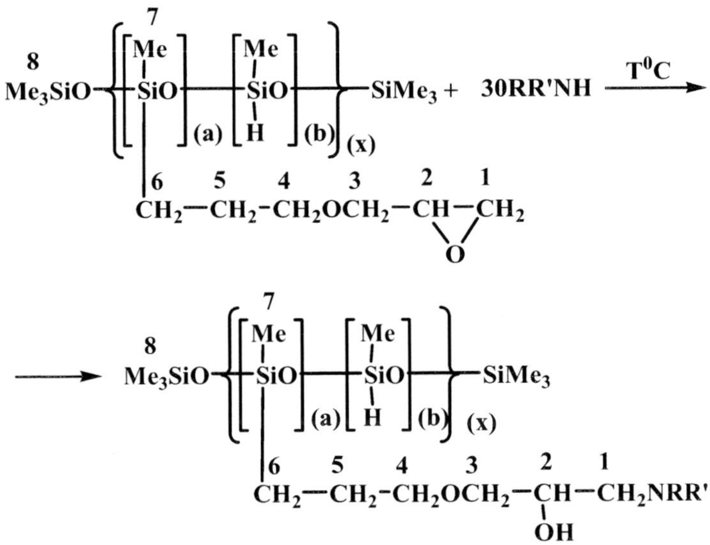

Scheme 3. Reaction of epoxyorganosiloxanes with secondary amines.

Where: [(a)+(b)](x)≈30. R=R'=C_2H_5 (X), R=Me, R'=C_6H_5 (XI), R=R'=C_4H_9 (XII).

The composition and structure of tertiary aminohydroxysiloxanes was established by means of elementary analysis, by finding the molecular masses by IR and NMR spectra data. In ^{1}H NMR spectra of compounds X÷XII one can observe resonance signals with chemical shifts 0.1-0.15 ppm for methyl protons at silicon (in 7 and 8 position), δ=0.5÷0.6 ppm (C^6H_2); additionally complicated signals with chemical shift δ= 0.9÷1.1 ppm for methyl protons in alkaline fragment (compounds X and XII); δ=1.5÷1.6 ppm (C^5H_2); δ=2.3÷2.7 ppm (NC^1H_2 and NCH_2); δ=2.9÷3.0 ppm (C^4H_2); δ=3.2 ppm (C^3H_2); δ=3.3 ppm (C^2H); δ=3.6 ppm (C^1H_2); δ=3.7 ppm (OH) and δ=4.2 ppm (≡Si-H) (see Figure 8 and 9 for compound X÷XII). Full identification of tertiary aminohydroxymethylsiloxanes was received, for example, of benzylaminohydroxysiloxane X by using COSY NMR spectra. In the ^{13}C NMR spectra of compound X, Figure 10 one can see the signals for carbon attached with hydroxyl groups at δ=67.13 ppm and signals for carbon attached

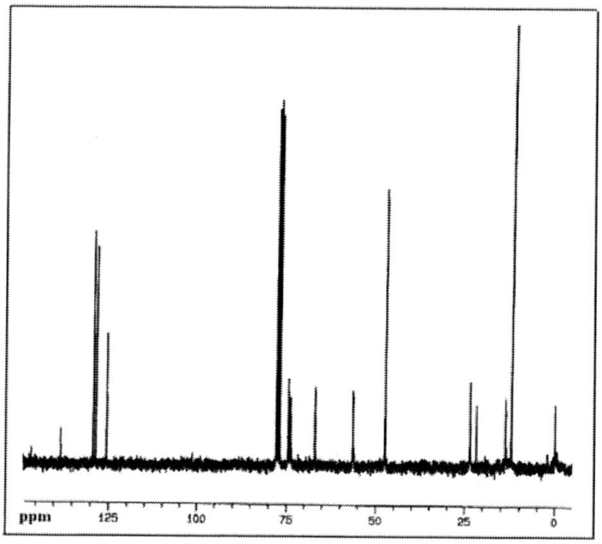

Figure 10. ^{13}C NMR spectrum of oligomer X.

with nitrogen at δ 47.57 and δ 56.50 ppm. In Figure 11 one can see the resonance signals on the surface. Where, the resonance signal at ~0.1 ppm, corresponds to methyl groups at 7 and 8 positions. The resonance signal at δ≈0.5; 1.5 and 3.4 ppm corresponds to methylene groups in 6, 5 and 4 positions.

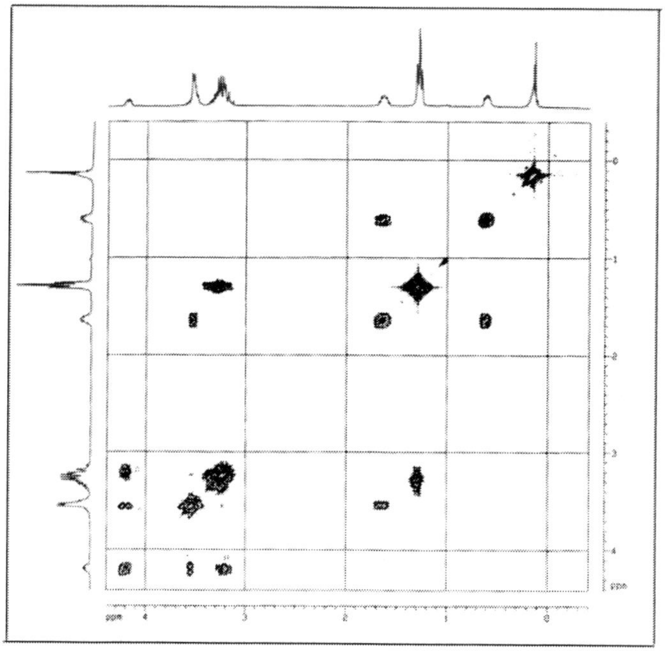

Figure 11. COSY NMR spectrum of oligomer X.

The triplet signal in the range δ=0.9÷1.1 ppm corresponds to methyl groups in the ethyl fragment. In the IR spectra of tertiary aminohydroxysiloxanes X ÷XII one can see the narrow absorption bands characteristic for secondary hydroxyl group at 3600 cm^{-1}, from which it is evident that the reaction of amines with epoxyorganosiloxanes mainly proceeds with the formation of α-adduct. The tertiary amines did not react with epoxy containing comb-type organosiloxanes because the tertiary amines did not have the active hydrogen groups which may enter the ring-opening reaction.

For the purpose of synthesis of siliconorganic quaternary ammonium salts the reaction of hydrogen bromide or acetic acid with aminohydroxyorganosiloxanes has been investigated. It was very interesting to test the stability of the siloxane chain and to determine that such strong acid as hydrogen bromide (or acetic acid) can break the linear siloxane chain. The reaction was carried out at 0^0C temperature to prevent the break of ≡Si-O-Si≡ backbone. The hydrogen bromide in equimolar amount was added and stirred for 4 hours and proceeds according to the following scheme 4.

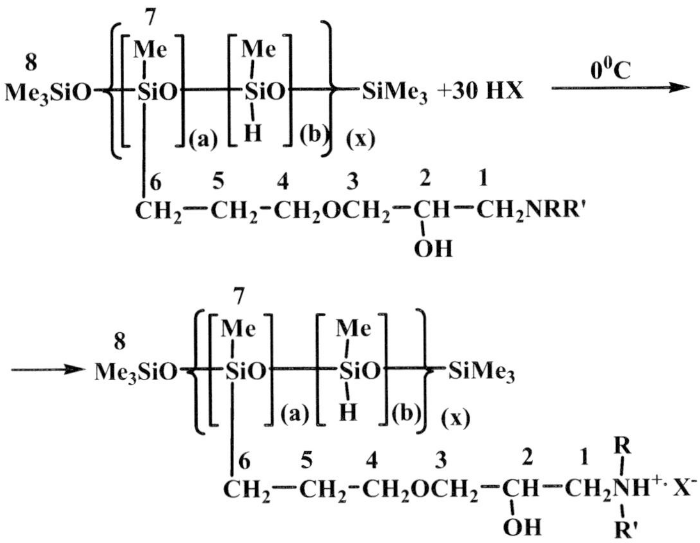

Scheme 4. Reaction of aminohydroxyorganosiloxanes with acids.

Where: [(a)+(b)](x)≈30; X=Br: R=Me, R'=C_6H_5 (XIII); R=R'=C_2H_5 (XIV). X=CH_3COO, R= R'=C_2H_5 (XV).

The synthesized salts are vitreous transparent colorless products, the compounds XIV and XV are soluble in water and the compound XIII in organic solvents such as CH_2Cl_2, chloroform and so on. By GPC analysis it was shown that during the reaction of aminohydroxysiloxanes with the above-mentioned acids break down reactions of ≡Si-O-Si≡ backbone do not take place (see table 1). The compounds XIII-XV were characterized by ^1H, ^{13}C NMR and IR spectra data. As it's seen from NMR spectra protonated compounds XIII-XV Figures12-15, one can see resonance singlet signals for methyl protons at silicon atom δ≈0.1-0.13 ppm (for hydrogen protons at 8 and 7 position accordingly), for protons in methylene groups with chemical shifts at δ≈0.5; 1.5 ppm; in the range 2.9÷3.6 ppm (6,5, 4 and 3 position).

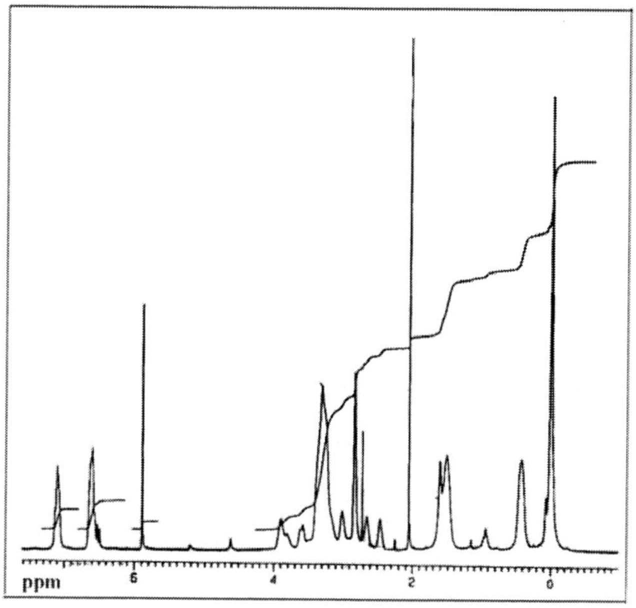

Figure 12. ^{1}H NMR spectrum of oligomer XIII.

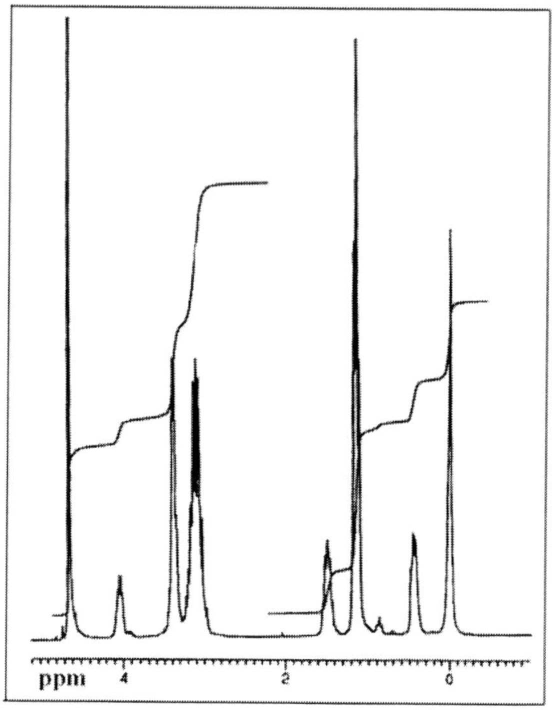

Figure 13. ^{1}H NMR spectrum of oligomer XIV.

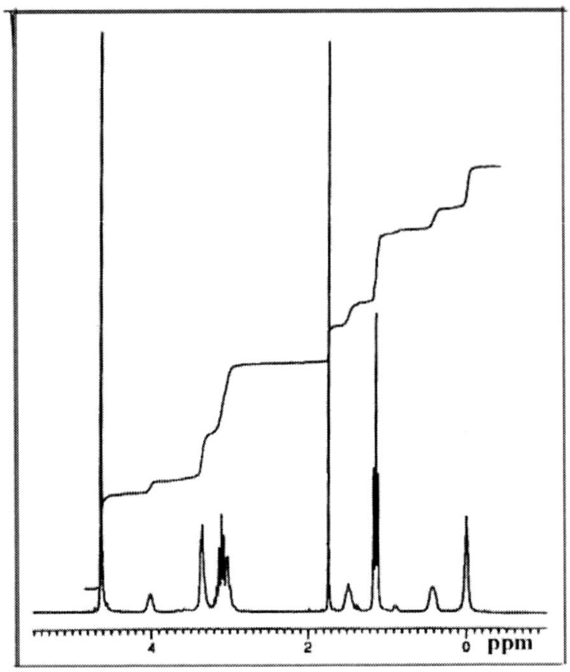

Figure 14. ^{1}H NMR spectrum of oligomer XV.

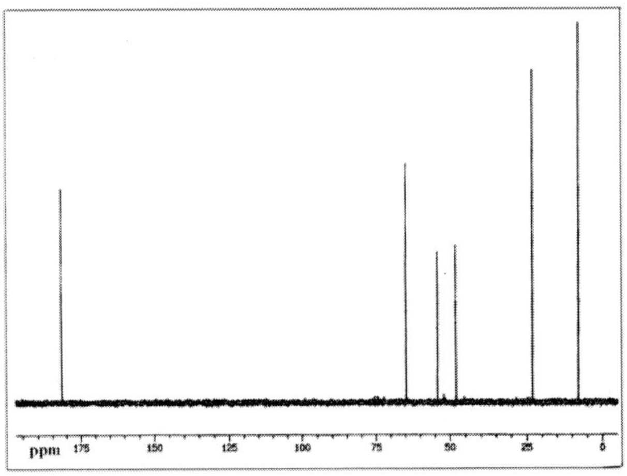

Figure 15. ^{13}C NMR spectrum of oligomer XV.

The comparison of chemical shifts for N-CH_2 - groups in tertiary and quaternary aminohydroxysilxanes, shows a significant shift in the low field for quaternary compounds.

Full identification of protonated aminohydroxymethylhydridesiloxanes COSY NMR spectra was received for compound XIV (Figure 16). On surface COSY NMR spectra in Figure 16 the resonance signals at $\delta \approx 0.1$ and $\delta \approx 0.13$ ppm correspond to methyl protons at 7 and 8 positions. The resonance broadened signals with chemical shift and with center at $\delta \approx 0.5$; 1.3; 1.7; 3.2 and 3.5 ppm correspond to protons of methylene groups in 6÷1 positions and in the ethyl fragment.

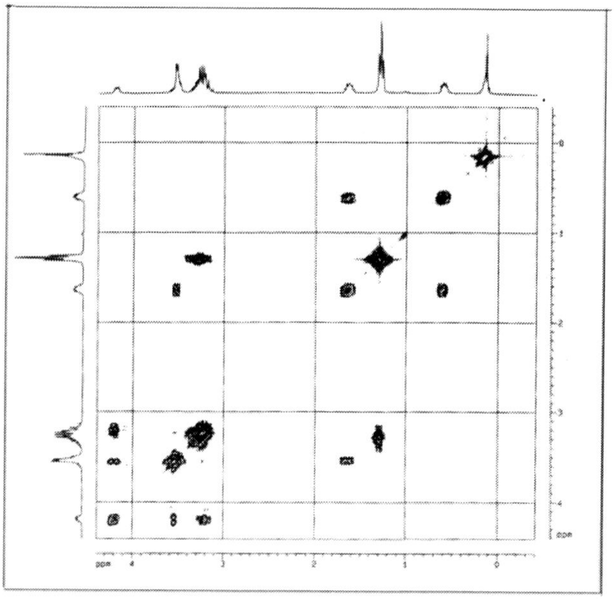

Figure 16. COSY NMR spectrum of oligomer XIV.

The large absorption band in the region 3200-3600 cm^{-1} corresponds to the associated hydroxyl and cationic nitrogen groups. The absorption bands in the region 1590-1650 cm^{-1} corresponds to NH_2^+ group.

For some synthesized aminohydroxylmethyl(hydride)siloxanes the roentgenographic investigtions have been carried out. As it is seen in Figure 17 the synthesized methylhydridesiloxane.

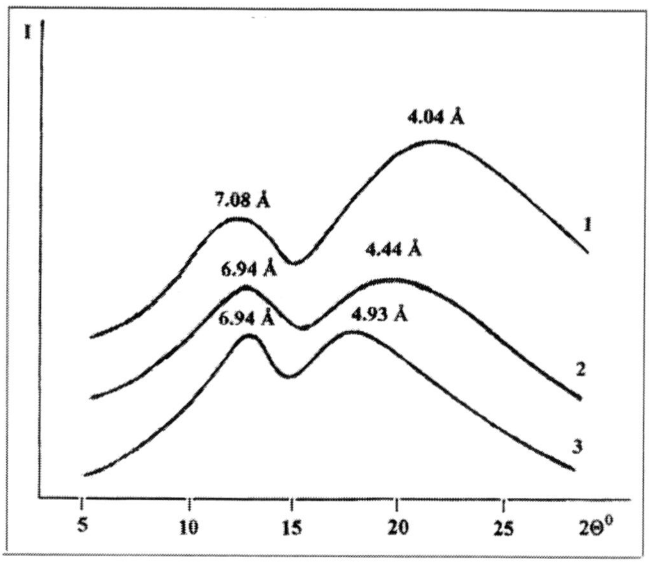

Figure 17. Diffractograms of comb-type organosiloxanes with aminohydroxy groups in the side chain. Where curve 1 corresponds to oligomer XIV, curve 2-XI and curve 3 - X.

Average molecular masses were determined by gel permeation chromatography with amino and hydroxyl groups in the side chain as well as protonated corresponding compounds which are represented as amorphous systems. On the diffractograms one can observe two maximums $2\theta^0=12.5^0$ and $2\theta=18.0^0\div22.0^0$. The main maximum $2\theta^0=12.5^0$ which corresponds to interchain distance, changes from $d_l=6.94$ Å (for compounds X and XI) up to $d_l=7.08$ Å, these increasing values of interchain distances correspond to protonated aminohydroxylmethylhydridesiloxane XIV.

The thermogravimetric investigation of some synthesized compounds in isothermal condition (200^0C) was carried out. As it is seen from Figure 18 (curves 1 and 2), for aminohydroxylsiloxanes X and XI mass losses for 10 h do not exceed 21-24%, while for protonated compound XIV for first the 3 h mass losses consist of about 74% (Figure 18, curve 3), then the destruction process proceeds slightly.

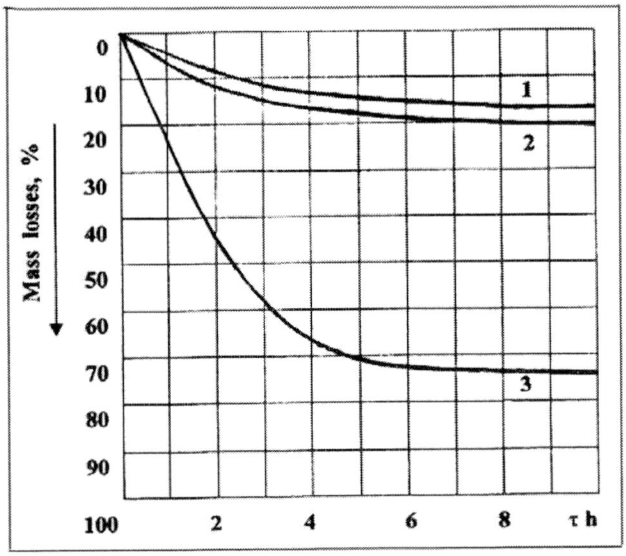

Figure 18. Thermogravimetric curves in isothermal (200^0C) conditions. Where curve 1corresponds to oligomer XI, curve 2 – X and curve 3 – XIV.

So, by us for the first time methylhydridesiloxane with hydroxyl and amino functional group in the side chain was synthesized, and by protonation of aminohydroxysiloxanes protonated aminohydroxysiloxanes were obtained.

References

[1] E.P. Plueddemann, G. Fanger. *J. Amer. Chem. Soc.,* 1959, 81, 2632.

[2] G. Greber, L. Metzinger. *Makrom.Chem.,* 1960, 39,189.

[3] 3. J.S. Riffle, I. Yilgor, C. Tran, G.L. Wilkes, J.E. McGrath, A.K. Bantia. "Epoxy Resin Chemistry II", *ACS Symp. Ser.,* 1983, 221, 21.

[4] D. Gagnebien, P.J. Madec, E. Marechal. *Eur. Polym. J.,* 1985, 21, 289.

[5] D.A. Girgvliani, L.M. Khananashvili, S.P. Davitian, A.R. Khachatrian, N.G. Tsomaia, A.O. Tonoyan. *J. Polym. Materials,* 1994, 25, 65.

[6] L.M. Khananashvili, I.Z. Bochorishvili, E.G. Markarashvili, N.I. Tsomaia. *Proceedings of the Academy of Sci. of Georgia, Chem.Series,* 1995, 20-21, 35 (in Russian).

[7] L.M. Khananashvili, N.I. Tsomaia, D.U. Murachashvili, D.A.Girgvliani, M.G. Karchkhadze, R.Sh. *J. Polym. Materials,* 1997, 37, 65.

[8] B.S. Rao, P.J. Madec, E. Marechal. *Makromol.Chem.*, Rapid Commun., 1986, **7**, 703.

[9] D.U. Murachashvili, L.M. Khananashvili, V.M. Kopilov, N.I. Tsomaia, M.I. Shkolnik. *Proceedings of the Academy of Sci. of Georgia, Chem. Series*, 1993, 19, 201 (in Russian).

[10] L.Khananashvili, G.Gurgenidze, M.Karchkhadze, O.Mukbaniani. XVI D. Mendeleev *Symposium in General and Applied Chemistry*, Saint-Petersburg, 1998, 319 (in Russian).

[11] O.V. Mukbaniani, L.M. Khananashvili, I.G. Esartia, S. Khaduri. *Intern Journal of Polym. Mater.*, 1994, 22 (1-4), 131.

[12] D. Murachashvili, L. Khananashvili, N. Tsomaia, V. Kopilov, M. Shkolnik, E. Khubulava. *Proceedings of the Academy of Sci. of Georgia, Chem. Series*, 1996, 22, 36 (in Russian).

In: Modification Reactions of Oligomethylhydridesiloxanes ISBN: 1-60021-365-0
Editors: O.Mukbaniani, et al. pp. 139-147

Chapter 9

SYNTHESIS OF CYCLOLINEAR ORGANOSILOXANE COPOLYMERS BY HYDROSILYLATION REACTION

O. Mukbaniani, T. Tatrishvili and N. Mukbaniani
I. Javakhishvili Tbilisi State University, I. Chavchavadze Ave., 1,
Tbilisi 380028, Georgia

ABSTRACT

The reaction of hydride addition of methylhydridesiloxanes to vinylorganocyclosiloxane, at the ratio of initial compounds 1:30, in the presence of platinum hydrochloric acid has been investigated. The hydride addition reaction order, activation energies and rate constants were found. The polyorganosiloxanes with organocarbocyclosiloxane fragments in the side chain were synthesized, and were completely soluble in organic solvents. ^{1}H and ^{13}C NMR, IR, thermogravimetric, gel-permeation chromatography, differential scanning calorimetry and X-ray methods were characterized by the synthesized oligomers.

Key words: organosiloxanes, hydride polyaddition, oligomers thermal-oxidative stability.

INTRODUCTION

Before our investigations [1, 2] there was no information in the literature about branched dimethylsiloxane copolymers containing diorganocyclosiloxane fragments, instead of the linear dimethylsiloxane links. From the literature [3] it is known that the main difference between oligo(dimethyl)methylsiloxanes with a branched structure and the equal viscosity of linear oligodimethylsiloxanes, is the absence of a tendency to crystallize above a definite branching content of methylsilsesquioxane links. The main reason of the suppression of the crystallization in oligo(dimethyl)methylsiloxanes is the break of regular structure of the linear siloxane chain by the insertion of methylsilsesquioxane links.

The most effective method of modification of linear organosilicon polymers is the insertion of various elements or groups with a different chemical nature in the structure of the macromolecular chain. As a result of the insertion in the dimethylsiloxane chain of different fragments there are changes in the physical-chemical properties [4]. The insertion of the cyclic fragments in the main linear dimethylsiloxane chain hinder the chain transfer reactions, which proceed with the release of D_n-type (where D = Me_2SiO) cycles during the thermal depolymerization, that raises the thermal-oxidative stability of the polymers [5].

Using heterofunctional condensation reactions, there are two possibilities of synthesis of organosiloxane copolymers with monocyclic fragments in the side chain. The first one is the reaction of heterofunctional condensation of dichlororganosiloxyorganocyclosiloxanes with α,ω-dihydroxydiorganosiloxanes and the second one is the modification reaction of functional group containing linear organosiloxanes by the insertion of cyclic fragments.

For the purpose of synthesis of the linear organosiloxane copolymers with cyclic fragments in the side chain the heterofunctional condensation of dichlororganosiloxycyclotri(tetra,penta)siloxanes and dichlororganosililcarbocyclotrisiloxane with dihydroxydimethylsiloxanes in the presence of pyridine has been investigated. It was shown that for short lengths of the linear dimethylsiloxane link (n≤4), the reaction of heterofunctional condensation proceeds both intermolecularly with formation of copolymers and intra-molecularly with formation of bicycloorganosiloxanes. The reaction proceeds according to the following scheme [6]:

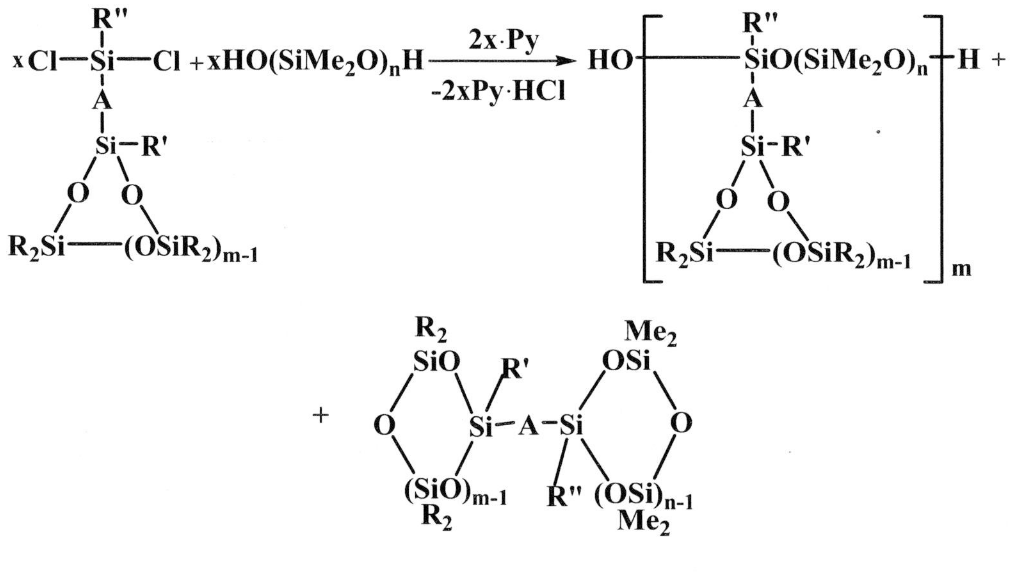

Where: R=R'=R"=Me, Ph; R≠ R'≠R"; A=O, C_2H_4; m=2÷4, n=2÷51.

As a result organosiloxane copolymers with regular arrangement of cyclic fragments in the side chain have been obtained.

One of the light methods of synthesis of organosiloxane copolymers with cyclic fragments in the side chain is the modification of the functional group containing methyl- and ethylhydridesiloxanes. From this modification reaction it must be noted that the dehydrocondensation reaction of ≡Si-H and ≡Si-OH containing organosiloxanes in the presence of alkaline catalysts proceeds with elimination of hydrogen without cleavage of ≡Si-0-Si≡ bonds [7].

It is known from the literature [8 - 12] that hydrosilylation is widely used for synthesis of cyclolinear copolymers. Discussed in this paper is the synthesis of organosiloxane copolymers with carbocyclosiloxane fragments in the side chain by hydride addition of methylhydridesiloxanes and methylhydridesiloxanedimethylsiloxane copolymer to vinyl-containing organocyclosiloxanes. It is shown that a hydride addition of dihydrideorganocyclosiloxanes to divinyl-containing organocyclosiloxanes in the presence of platinum hydrochloric acid does not cause cyclosiloxane ring scission [8 - 12].

EXPERIMENTAL PART

The organic solvents were purified by drying and distillation. The initial vinylorganocyclosiloxanes were synthesized using well-known methods [6, 13]. The IR spectra of all the samples were taken on an IR-20 spectrophotometer, while the NMR spectra on a Perkin-Elmer R-32 instrument at an operating frequency of 90 MHz in a solution of CCl_4 and dioxane with the internal standard HMDS or TMS and on JNOL-FX-100 spectrometer in a solution of deuterium chloroform, the line of which was used as a reference. ^{29}Si NMR spectra were taken on a WP-200 SY (Bruker) instrument at an operating frequency of ^{29}Si - 39.77 MHz.

The viscosities of copolymers were investigated in toluene solutions at 25 °C. Wide angle X-ray diffractograms were taken on a "DRON-2" (Burevestnik, Saint-Petersburg, Russia) instrument using A-CuK_α radiation. The values were measured without a filter, and the angular velocity of the motor was ≈*2 deg/min.*

Differential scanning calorimetric analysis was performed on a Perkin-Elmer DSC-7. Heating and cooling scanning rates were ~5 ^{0}C/min.

Thermogravimetric investigations were carried out on a Paulic-Paulic-Erdey derivatograph, model MOM-102. The test conditions were: temperature rise rate v ≈5 deg/min, in the open area.

RESULTS AND DISCUSSION

The main reason for the investigation was to create convenient methods for the synthesis of organosiloxane copolymers with monocyclosiloxane fragments in the side chain. To establish the influence of the cyclic fragment on the properties of copolymers the initial difunctional organodichlorosiloxyorganocyclosiloxanes and a new class of cyclolinear copolymers, with monocyclic fragments in the side chain were synthesized.

For the purpose of synthesizing carbosiloxane copolymers with cyclic fragments in the side chain, a hydride addition of methylhydridesiloxanes to vinylorganocyclosiloxanes in the presence of catalysts was studied. Preliminary heating up of 1-vinyl-1-methyltetraphenylcyclotrisiloxane (at concentration C = 0.02628 mol/l) during 8 hours in the presence of platinum hydrochloric acid (9×10-5 g per 1 g of reagents) in the temperature range of 80 - 100°C has not induced polymerization of vinylorganocyclosiloxanes, and no changes in ^{1}H NMR spectra for methyl protons have been detected. Hydrosilylation of methylhydridesiloxane (n=35) with 1-vinyl-1-methyltetraphenylcyclotrisiloxane at 1:35 ratio of the initial reagents at 80, 90 and 100°C was studied.

Figure 1 shows that hydrogen conversion is increased from 85% to 95% with a temperature rise from 80 to 100°C. Hydrosilylation is intensive during the initial 50 - 60 min and after 80 min becomes negligible. Note also that the reaction is incomplete, and substitution of all active ≡Si-H groups by cyclotrisiloxane fragments is not observed. This may be explained by the steric effect induced by organocyclosiloxane fragments during formation of the macromolecular chain. Complete hydrosilylation is observed at 1:17.5 ratios of the initial reagents. Hydride addition of α,ω-bis(trimethylsiloxy)methylhydridesiloxane to vinyl containing organocyclosiloxanes proceeds in accordance with the general scheme as follows:

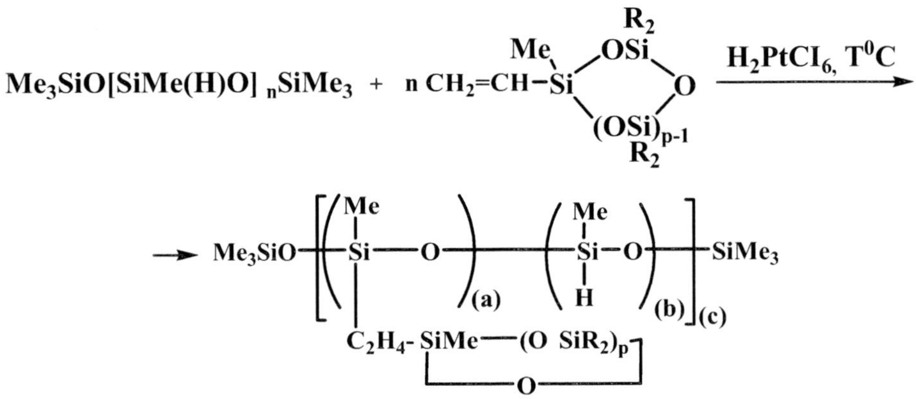

Where: [(a) + (b)](c) = n≈35; R=Ph, p = 2 – I^1(80^{0}C), I^2 (90^{0}C), I^3 (100^{0}C), at 1:35 ratio. I^4 (100^{0}C), at 1:17.5 ratio. R = Me, p=3 – II (90^{0}C), p=4 – III (90^{0}C).

Synthesized copolymers represent light-yellow liquids or solids, depending on substituted groups, and are well soluble in ordinary organic solvents, with η_{sp} = 0.05 – 0.07. Structure and composition of synthesized copolymers were determined by elementary and functional analysis and from IR and NMR spectral data. Some physical- chemical properties of the comb-shaped methylsiloxane copolymers with carbocyclosiloxane fragments in the side chain are shown in table 1.

IR spectra of copolymers display absorption bands typical of unreacted ≡Si-H bond. They also display new absorption bands in the range of 2160 –2170 cm^{-1}, which is typical of fan oscillations of CH_2 group and for asymmetrical valence oscillations of ≡ Si-O-Si≡ bonds at 1020 and 1075 cm^{-1}, and typical of both linear siloxane units and siloxane bonds in cyclotri- and –cyclotetrasiloxane rings. In copolymers with cyclotrisiloxane fragments in the side chain, absorption bands typical of asymmetrical valence oscillations of ≡ Si-O-Si≡ bonds in linear dimethylsiloxane chain and cyclotrisiloxane ring overlap.

Table 1. Some physical-chemical properties and yields of comb-shaped methylsiloxane copolymers with organocarbocyclosiloxane fragments in the side chain.

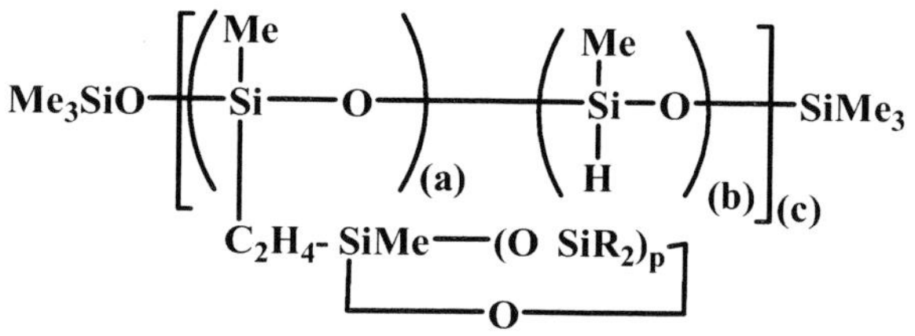

№	Yield, %	*p*	*R*	T_{react}, ^0C	Initial reagent ratio.	η_{spec}*	d_1,Å	T_g ^0C	H_2 conversion, %
I^1	93	2	*Ph*	*80*	*1:35*	0,05	10,30	-	85
I^2	94	2	*Ph*	*90*	*1:35*	0,06	-	-	90
I^3	94	2	*Ph*	*100*	*1:35*	0,07	10,30	+12 ÷+15	95
I^4	96	2	*Ph*	*100*	*1:17.5*	0,05	9,80	0-+3	100
II	96	3	*Me*	*90*	*1:35*	0.05	8.45	-94	92
III	95	4	*Me*	*90*	*1:35*	0.05	8.76	-110	90

*In 1% toluene solution at 25^{0}C.

^{1}H NMR spectrum of copolymer I^3 (table 1) displays broadened signals for methyl protons (3H) in -O-SiMe-O- fragment with chemical shift δ =0.27 ppm and methyl protons in the backbone with chemical shift δ = 0.10 ppm. For copolymers II and III (table 1), there is a broadened singlet reflex (4H) with chemical shift δ = 0.72 ppm, typical of $-CH_2$-CH_2- fragment and indicating the Farmer rule during hydrosilylation, which is also observed. Singlet reflexes from methyl protons at 0.13 and 0.15 ppm, typical of cyclic fragments, are detected.

Figure 1 indicates a hydrogen conversion increase from 85% to 95% with temperature rise from 80°C to 100°C. Hydrosilylation proceeds intensively during the initial 2 hours. Figure 2 shows time dependence of reverse concentration. The second order of hydrosilylation is detected. Hydrosilylation rate constants were calculated: $k_{80^0C} = 5.5431\times10^{-1}$, $k_{90^0C} = 7.5355\times10^{-1}$, $k_{100^0C} = 1.00$ l/mol·s.

Figure 3 shows dependence of the hydrosilylation rate constant logarithm, lg*k*, on reverse temperature, and activation energy was calculated, which is equal to $E = 31.6$ kJ/mol.

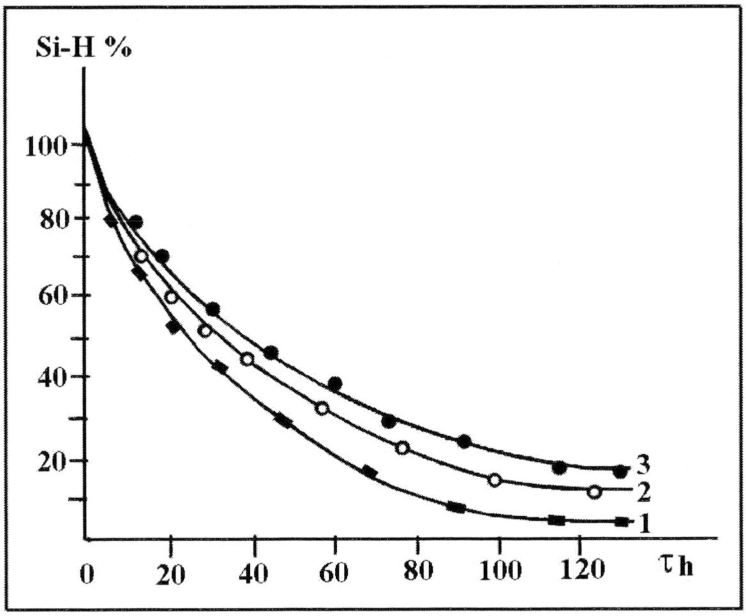

Figure 1. Hydrogen conversion in hydride addition of α,ω–bis(trimethylsiloxy)methylhydridesiloxane to 1-vinyl-1-methyltetraphenylcyclotrisiloxane: 1 - at 100°C; 2 – at 90°C; 3 – at 80°C.

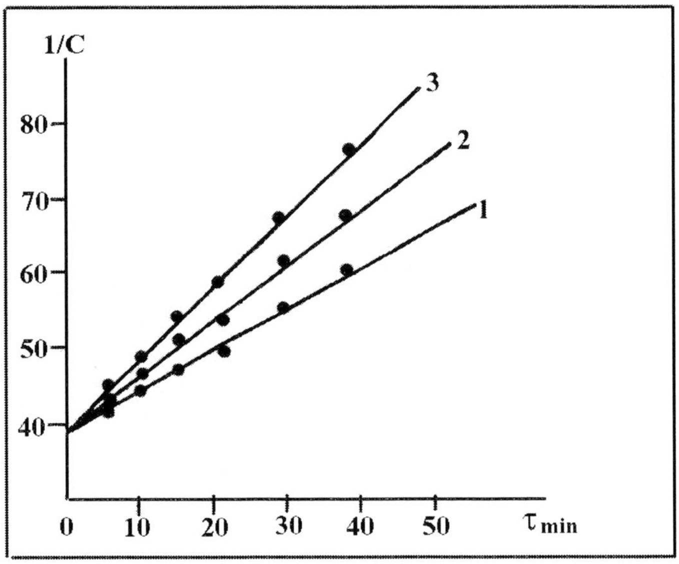

Figure 2. Time dependence of reverse concentration for hydride addition of methylhydridesiloxane to 1-vinyl-1-methyltetraphenylcyclotrisiloxane at 1:35 ratio of initial reagents: 1 - at 80°C; 2 - at 90°C; 3 - at 100°C.

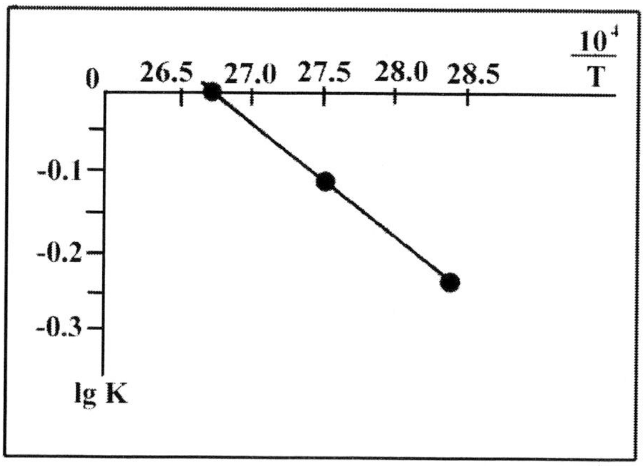

Figure 3. Dependence of the reaction rate constant logarithm of hydride addition of methylhydridesiloxane to 1-vinyl-1-methyltetraphenylcyclotrisiloxane on reverse temperature at 1:35 ratio pf initial reagents.

X-ray diffraction studies show that synthesized copolymers are single phase amorphous products and interchain distance changes in the range d_1= 10.30 – 8.45 Å. At phenyl substituted groups for copolymer I^3 (table 1) interchain distances is equal to d_1 = 10.30 Å. The interchain distance decreases with the number of cyclosiloxane fragments in the side chain (copolymer I^4, table 1). With an increase of the volume of cyclic fragment in the case of methyl substituted groups copolymer II and III the value of interchain distances slightly increases.

Figure 4 shows differential scanning calorimetric curves of synthesized copolymers. In accordance with data from Figure 4 and table 1 copolymer I^3 possesses a higher glass transition temperature, compared with copolymers II and III (table 1), which is explained by the presence of phenyl framing groups in the cyclic fragments at silicon atoms. The glass transition temperature is decreased with the quantity of cyclic fragments (refer to table 1).

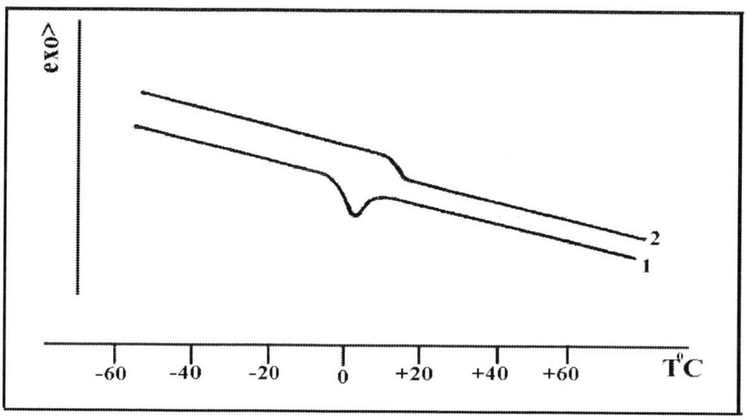

Figure 4. DSC patterns for oligomers with cyclic fragments in the side chain: 1 - oligomer I^3; 2 - oligomer I^4.

Substitution of the methylcyclotetrasiloxane fragment by the methylcyclopentasiloxane one decreases the glass transition temperature T_g, by 13 - 15°C, which coincides with the previously published data [6].

The thermogravimetric investigation of synthesized copolymers was carried out. From TGA it can be seen that the replacement of some of the methyl groups by the phenyl ones in a cyclic fragment results in a rise of decomposition of starting temperature up to 280-300 ^{0}C .This is associated with a high resistance of phenyl groups to oxidation as well as with their inhibiting effect on oxidation of methyl groups [14]. The replacement of an oxygen bridge by an ethylene group results in a slight decrease of thermal-oxidative stability of the copolymer [6].

For phenyl containing copolymers, 5% mass losses are observed in the temperature region 280-300°C; and with an increasing temperature the loss of mass keeps on increasing. The main destruction process proceeds over the range 300-600°C, and above 700°C the curves of mass losses do not change. For copolymers with methylcyclosiloxane fragments as a pendant group, 5% mass losses can be observed in the temperature region 230-250 °C. Introduction of organosiloxane cyclic fragments in the side chain of the methylsiloxane backbone induces a considerable decrease of the degradation process of such copolymers by depolymerization mechanism at 320 - 340°C, whereas unblocked PDMS is depolymerized by almost 95% at these temperatures.

Comparison of thermal-oxidative stability of copolymers with organocyclosiloxane fragments in the side chain and copolymers containing cyclotetra-, -penta-, and -hexasiloxane fragments along the main chain [15-18], showed approximately the same thermal-oxidative stability, because the insertion of cyclic fragments in the side chain hinders the chain transfer reactions that proceed with release of the Dn-type cycles.

Therefore, it is evident that cyclic fragments in the side chain in the case of destruction behave like copolymers with cyclic fragments along the dimethylsiloxane chain.

Thus, that was the first time, when behavior of methylsiloxane oligomers and copolymers with various cyclic fragments in the side chain similar to dimethylsiloxane copolymers with cyclic fragments in the backbone was indicated [8].

REFERENCES

[1] O. Mukbaniani, S.M. Meladze, G. Zaikov, L. Khananashvili. *Polymer Year Book*, 2003, 6, 141.

[2] O.V. Mukbaniani and G.E. Zaikov. The book, *New Concepts in Polymer Science*, «Cyclolinear Organosilicon Copolymers: Synthesis, Properties, Application». Printed in The Netherlands, ///VSP///, Utrecht, Boston – 2003, 499.

[3] M.V. Sobolevski, *Oligoorganosiloxanes: Properties, Production and Application, Khimia*, Moscow, 1985 (in Russian).

[4] V.P. Mileshkevich. Kauchuk I Rezina, 1978, 6, 4 (in Russian)

[5] T.H. Tomas and T.C. Kendrick. *J.Polym.Sci.*, 1969, 2A(7), 537.

[6] O.V. Mukbaniani, S.M. Meladze, I.G. Esartia, I.I.Tverdokhlebova, L.M. Khananashvili. *J. of Applied Polymer Sci*, 1999, 74, 583.

[7] O.V. Mukbaniani, L.M. Khananshvili, I.G. Esartia, S.D. Khaduri. *Int. J. of Polymeric Materials* 1994, 24(1), 131.

[8] L.M. Khananshvili, O.V. Mukbaniani, N.A. Koyava, G.V. Porchkhidze, and Yu.I. Tolchinski. *Int. J. Polym. Mater.*, 1992, 17, 113.

[9] M.G. Karchkhadze, N.O. Mukbaniani, L.M. Khananashvili, S.M. Meladze, N.G. Kvelashvili, and T.P. Doksopulo. *Int. J. Polym. Mater.*, 1998, 41, 89.

[10] N.O. Mukbaniani, M.G. Karchkhadze, A.Sh. Samsonia, R.Sh. Tkeshelashvili, and L.M. Khananashvili. *Bulletin of the Georgian Academy of Sciences,* 1999, 160(1), 84 (in Russian).

[11] A.A. Zhdanov, T.A. Pryakhina, T.V. Strelkova, R.I. Afonina, and V.M. Kotov. *Vysokomol. Soedin.*, 1993, 35(5), 475 (in Russian).

[12] O.V. Mukbaniani, L.M. Khananashvili, M.G. Karchkhadze, R.Sh. Tkeshelashvili, and N.O. Mukbaniani. In: *Synthesis and Properties of Polymers,* Ed. G.E. Zaikov, Nova Science Publ., Inc. Commack, 1996, 89.

[13] S.M. Meladze, O.V. Mukbaniani, L.M. Khananshvili. *Zhurn. Obshch. Khimii* 1981, 51(7), 1624 (in Russian)

[14] A.F.Bulkina, V.S. Papkov, A.A. Zhdanov and K.A. Andrianov, *Vysokomol. Soed.*, 1972, 20A, (1), 70 (in Ruissian).

[15] K.A. Andrianov, A.I. Nogaideli, G.L. Slonimsky, V.Yu. Levin, N.N. Makarova, Yu.P. Kvachev and O.V. Mukbaniani. *Vysokomol. Soed.*, 1976, 18B(5), 359 (in Russian).

[16] S.M. Meladze, O.V. Mukbaniani and L.M. Khananashvili. *Vysokomol. Soed.,* 1984, 26B(4), 250 (in Russian).

[17] N.A. Koiava, O.V. Mukbaniani and L.M. Khananashvili. *Vysokomol. Soed.*, 1985, 27A(4), 2261 (in Russian).

[18] V.A. Achelashvili, O.V. Mukbaniani, L.M. Khananashvili and G.I. Sturua. *Proceedings of the Georgian Academy of Sciences,* (Series of collection of articles), 1988, 53 (in Russian).

In: Modification Reactions of Oligomethylhydridesiloxanes ISBN: 1-60021-365-0
Editors: O.Mukbaniani, et al. pp. 149-162

Chapter 10

Modification Reactions of Methylhydrosiloxanes with Tricyclodecadiene

O. Mukbaniani[1], G. Zaikov[2], T. Tatrishvili[1], N. Mukbaniani[1], and Kh.Koberidze[1]

[1] Ivane Javakhishvili Tbilisi State University, Ilia Chavchavadze Avenue, 1, Tbilisi 0128,Georgia

[2]Institute of Biochemical Physics Russian Academy of Sciences 4 Kosygin Street, Moscow 119991 (Russia)

Abstract

The reaction of hydride polyaddition of methylhydridesiloxane and methylhydridesiloxane-dimethylsiloxane oligomers to an unsaturated carbocyclic compound - tricyclodecadiene, at the various ratios of initial compounds, in the presence of platinum hydrochloric acid has been investigated. It was shown that the hydrosilylation reaction proceeds both to "1-2" and "9-10" position and by intermolecular with formation of branching systems. The polyorganosiloxanes with tricyclodecenyl fragment in the side chain, completely soluble in organic solvents were synthesized. For full characterization of hydride addition of methylhydridesiloxane to tricyclodecadiene, by quantum-chemical half empiric AM1 method, for all initial, intermediate and final products, in the modelling reaction of hydrosilylation of methyldimethoxysilane to tricyclodecadiene, the heats of formations (ΔH_f), change of energy (ΔH) of the system depending on the change of distance ($R_{C\text{-}Si}$) between ≡C-Si≡ bonds, as well as the charge values (q) on the atoms, dipole moments (μ) and bonds orders (Pij) are calculated. It was concluded, that the course of hydride addition of modelling reaction of methyldimethoxysilane to tricyclodecadiene energetically is more favourable by 9.10-addition. The hydride polyaddition reaction order, activation energies and rate constants were found. The synthesized oligomers were characterized by ^1H and ^{13}C NMR, FTIR, thermogravimetry, gel permeation chromatography, differential scanning calorimetry and wide angle X-ray methods.

Key words: organosiloxanes, hydride polyaddition, oligomers, thermal-oxidative stability.

INTRODUCTION

The properties of siliconorganic polymers depend on the structure of a macromolecular chain and of the nature of the surrounding organic groups at silicon atom [1].

Therefore it's evident that effective methods of synthesis of a new type siliconorganic co-polymers with complex's valuable properties attach a great importance. Particularly the synthesis of polyorganosiloxanes with un-saturating surrounding groups at the silicon atom, for receiving of siliconorganic rubbers [2] and for modification of carbo chain polymers [3] is very interesting.

Macromolecular grafting is a synthetic approach that involves the functionalization of a preformed polymer backbone containing reactive groups with an appropriate reagent. There is an increasing interest in catalytic grafting of polyhydrosiloxane as to an alternative route for the synthesis of organic–inorganic hybrid polymers, owing to the complexities involved in direct generation of well-defined linear hybrid polymers [4]. Besides the process of equilibration, the hydrosilylation process is the most widely used method of preparing organofunctional polymethylsiloxanes starting from poly(methylhydro)siloxanes [5, 6]. From this point of view the chemical modification reactions are very interesting. In the literature the modification reactions of the linear polymethylvinylsiloxanes with cyclopentadiene groups is known [7]. In this case the synthesized polymers are characterized with a higher reaction ability for vulcanization and thermal-oxidative stability.

From the literature [8] it's known, that electrophilic addition of monomer type methylhydrochlorosilanes to tricyclodecadiene in the presence of a Spier catalyst proceeds both to 9.10 and 1.2 addition depending on the nature and amount of organic groups at silicone atoms, according to the following scheme:

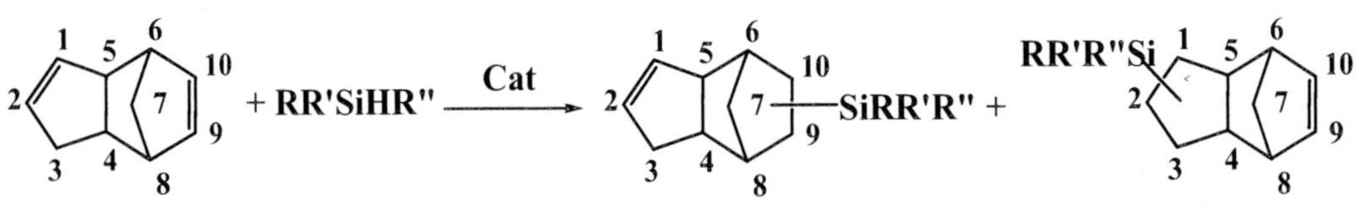

Where: R=R'=R"=Alk, Ar; R≠R'≠R"≠Alk, Ar; R"=CI.

By the authors [9] it was shown that using such catalysts as $Rh_4(CO)_{12}$ or $Co_2(CO)_8$ in this reaction the hydrosilylation proceeds selective mainly via 9,10-addition, without obtaining isomeric adducts.

2. Experimental Part

2.1. Materials and Techniques

The starting materials for the synthesis of thermo reactive comb-type methylorganosiloxane oligomers with functional groups in the side chain were α,ω–bis(trimethylsiloxy)methylhydrosiloxane and tricyclodecadiene.

The initial α,ω–bis(trimethylsiloxy)methylhydridesiloxane with degree of polymerization ~$m \approx 30$, 53 and α,ω–bis(trimethylsiloxy)methylhydridesiloxane-dimethylsiloxane oligomer with degree of polymerization ~$m \approx 33$, $n \approx 23$ tricyclodecadiene and catalyst H_2PtCl_6 were obtained from Sigma-Aldrich and used as received. The organic solvents were cleaned by drying and distillation.

The organic solvents were cleaned by drying and distillation. The purity of the initial tricyclodecadiene was controlled by its boiling temperature and refraction indexes. The purity of the starting compounds was controlled by gas-liquid chromatography "LKhM-8 MD", (Russian model) phase SKTF-100 (10%), the NAW chromosorb, carrier gas He and the 2M column.

FTIR spectra were obtained on the Nicolet Nexus 470 with MCTB detector. ^1H and ^{13}C spectra were recorded on AC-250 spectrometer at 250 MHz in the solution of deuterium solvents as solvent and internal standard.

Thermogravimetric analysis was performed on a Perkin-Elmer TGA-7, the thermogravimetric analyser at ~10 ^{0}C/min. Differential scanning calorimetric analysis was performed on a Perkin-Elmer DSC-7. Heating and cooling scanning rates were ~5 ^{0}C/min.

Gel-permeation chromatography (GPC) investigations were performed using a Waters Model 6000A chromatograph with an R 401 differential refractometer detector. The column set comprised 10^3 and 10^4 Å Ultrastyragel columns (Waters, USA). Sample concentration was approximately 3% by weight in toluene; a typical injection volume for the siloxane was 5 μL. Standardization of the GPC was accomplished by the use of styrene or polydimethylsiloxane standards with the known molecular masses.

Wide-angle X-ray diffractograms were taken on a DRON-2 (Burevestnik, Saint-Petersburg, Russia) instrument using A-CuK_α radiation. The values were measured without a filter, and the angular velocity of the motor was ~2 deg/min.

2.2. Hydride addition of methylhydridesiloxane to tricyclodecadiene (TCD)

The hydrosilylation was carried out in the three-necked flask equipped with stirrer and reflux condenser with a calcium chloride tube. The initial reagents, α,ω–bis(trimethylsilyl)-methylhydridesiloxane and tricyclodecadiene were placed in the flask and thermostated in an oil bath until constant temperature was achieved. As a catalyst, 0.1 M solution of platinum hydrochloric acid in tetrahydrofuran ($5 \div 10 \times 10^{-5}$ g per 1 g of the starting substance) was used. The reaction was performed in the argon atmosphere in the presence of toluene at temperature region 60÷90°C. Then the reaction mixture was connected to the vacuum (P=1÷2 mmHg) and the toluene and unreacted TCD was removed in the temperature range 40-50^{0}C under the vacuum.

3. Results and Discussion

In literature there is some information of receiving tricyclodecenil containing alkyl siloxanes by hydrosilylation of TCD by hydrideorganochlorosilanes in the presence of platinum hydrochloric acid [7], in the presence cobalt and rhodium carbonyls [8]. There are known polycondensation type organo-inorganic copolymers with tricyclodecenil groups in the chain [9]. These above mentioned copolymers with unsaturated fragments in the chain by a practical stand-point are very interesting systems, because they contain unsaturated bonds, which give us the possibility to change the properties of obtained siliconorganic elastomers in a wide range.

Preliminary heating of initial compounds in the temperature range of 60÷90^{0}C in the presence of catalyst, showed that in these conditions polymerization of tricyclodecadiene, break in siloxane backbone, or elimination of methane do not take place. Besides, there are no changes in the NMR and FTIR spectra of TCD and methylhydridesiloxane. By gas-liquid chromatography it was established that the polymerization of TCD in this condition does not proceed.

For the purpose of synthesis of the linear siliconorganic elastomers with tricyclodecenil fragments in the side chain, the modification reactions of α,ω-bis(trimethylsiloxy)methyl-hydridesiloxane (n≈ 30,53) and α,ω-bis(trimethylsiloxy)methylhydridesiloxane-dimethylsiloxane with TCD were studied. The hydrosilylation reactions were studied at 1:30, 1:53 and 1:25 molecular ratio of initial compounds in the presence of platinum hydrochloric acid as a catalyst. As it's seen from the molecular structure of TCD:

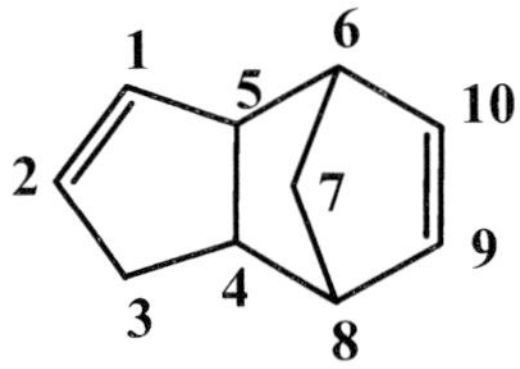

it contains two unsaturated bonds in "1-2" and "9-10" position. From literature data [8,9] during hydrosilylation of TCD in case of a big amount of electro acceptoric groups containing organochlorosilanes (methyldichlorosilane) hydride addition at the double bond proceeds to "9-10" position. At a small amount of electro acceptoric groups (in the case of trimethylchlorosilane) the addition proceeds both to "1-2" and "9-10" position. In the presence of electro donor groups (in the case of trimethylsilane) the hydrosilylation mainly proceeds to "1-2" position. It was possible that in our case the reaction must be proceeding to "1-2" position. As it's known from literature the electrophilic addition of ≡Si-H bond to TCD must proceed with the formation of a transitional complex. The stability of the complex at "9-10" position because steric difficulty is lower, than the stability of the complex to "1-2" position. As a result of this the addition mainly must proceed to "9-10" position.

In the ^{1}H NMR spectra of TCD, Figure 1, one can observe the multiplet resonance signals for unsaturated bonds with center δ=5.50 ppm and δ=5.95 ppm characteristic of unsaturated bonds. One can observe for protons in methylene groups at 7a and 7b position with center δ=1.55 ppm and δ=1.35 ppm (J_{hemin}=14 Hz) δH($3_{a,e}$)=2.20 ppm.

Table 1. ^{1}H NMR resonance signals of TCD and oligomers (II, III).

C-H bond	H-1	H-2	H-3	H-4	H-5	H-6	H-7	H-8	H-9	H-10
TCD (ppm)	5,95	5,95	2,20	1,67	1,67	2,22	1,40	2,17	5,50	5,50
II (ppm)	5,95	5,95	2,20	1,70	1,70	1,80	1,30	1,80	-	-
III (ppm)	5.96	5,96	2,20	1,70	1,70	1,80	1,30	1,80	-	-

In the ^{1}H NMR spectra of hydrosilylation product II one can observe the resonance of unimportant signals for protons in the unsaturated fragment at "9-10" position, and mainly unchanged multiplet resonance signals at δ=5.3÷5.7 ppm, with integral intensity ~1:10. So, from ^{1}H NMR spectra of hydrosilylation product it's evident that the hydride addition mainly proceeds at "9-10" position and about 10% in "1-2" position.

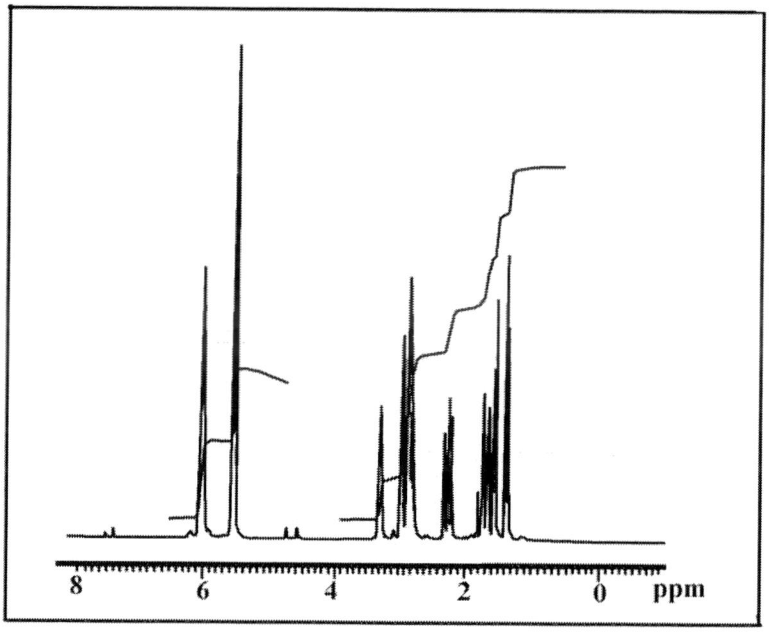

Figure 1. ^{1}H NMR spectrum of tricyclodecadiene.

The above-mentioned conclusion is in conformity with quantum-chemical calculation. By Quantum-chemical semi empiric method AM1 (Austin Model) the electronic characteristics: the charges on the atoms and bonds orders were calculated. The obtained results are presented on the diagram 1.

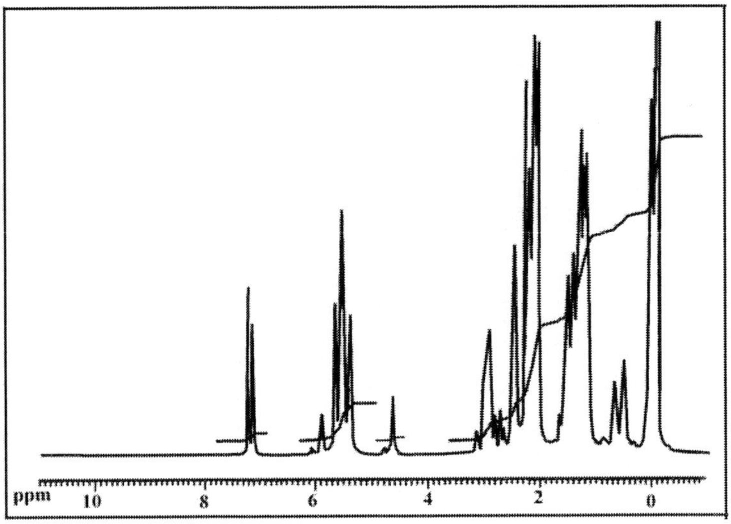

Figure 2. ^{1}H NMR spectrum of α,ω-bis(trimethylsiloxy)methyltricyclodecenylsiloxane II.

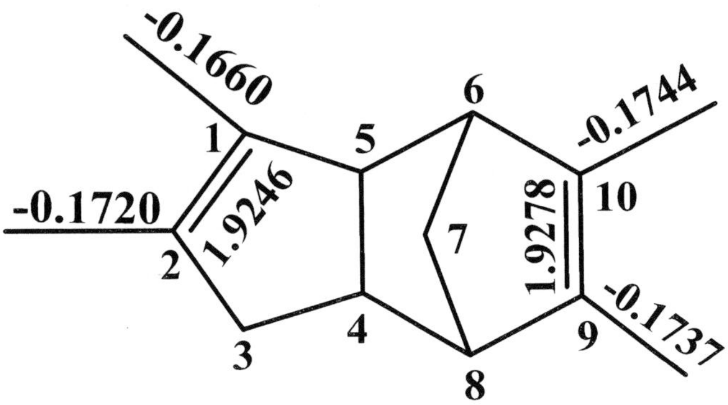

Diagram 1. The charges on the atoms and bond orders in TDC molecule.

As it's seen from the diagram, there is a higher value of the charge observed on the carbon atoms in "1-2" and "9-10" position. This indicates that these carbon atoms are reactionable centres for electrophilic addition. Also, bond orders in "1-2" and "9-10" position are characterized with higher values. For example $P_{1,2}(\pi)$=1.9246 and $P_{9,10}(\pi)$=1.9278 (where index π showed bond orders of π components). Because $P_{9,10}(\pi)>P_{1,2}(\pi)$, it's possible to predict that the addition is more possible with participation of $P_{9,10}(\pi)$ electrons. This conclusion does not accept the addition in "1-2" position.

The hydrosilylation reaction was studied at the constant temperature, in the range 60÷90^{0}C. It was established that during the reaction in melt condition, at beginning stages of addition, three dimensional systems were obtained, which may be explained by intermolecular reactions in macro chain. For the purpose of synthesis of soluble organosiloxanes the reactions were carried out in an absolute toluene solution (C=0.1017 mol/l). The course of the reaction was observed by the decrease of the amount of active ≡Si-H groups (Figure 3).

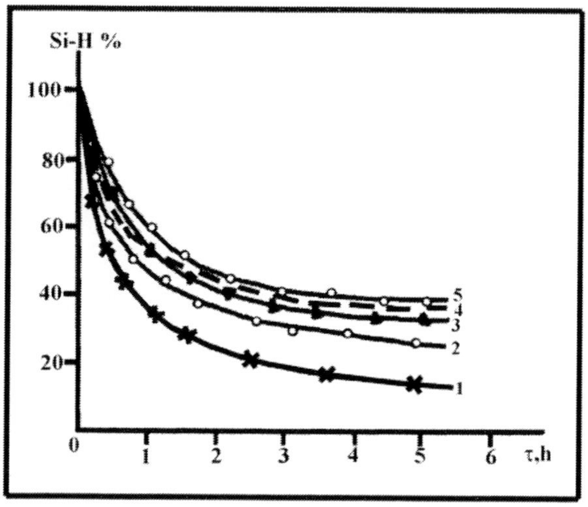

Figure 3. Dependence of changes of active Si-H% groups on the time upon of hydride addition of α,ω-bis(trimethylsiloxy)methylhydridesiloxane to tricyclodecadiene, where curve 1 is at 90^{0}C, curve 3 is at 80^{0}C, curve 4 is at 70^{0}C and curve 5 is at 60^{0}C. Curve 2 corresponds to hydride addition of α,ω-bis(trimethylsiloxy)methylhydridesiloxane-dimethylsiloxane to tricyclodecadiene at 90^{0}C.

It was shown that with the increase of temperature the depth of hydrosilylation increases. At 60^{0}C the conversion of active ≡Si-H groups after 5 h is about 61% (Figure 3, curve 4), at 90^{0}C the conversion is equal to 85% (Figure 3, curve 1). In the case of methylhydridesiloxane-dimethylsiloxane oligomer the conversion of active ≡Si-H groups at 90^{0}C is ~75%.

The hydrosilylation proceeds vigorously during the first 90-100 min, after that the conversion of active ≡Si-H bonds increases slightly. On the basis of experimental investigations it's possible to assume that hydrosilylation mainly proceeds with obtaining various linked branching systems, according to the following scheme 1:

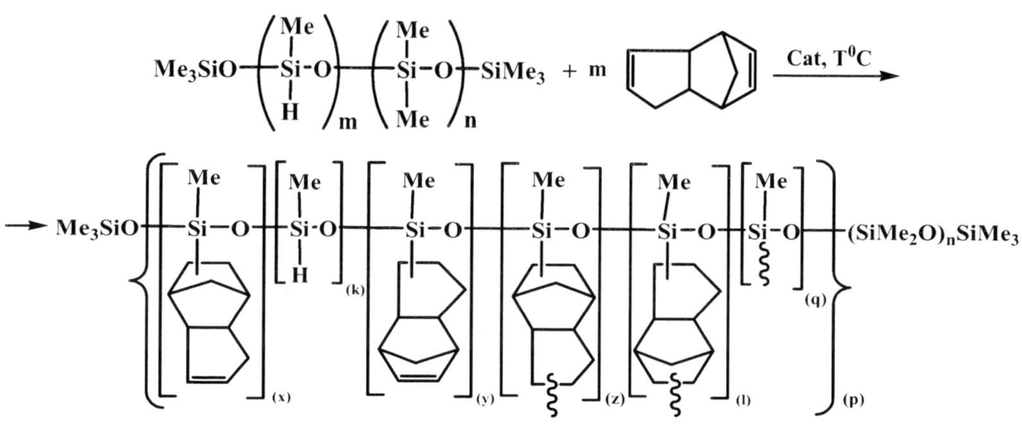

Scheme 1. Hydrosilylation reaction of α,ω-bis(trimethylsiloxy)methylhydridesiloxane to tricyclodecadiene.

Where: [(x)+(k)+(y)+(z)+(l)+(q)](p)=m≈30, n=0: 90^0 C(I) m≈53, 60^{0}C(II^1); 70^{0}C(II^2), 80^{0}C(II^3), 90^{0}C (II); [(x)+(k)+(y)+(z)+(l)+(q)](p)=m≈30, m≈33, n=23: 90^{0}C (III).

It was established that with the rise of temperature, cross-linked polymers are obtained. During hydrosilylation at 90^{0}C at 85% conversion of active ≡Si-H bonds about 33-35% insoluble polymers were obtained, this may be explained by the secondary interchain hydrosilylation reactions on the bases of the catalyst in the reaction system, according to the scheme 2:

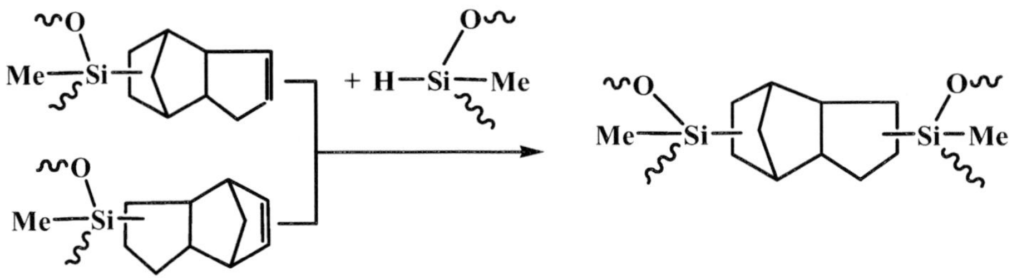

Scheme 2. Possible scheme of inter-molecular hydrosilylation reaction.

Consequently in TCD, -CH=CH- bond in "1-2" position as an unsaturated bond in "9-10" position is characterized with a higher reaction ability reference to hydrosilylation. Therefore for reception of soluble copolymers hydrosilylation reaction stepped up to achieve 75-80 % of conversion of active ≡Si-H groups. Then the reaction mixture was connected to the vacuum (P=1÷2 mmHg) and the toluene and un-reacted TCD was removed in the temperature range 40-50^{0}C under the vacuum up to constant weight. The synthesized copolymers are transparent, opalescent viscous products well soluble in ordinary organic solvents with η=0.07÷0.13. Some physical-chemical properties of oligomers are presented in table 2.

Table 2. Physical-chemical properties of methyl(tricyclodecenyl)(hydride)siloxanes.

No	Yield, %	React. Temp-re 0C	η*sp	Tg, 0C	$\bar{M}$ωx10-3	P, polydisp	5%, mass losses	d1, Å
I	85	90	0.08	-1	7.8	2.6	240	9.10
I1	65	-	0.13	+8	82.1	21.8	300	9.33
II1	83	60	0.07	-	-	-	-	-
II2	75	70	0.08	+2	7.8	1.8	-	-
II3	70	80	0.09	-	-	-	-	-
II	67	90	0.09	+6	10.3	2.3	270	9.33
III	70	70	0.11	-15	-	-	250	8.63

*In 1% solution of toluene at 25^{0}C.

In the FTIR spectra of synthesized oligomers one can observe absorption bands characteristic for -CH=CH- bonds at 1600 and 3550 cm^{-1}, the absorption bands at 2165, 1275 and 1020 cm^{-1}, characteristic for ≡Si-H, ≡Si-Me and ≡Si-O-Si≡ linear bonds accordingly.

For full characterization of hydride addition of methylhydridesiloxane to tricyclodecadiene, by quantum-chemical semi empiric AM1 method [9], for all initial, intermediate and final products, in the modeling reaction of hydrosilylation of methyldimethoxysilane to tricyclodecadiene, the heats of formations (ΔH_f), change of energy (ΔH) of the system depending on the change of distance ($R_{C\text{-}Si}$) between ≡C-Si≡ bonds, also the charges values (q) on the atoms, dipole moments (μ) and bonds orders (Pij) are calculated.

For the beginning the modelling reaction of a hydride addition of $[Me(MeO)_2SiH]$ methyldimethoxysilane to tricyclodecadiene to 9,10-direction was considered. The model reaction proceeds with the formation of compound IV according to the following scheme 3.

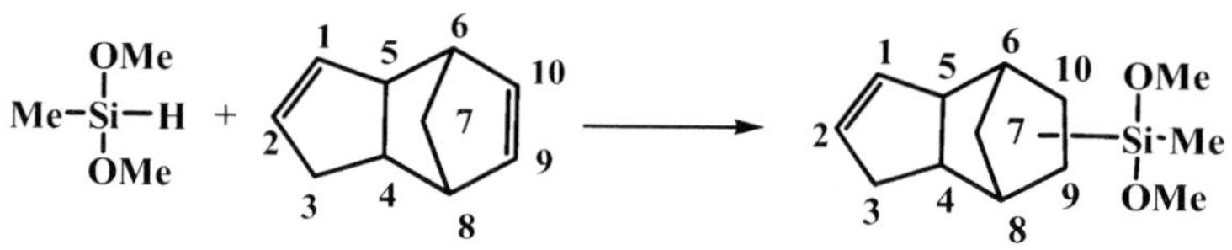

IV

Scheme 3. Modelling reaction of hydride addition of methyldimethoxysilane $[Me(MeO)_2SiH]$ to tricyclodecadiene 9.10-addition.

The distances between the double bonded carbon atoms (C_{10}, C_9) and silicon atoms in tricyclodecadiene was taken about 1.0 Å more than the bond, expected in a product.

The change in step of distances ($R_{C_{10}\text{-}Si}$) between an atom of silicon and a double bonded carbon atom (C_{10}) was ~0.05 Å. Dependence of change of energy (ΔH) of the system on the distances ($R_{C_{10}\text{-}Si}$) between the silicon and carbon atoms is presented on the Figure 4.

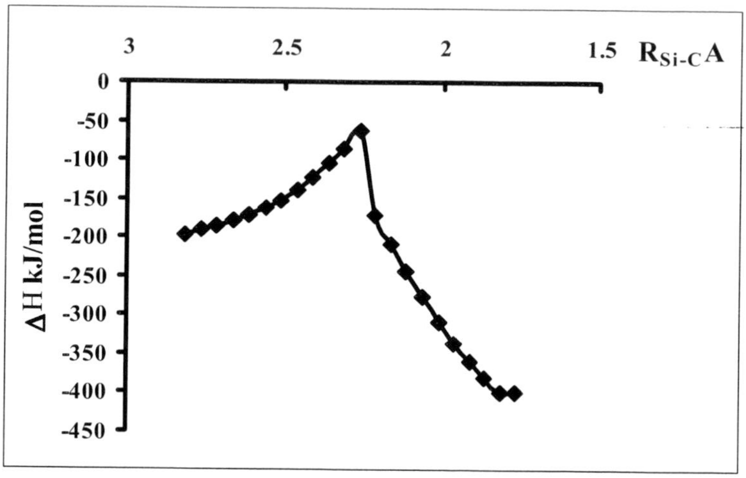

Figure 4. Dependence of changes of energy (ΔH) on distances ($R_{C_{10}\text{-}Si}$) between silicon and carbon atoms during of modelling hydrosilylation reaction of methyldimethoxysilane to tricyclodecadiene (9.10-addition).

As it is seen from Figure 4, at approximation of the silicon to the carbon atom (C_{10}) up to $R_{C_{10}\text{-}Si}$=2.26 Å distances the energy of system rises. At the same time the bond order between carbon C_{10} and C_9 atoms in a molecule of tricyclodecadiene decreases from 1.909 up to 1.699. The bond order between the hydrogen and silicon atoms also decreases ($R_{Si\text{-}H}$=0.922-0.775) and in the modelling compound methyldimethoxysilane the process of formation of new bonds ($R_{C_{10}\text{-}Si}$=0.015-0.105 and $P_{C_9\text{-}H}$= 0.0006-0.063) is observed. In case of a distance up to 2.21 Å the energy of the system sharply decreases, the double C=C bond passes in ordinary C-C bond ($P_{C_{10}\text{-}C_9}$=0.991), and as to the bond order between the silicon and carbon atom C_{10}, it reaches to 0.798. Thus the hydrogeis completely eliminated from silicon ($P_{Si\text{-}H}$=0.013) and attaches to the carbon atom C_9 ($P_{C_9\text{-}H}$=0.929). The same results have been obtained during hydride addition of silicon and double bonded C_9 carbon atom.

With consecutive reduction of distances the energy of the system monotonously decreases and with strengthening formed new bonds. The heat of formation of the product IV of hydride 9.10-addition (ΔH= -399.129 kJ/mol) is in good correlation with calculated heat of formation (ΔH= -399.012 kJ/mol) of compound IV obtained by hydride addition of methyldimethoxysilane to tricyclodecadiene at 9.10-direction. Activation energy of 9.10-addition of the modelling reaction is equal to E_{act}=133.21 kJ/mol.

The 1.2-addition of the modelling compound methyldimethoxysilane to tricyclodecadiene is considered also. The modelling reaction of 1.2-addition proceeds with the formation of compound V according to the scheme 4:

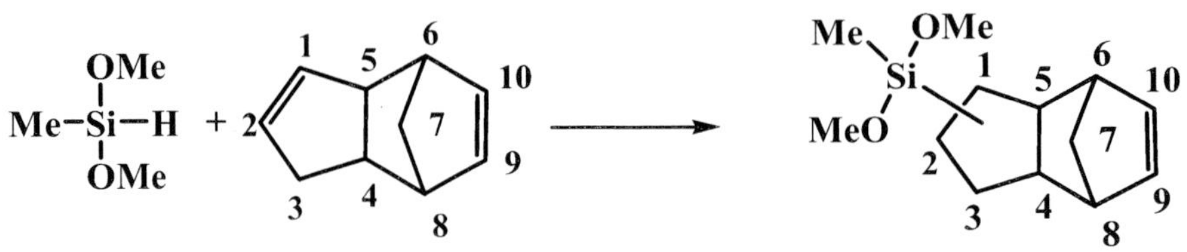

V

Scheme 4. Modelling reaction of hydride addition of methyldimethoxysilane [Me(MeO)$_2$SiH] to tricyclodecadiene 1.2-addition.

In Figure 5 the dependence of change of energy (ΔH) of the system on the distances ($R_{C_1\text{-}Si}$) between the silicon and carbon atom is presented. Similarly the above-mentioned distance between the double bonded carbon atom (C_1) and silicon was taken about 1.0 Å more than the bond, expected in a product. The change in pace of distances ($R_{C_1\text{-}Si}$) between the atom of silicon and the double bonded carbon atom (C_1) was about 0.05 Å.

As it is seen from Figure 8, at approximation of the silicon to the carbon atom (C_1) up to $R_{C_1\text{-}Si}$=2.30 Å distances the energy of the system rises. At the same time the bond order between of carbon C_1 and C_2 atoms in a molecule of tricyclodecadiene decreases from 1.901 up to 0.978. The bond order between hydrogen and silicon atoms also decreases ($R_{Si\text{-}H}$=0.917-0.884) and in the modelling compound methyldimethoxysilane, the process of formation new bonds ($R_{C_1\text{-}Si}$=0.0085-0.853 and $P_{C_2\text{-}H}$= 0.0009-0.011) is observed. In the case of distances up to 2.25 Å the energy of the system sharply decreases, the double C=C bond passes in an ordinary C-C bond ($P_{C_1\text{-}C_2}$=0.980), and as to the bond order between the silicon and carbon atom C_1, that it reaches to 0.853

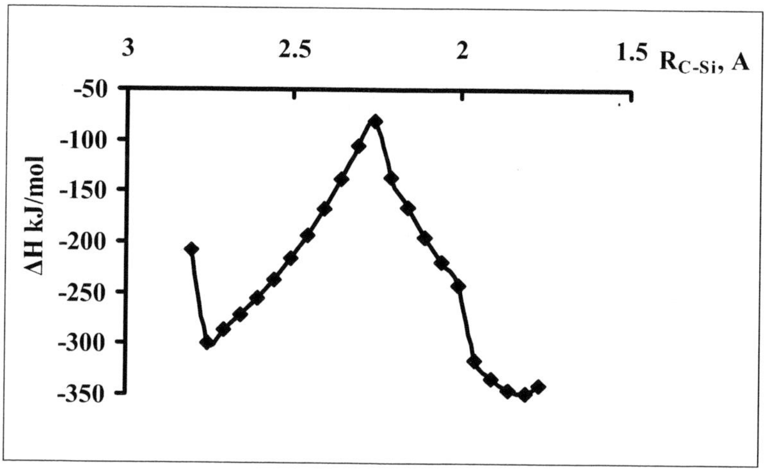

Figure 5. Dependence change of energy (ΔH) on distance ($R_{C_1\text{-}Si}$) between silicon and carbon atoms during modelling hydride 1.2-addition of methyldimethoxysilane to tricyc-lodecadiene.

Thus the hydrogen is completely eliminated from silicon ($P_{Si\text{-}H}$=0.043) and attaches to the carbon atom C_2 ($P_{C_2\text{-}H}$=0.887). The same results have been obtained during hydride addition of silicon and a double bonded C_2 carbon atom.

With consecutive reduction of distances the energy of the system monotonously decreases and with strengthening forms new bonds. The heat of formation of the product V of hydride 1.2-addition (ΔH=-399.129 kJ/mol) is in good correlation with the calculated heat of formation (ΔH=-399.012 kJ/mol) of compound V obtained by hydride addition of methyldimethoxysilane to tricyclodecadiene at 1.2-direction. Activation energy of 1.2-addition of the modelling reaction is equal to E_{act}= 218.67 kJ/mol.

Thus in a comparison of the Figures 4 and 5, and also energy of activation (the character of curve of change of energy and the value of bond orders) it can be concluded, that the course of the modelling reaction of hydride addition of methyldimethoxysilane to tricyclodecadiene energetically is more favourable by 9,10-addition, which is in agreement with NMR spectral data.

In Figure 6 the dependence of reverse concentration of the reactant on the time is presented. One can see that at the beginning stages the hydrosilylation is the reaction of the second order. The hydrosilylation reaction rate constants at various temperatures were calculated: k_{60^0C}= 1.051x10^{-1}, k_{70^0C}=1.405x10^{-1}, k_{80^0C}=1.944x10^{-1} l/mol·s.

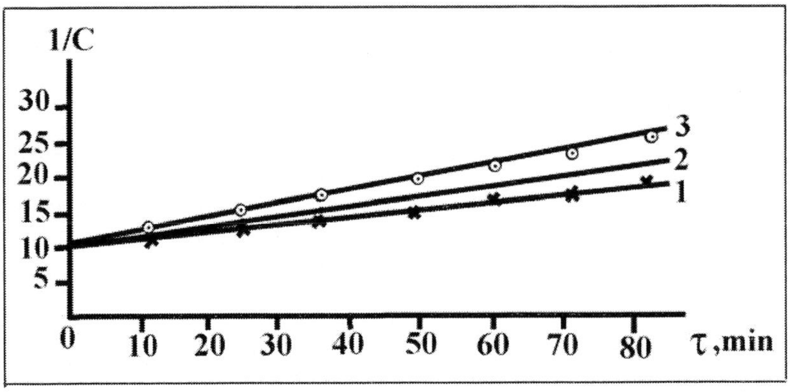

Figure 6. Dependence of reverse concentration on the time, where curve 1 is at 60^0C, curve 2 is at 70^0C and curve 3 is at 80^0C.

From the dependence of the logarithm of the hydrosilylation reaction rate constants on the reverse temperature the activation energy of the hydrosilylation reaction of methylhydridesiloxane to tricyclodecadiene was calculated ~E=32.3 kJ/mol.

The synthesized oligomers were studied by GPC method. Figure 7 shows monomodal molecular weight distribution of oligomer I, $\overline{M}_{\omega}\approx 7.8x10^3$ and polydispersity is equal to P≈2.6. Whence it follows, that in case of full hydrosilylation $\overline{M}_{\omega}$ does not exceed $6x10^3$. Increased molecular weight once again proved that the hydrosilylation reaction proceeds to macromolecular interchain hydride addition.

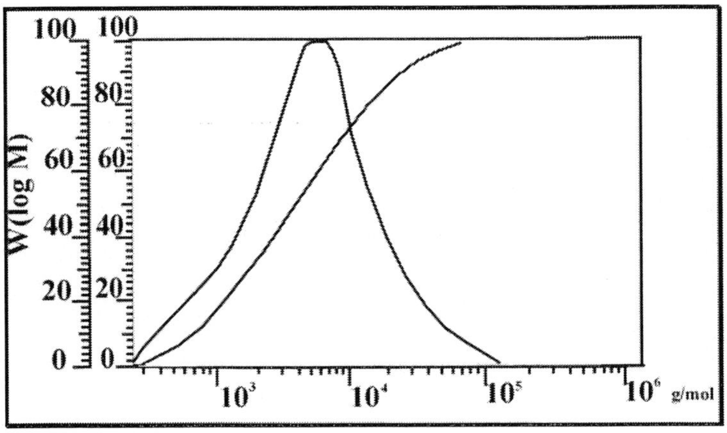

Figure 7. Gel permeation chromatographic curves of oligomer I.

The oligomer I (30% toluene solution) was heated additionally for 4-5 hours at 80-90^0C temperature interval, at this time the value of active ≡Si-H group's decreases from 18% up to 3%. After filtration the toluene was removed at 40-50^0C in a vacuum (P=1÷2 mmHg) up to constant mass and transparent yellow oligomer I^1 (65%) was received, which is soluble in ordinary organic solvents.

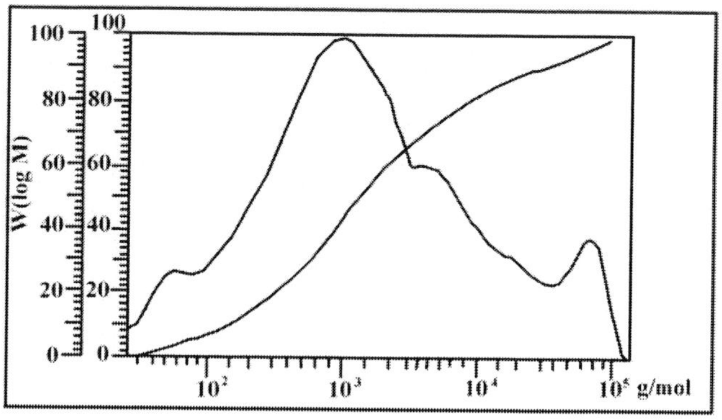

Figure 8. Gel permeation chromatographic curves of oligomer I^1.

By GPC analysis Figure 8, showed that during additional heating the macromolecular intermolecular hydride addition takes place and polydispersity of oligomer I^1 with $\bar{M}_\omega \approx 8.3x10^4$ and $P \approx 21.8$ were received.

The TGA analysis of methylsiloxane oligomers containing tricyclodecenil in the side chain was carried out and it was shown that they are characterized by a higher thermal oxidative stability than the trimethylsiloxy group with blocked polydimethylsiloxanes. As it's seen from Figure 9, 5% mass losses for obtaining for oligomers II^2 and III is observed in the range 290-300^{0}C. The main destruction process proceeds in the 350-700^{0}C temperature range, after 700^{0}C mass losses and not observed. The oligomer III is characterized by lower thermal-oxidative stability, than oligomer II^2.

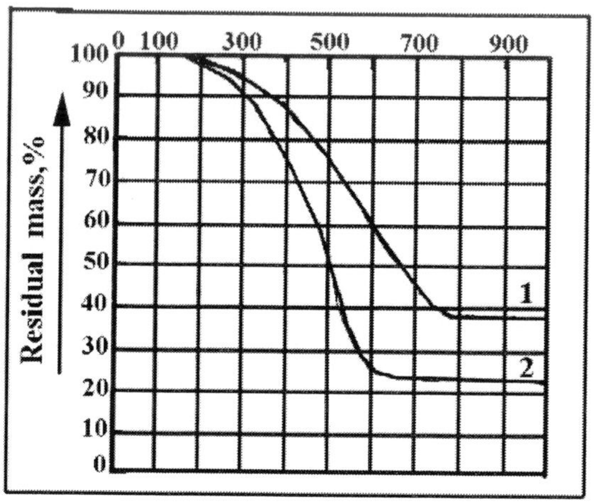

Figure 9. Thermogravimetric curves of oligomers, where curve 1 is for oligomer II^2 and curve 2 is for oligomer III.

By DSC analysis (Figure 10) it was shown that the first temperature transition for oligomer I, which corresponds to glass transition temperature (T_g) observed at ~-1^{0}C and for oligomer I^1 is observed at ~+8^{0}C. With an increase of the amount of tricyclodecenil fragments in the linear siloxane chain (with an increase of the depth of hydride addition) the T_g of oligomers rises (see Figure 9, curves 1-4). For oligomer III glass transition temperature is equal to ~ -15^{0}C.

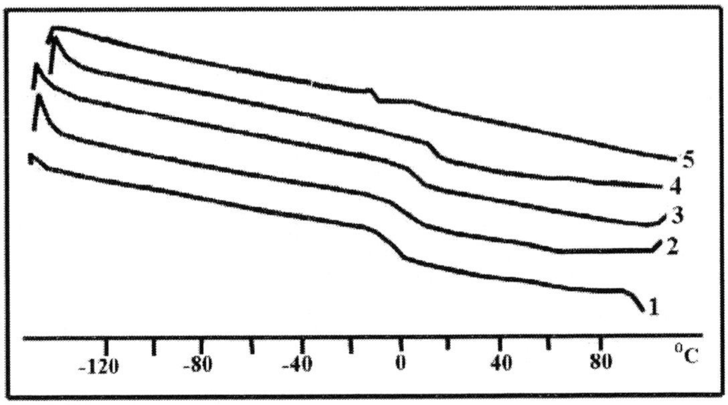

Figure 10. DSC curves of oligomers, where curve 1 is for oligomer I, curve 2 is for oligomer II^1, curve 3 is for oligomer II, curve 4 is for oligomer I^1 and curve 5 is for oligomer III.

The tricyclodecenil containing methylhydridesiloxane oligomers are amorphous one phase systems. With an increase of the amount of tricyclodecenil fragments in the dimethylsiloxane chain the value of interchain distances d_l slightly increases.

The synthesized tricyclodecenil and hydride containing linear methyl siloxanes are interesting products, because they contain reactionable functional groups which easily enter in the intermolecular addition reaction at the expense of in system existence platinum hydrochloric acid, that gives us the possibility to use such oligomer to prepare siliconorganic rubbers.

REFERENCES

[1] O.V. Mukbaniani, L.M. Khananashvili. Intern. *J. Polymer. Materials,* 1994, 27, 31.

[2] G.N. Aneli, Kh.E. Koberidze, O.V. Mukbaniani, M.G. Karchkhadze, L.M. Khananashvili. *Russian Polymer News*, 1998, 3, 7.

[3] N.M. Galashina. *Vysokomol. Soed.,* 1994, 36A, 640 (in Russian).

[4] 4. B.P.S. Chauhan, J.S. Rathore, N. Glloxhani. *Appl. Organometal. Chem.,* 2005, 19, 542.

[5] M. P. McGrath, E. D. Sall, S. J. Tremontm. *Chem. Rev.,* 1995, 95, 381.

[6] B. Marciniec. In Comprehensive Handbook on Hydrosilylation, Pergamon Press, Oxford, 1992.

[7] J.N. Aneli, O.V. Mukbaniani, M.G. Karchkhadze, Kh.E. Koberidze, R.Sh. Tkeshelashvili, L.M. Khananashvili. *Georgian Patent,* N1348101-96.

[8] 8. A.I. Nogaideli, L.M. Khananashvili, L.I. Nakaidze, T.V. Chogovadze, V.S. Tskhovrebashvili, R.V. Volkova, G.A. Nogaideli. *Dokl. Akad. Nauk GSSR*, 1978, 92, 341 (in Russian).

[9] 9. G.K.I. Magomedov, O.V. Shkolnik, L.M. Khananashvili, L.I. Nakaidze, M.G. Katsitadze. *Zhurn. Obshch. Khimii,* 1988, 59, 104 (in Russian).

In: Modification Reactions of Oligomethylhydridesiloxanes ISBN: 1-60021-365-0
Editors: O.Mukbaniani, et al. pp. 163-178

Chapter 11

HYDRIDE ADDITION OF METHYLHYDRIDSILOXANES TO CYCLOHEXA-1,3-DIENE

***O. Mukbaniani*[1], *G. Zaikov*[2], *T. Tatrishvili*[1], *Kh. Koberidze*[1], *and S. Meladze*[1]**

[1] I. Javakhishvili Tbilisi State University, I. Chavchavadze Ave., 1, Tbilisi 380028, Georgia

[2]Institute of Biochemical Physics Russian Academy of Sciences 4 Kosygin Street, Moscow 119991 (Russia)

ABSTRACT

The reaction of hydride polyaddition of methylhydridesiloxanes to cyclohexa-1,3-diene, at the ratio of initial compounds 1:30, in the presence of platinum chlorohydric acid has been investigated. The hydride polyaddition reaction order, activation energies and rate constants were found. The polyorganosiloxanes with cyclohexa-1,3-diene fragments in the side chain, which are completely soluble in organic solvents were synthesized. By quantum-chemical half-empirical method AM1 the electronic characteristics of cyclohexa-1,3-diene: the charges on the carbon atoms and bonds orders were calculated. For full characterization of the hydride addition of methylhydridesiloxane to cyclohexa-1,3-diene and by quantum-chemical half empiric AM1 method, for all initial, intermediate and final products, in the modelling reaction of hydrosilylation of methyldimethoxysilane to 1,3-cyclohexa-diene, the heats of formations (ΔH_f), change of energy (ΔH) of the system depending on the change of distance (R_{C-Si}) between ≡C-Si≡ bonds, also the charges values (q) on the atoms, dipole moments (μ) and bonds orders (Pij) are calculated. It was concluded, that the course of hydride addition of modelling reaction of methyldimethoxysilane to cyclohexa-1,3-diene energetically is more favourable by 1,4-addition. ^{1}H and ^{13}C NMR, FTIR, thermogravimetric, gel-permeation chromatography, differential scanning calorimetry and X-ray methods were used to characterize the synthesized oligomers.

Key words: organosiloxanes, hydride polyaddition, oligomers thermal-oxidative stability.

INTRODUCTION

The properties of siliconorganic polymers aside from other factors depend on the structure of a macromolecular chain and on the nature of the surrounding organic groups at the silicon atom [1].

The effective methods of synthesis of the new type siliconorganic copolymers with complex valuable properties attach a great importance. Particularly interesting is the synthesis of polyorgano-siloxanes with unsaturated surrounding groups at silicon atom, for receive siliconorganic rubbers [2] and modifiers for carbo chain polymers [3] which is very important.

From this point of view the chemical modification reactions are very interesting. In the literature the modification reactions of the linear polymethylvinylsiloxanes with norbornene groups are known [4]. In this case the synthesized polymers are characterized with a higher reaction ability for vulcanisation and thermal-oxidative stability. By the catalytic dehydrocondensation of polymethylhydridesilones with hydroxyorganocyclosiloxanes in the presence of catalysts the polyorganosiloxanes with cyclic fragments in the side chain were synthesized [5]. By hydrosilylation reactions of methyl(eth-yl)hydridesiloxane with vinylorganocyclosiloxanes in the presence of platinum hydrochloric acid polyorganosiloxanes with carboorganocyclosiloxane fragments in the side chain were synthesized [1,6]. In both cases the insertion of reactionable cyclic fragments in the side chain leads to the rise of thermal-oxidative stability and glass transition temperature of copolymers.

By a polymerisation reaction of methyltricyclodecenylcyclotri(tetra)siloxanes in the presence of catalysts, methylsiloxane copolymers with unsaturated tricyclodecenyl groups in the chain was obtained [7].

By a hydrosilylation reaction of tricyclodecadiene with α,ω-bis(trimethylsiloxy)-methylhydridesilo-xanes in the presence of platinum hydrochloric acid or platinum on the carbon as a catalyst, various linked methylsiloxane oligomers with tricyclodecenyl fragments in the chain were synthesized [8], according to the following scheme:

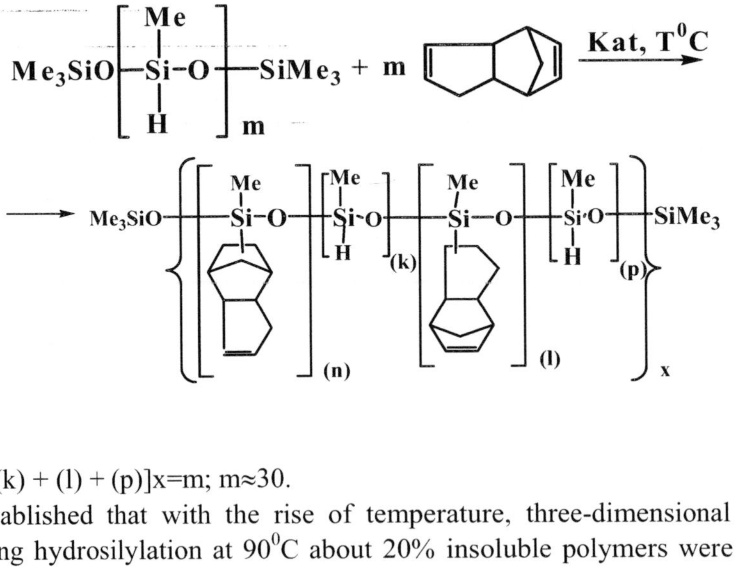

Where: [(n) + (k) + (l) + (p)]x=m; m≈30.

It was established that with the rise of temperature, three-dimensional polymers are received. During hydrosilylation at 90^0C about 20% insoluble polymers were obtained, this may be explained by the secondary interchain reactions on the bases of a catalyst in the reaction system. Soluble oligomers were obtained in case of 80% conversion of active ≡Si-H bonds.

EXPERIMENTAL PART

The purity of initial dicyclopentadiene was controlled by its boiling temperature and refractive index. The organic solvents were cleaned by drying and distillation. The purity of the starting compounds was controlled by gas-liquid chromatography "LKhM-8-MD" (Russian model), phase SKTF-100 (10%), the NAW chromosorb, carrier gas He and the 2M column.

FTIR spectra were obtained on Nicolet Nexus 470 with MCTB detector, while the ^{1}H and ^{13}C nuclear magnetic resonance (NMR) spectra were recorded on a Bruker AC-250 spectrometer at 250 MHz in the solution of deuterium solvents as a solvent and internal standard.

The starting materials for the synthesis of the linear siloxanes with cyclohexenyl fragments in the side chain were α,ω-bis(trimethylsiloxy)methylhydridesiloxanes with degree of polymerization $n \approx 30$ and cyclohexa-1,3-diene.

Gel-permeation chromatography (GPC) measurements were carried out by using toluene or THF as an eluent and IR detector. Standard polystyrenes or polydimethylsiloxanes were used for calibration. Gel permeation chromatography was carried out with the use of a Waters Model 6000A chromatograph with an R 401 differential refractometer detector. The column set comprised 103 and 104 Å Ultrastyragel columns. Sample concentrations were approximately 3% by weight in toluene, and typical injection volume for the siloxane was 5μL. Standardization of the GPC was accomplished by use the of styrene or PDMS standards with the known molecular weight.

Diffractograms were taken on a "Dron-2" instrument (Russian model), A-Cuα values were measured without a filter, the angular velocity of the motor ω=2 dg/min.

The thermogravimetric investigations were carried out on a "Paulic-Paulic-Erdey" derivatograph model MOM-102 (Hungary). The test conditions were: temperature rise rate ν=5 deg/min, in the open area.

RESULTS AND DISCUSSION

In the literature there is no information about cyclohexenyl side group containing linear methylsilo-xanes. The modification reactions of α,ω-bis(trimethylsiloxy)methylhydride-siloxanes ($n \approx 30$) with cyclohexa-1,3-diene for the purpose of synthesis of the linear siliconorganic elastomers with cyclo-hexenyl side groups were studied. The hydrosilylation reactions were studied at a 1:30 molecular ratio of initial compounds in the presence of platinum chlorohydric acid as a catalyst. Preliminary heating of initial compounds in the temperature range of $60 \div 90^{0}C$ in the presence of catalyst showed that in these conditions no changes of initial compounds take place. Besides, there are no changes in the NMR and FTIR spectra of cyclohexa-1,3-diene and methylhydridesiloxanes. By gas-liquid chroma-tography it was established that the polymerisation of cyclohexa-1,3-diene in this condition does not take place.

As it's seen from the structure of cyclohexa-1,3-diene the hydrosilylation may proceed to one double bond with preservation of the second one. But another side hydride addition may be realized in "1-2" direction or as a conjugated diene system in "1-4" direction (Scheme 1).

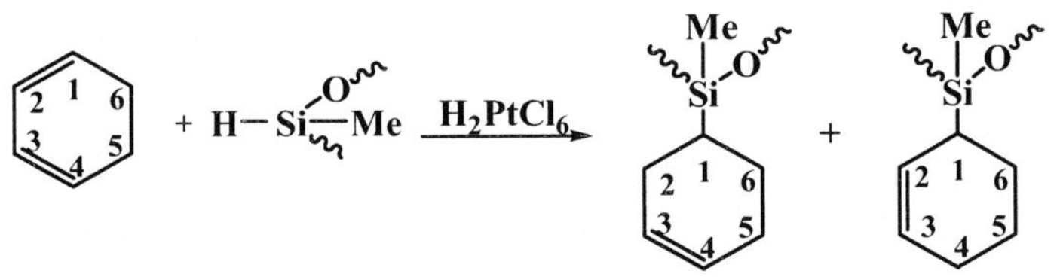

Scheme 1. Possible schemes of hydrosilylation reactions of cyclohexa-1,3-diene.

By quantum-chemical half-empirical method AM1 [9] the electronic characteristics of cyclohexa1,3-diene: the charges on the carbon atoms and bonds orders were calculated.

The obtained results are presented on the molecular diagram 1:

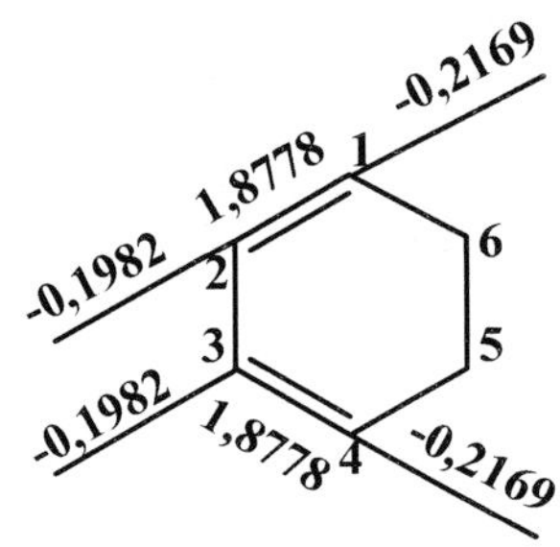

Diagram 1

As it's seen from the molecular diagram, the higher value of the charge is observed on the carbon atoms in 1 and 4 positions. This indicates that these carbon atoms are more reactionable centres for electrophilic addition. Because of that the hydride addition may be realized with more probability in "1-4" position, than 2 or 3 position ($q_{1.4}\approx-0.2169$, $q_{2..3}\approx-0.1982$). The ratio of atom charges values $q_{1.4}/q_{2.3}\approx1{,}09$, which is peculiarly in agreement with NMR spectra data. Besides bond orders in "1-2" and "3-4" position, they are characterized with the same values $P_{1.2}=P_{3.4}\approx1.8778$, which indicate that during hydrosilylation decisive rule execute the charges values on the atoms.

It was established that during the hydrosilylation reaction in melt condition, at the beginning stages of hydrosilylation (20% of the conversion of active ≡Si-H groups) three-dimensional systems were obtained, which may be explained not only by intramolecular reactions, but also by intermolecular reactions of the macro-chain (Scheme 2).

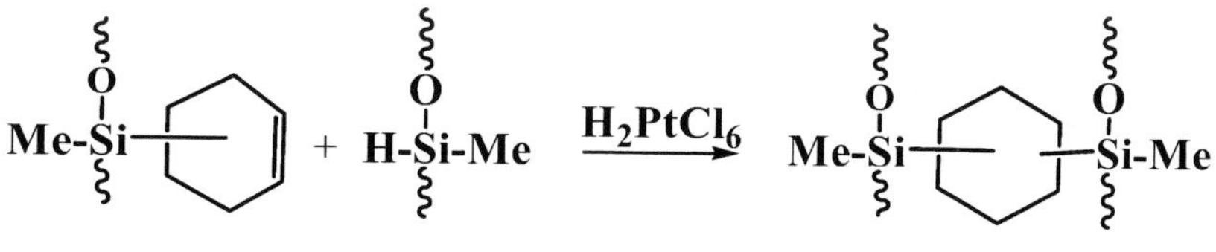

Scheme 2. Possible scheme of inter-molecular hydrosilylation.

For the purpose of synthesis of soluble organosiloxanes the hydrosilylation reaction was carried out in an absolute toluene solution (C≈0.1155 mole/l). The course of the reaction was observed by a decrease of the amount of active ≡Si-H groups. As it's seen from Figure 1, with the rise of temperature the depth of hydrosilylation (conversion of ≡Si-H groups) reaction increases.

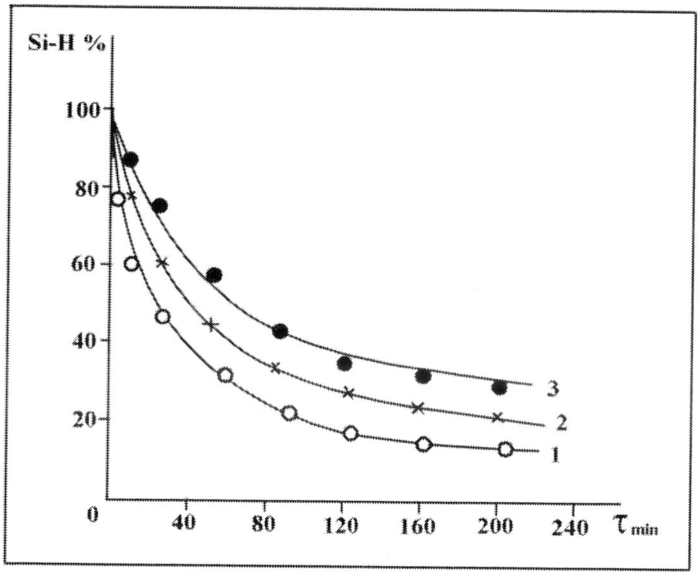

Figure 1. Dependence of changes of active ≡Si-H% groups on the time during hydride addition of α,ω-bis(trimethylsiloxy)methylhydridesiloxanes to 1,3-cyclo-hexadiene. Where curve 1 is at 80^{0}C, curve 2 is at 70^{0}C and curve 3 is at 60^{0}C.

After 3h hydrosilylation at 60^{0}C temperature the conversion of active ≡Si-H bonds is about 69% (Figure 1, curve 3). With the rise of the temperature the conversion of active ≡Si-H bonds increases and at 80^{0}C temperature it consists of about 85% (curve 1). It must be denoted that at higher stages of hydrosilylation, three-dimensional systems were also obtained. For the purpose of synthesis of soluble systems the hydrosilylation reaction put an end at ~80% depth of ≡Si-H conversion.

Consequently, the hydrosilylation reaction of methylhydridesiloxanes to cyclohexa-1,3-diene proceeds according to the following scheme 3.

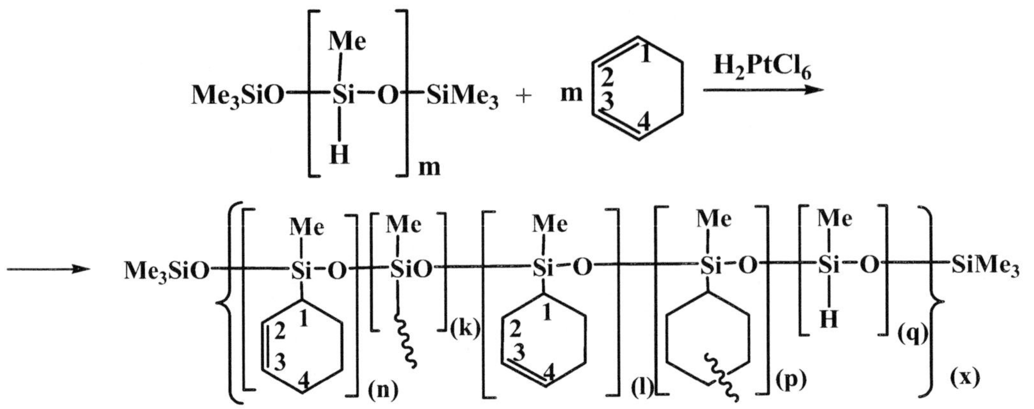

Scheme 3. Hydride addition of methylhydridesiloxanes to cyclohexa-1,3-diene.Where: [(n)+(k)+(l)+(p)+(q)](x)≈30, I[1] (60°C); I[2] (70°C); I (80°C).

The synthesized siliconorganic oligomers are colorless viscous products well soluble in ordinary aromatic type organic solvents with $\eta_{sp} \approx 0.8 \div 0.9$. Structure and composition of oligomers were provided by elementary analysis, by the finding of molecular masses, by FTIR and NMR spectral data. Elementary composition yields and some physical-chemical properties of synthesized oligomers are presented in table 1.

Table 1. Elementary composition, yields and some physical-chemical properties of oligomers.

Oligo-mer №	Yield, %	Reaction tempera-ture, °C	η_{sp}*	T_g, °C	d_1, Å	5%-mass losses	$\bar{M}_\omega x10^{-3}$	Elementary** composition, %		
								C	H	Si
I[1]	94	60	0,08	-14	8.40	230	-	*59.42* *58.95*	*8.66* *8.43*	*20.54* *21.02*
I[2]	94	70	0.08	-	-	-	-	*59.42* *59.21*	*8.66* *8.51*	*20.54* *20.94*
I	95	80	0.09	-10	8.43	230	48	*59.42* *59.30*	*8.66* *8.35*	*20.54* *20.77*

*At 25°C temperature, in 1% solution of toluene. **Overline there is calculated values and under-line - experimental values.

In the FTIR spectra of oligomers one can see characteristic bands for asymmetric valence oscillation of the linear ≡Si-O-Si≡ bonds at 1020 cm^{-1}. In the spectra one can observe characteristic bands for ≡Si-Me, -$SiMe_3$ and ≡Si-H bonds at 1275, 840 and 2170 cm^{-1} accordingly.

In the ^1H NMR spectra (Figure 2) of synthesized oligomer I one can see the singlet signals for methyl protons with chemical shift with centre $\delta \approx 0.1$ ppm, in the spectra one can also see widened singlet signals for methyl protons in fragment -CH=CH- with chemical shift δ≈6.1 and δ≈5.5 ppm. From spectra it is evident that the hydrosilylation proceeds to "1-4" direction also to "1-2" direction, exceed "1-4" direction and the ratio is ~1,6:1.

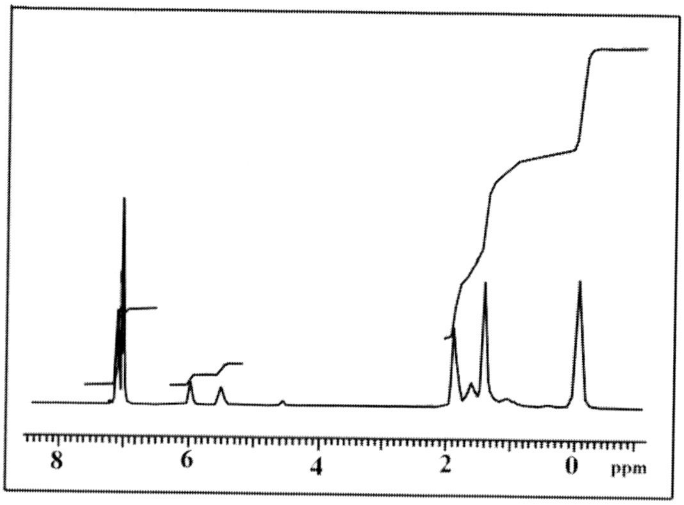

Figure 2. ^{1}H NMR spectra of α,ω-bis(trimethylsiloxy)methylhexenylsiloxane I.(Solvent and standard $CDCl_3$).

In the ^{13}C NMR spectra (Figure 3) of oligomer I one can observe for "1-4" and "1-2" hydride addition signals group with chemical shift: for "1-4" addition δ≈129.6; δ≈130.4 ppm and for "1-2" addition δ ≈126.7 and δ ≈139.9 ppm. Accordingly chemical shifts are easily given from the saturated section characterising for "1-4" addition at 23.7(C-1); 29.2(C-5); 27.8(C-4); 130.4(C-2); 23.7(C-1) and for "1-2" addition 24.1(C-6); 28.8(C-5); 129.6(C-4), 126.7(C-3); 22.8(C-2); 27.0(C-1).\

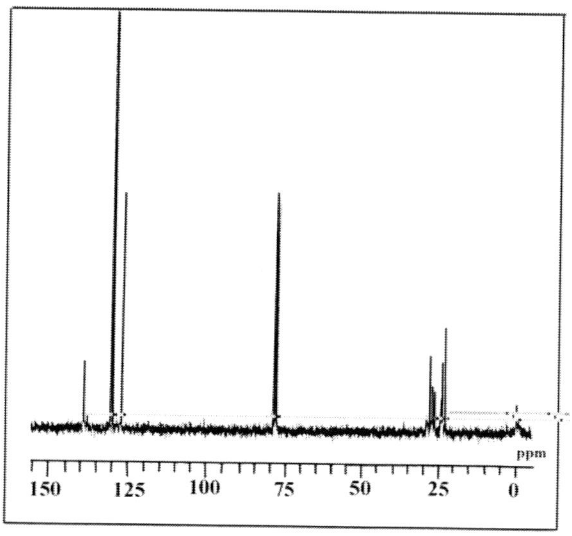

Figure 3. ^{13}C NMR spectra of α,ω-bis(trimethylsiloxy)methylhexenylsiloxane I. (Solvent and standard $CDCl_3$).

For full characterization of hydride addition of methylhydridesiloxane to cyclohexa-1,3-diene, by quantum-chemical semi empiric AM1 method [9], for all initial, intermediate and final products, in the modelling reaction of hydrosilylation of methyldimethoxysilane to cyclohexa-1,3-diene, the heats of formations (ΔH_f), change of energy (ΔH) of the system depending on the change of distance ($R_{C\text{-}Si}$) between ≡C-Si≡ bonds, also the charges values (q) on the atoms, dipole moments (μ) and bonds orders (Pij) are calculated (see tables 2 and 3).

For the beginning the modelling reaction of hydride addition of methyldimethoxysilane [$Me(MeO)_2SiH$] to cyclohexa-1,3-diene to 1,2-direction was considered. The model reaction proceeds with the formation of compound II according to the following scheme 4.

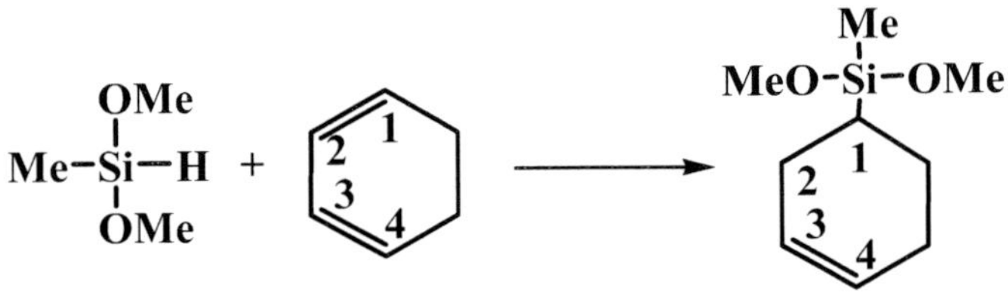

II

Scheme 4. Modelling reaction of hydride addition of methyldimethoxysilane [$Me(MeO)_2SiH$] to cyclohexa-1,3-diene to 1,2-direction.

The distances between the double bonded carbon atom (C_1) and silicon atoms in cyclohexa-1,3-diene was taken about 1,0 Å more than the bond, expected in a product. Figure 4 showed the starting position and numeration of atoms of compounds in the modelling hydrosilylation reaction of methyldi-methoxysilane to cyclohexa-1,3-diene:

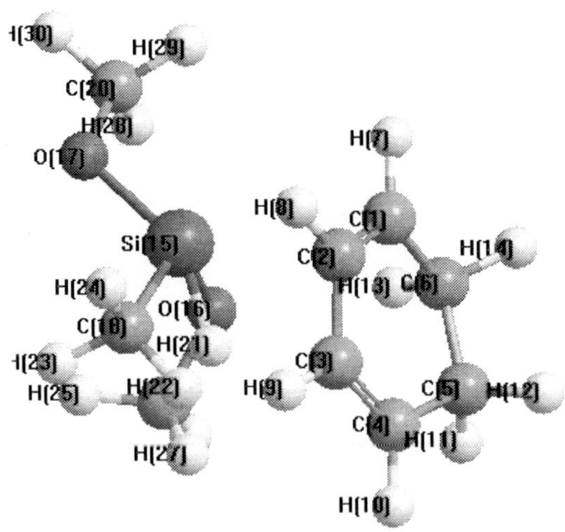

Figure 4. Starting position and numeration of atoms of compounds in modelling 1,2-hydrosilylation reaction of methyldimethoxysilane to cyclohexa-1,3-diene.

The change in step of distances (R_{C_1-Si}) between the atom of silicon and double bonded carbon atom (C_1) was 0,05 Å. Dependence of change of energy (ΔH) of the system on the distances (R_{C-Si}) between silicon and carbon atoms is presented in Figure 5.

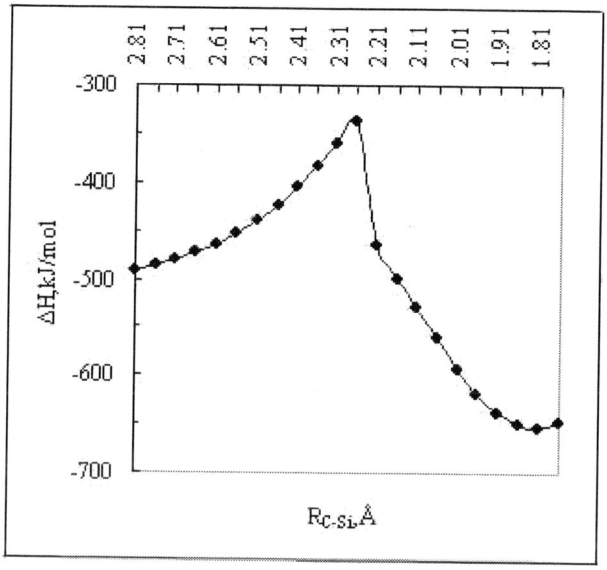

Figure 5. Dependence of changes of energy (ΔH) on distances (R_{C_1-Si}) between silicon and carbon atoms during of modelling hydrosilylation reaction of methyldimethoxysila-ne to cyclohexa-1,3-diene (1,2-addition).

As it is seen from Figure 5, at approximation of the silicon to the carbon atom (C_1) up to R_{C-Si}=2.26 Å distances, the energy of the system rises. At the same time the bond order between atoms of carbon C_1 and C_2 in a molecule of cyclohexa-1,3-diene decreases from 1.855 up to 1.648. The bond order between hydrogen and silicon atoms also decreases (R_{Si-H}=0.907-0.739) and in the modelling compound methyldimethoxysilane the process of formation of new bonds (R_{C_1-Si}=0.013-0.106 and P_{C_2-H}= 0.001-0.073) is observed. In case of distances up to 2.21 Å the energy of the system sharply decreases, the double C=C bond passes in ordinary C-C bond ($P_{C_1-C_2}$=0.996), and the bond order between the silicon and carbon atom C_1 reaches to 0.794. Thus the hydrogen is completely eliminated from the silicon (P_{Si-H}=0.011) and attaches to the carbon atom C_2 (P_{C_2-H}=0.912).

With consecutive reduction of distances the energy of the system monotonously decreases and with strengthening formed new bonds. The heat of formation of the product II of hydride 1,2-addition (ΔH=-654.04 kJ/mol) is in a good correlation with the calculated heat of formation (ΔH=-653.94 kJ/mol, see table 2) of compound II obtained by hydride addition of methyldimethoxysilane to cyclohexa-1,3-diene at 1,2-direction. Activation energy of 1,2-addition of the modelling reaction is equal to E_{act}=155.16 kJ/mol. In Figure 6 one can see the final structure of the modelling 1,2-addition of methyldimethoxysilane to cyclohexa-1,3-diene.

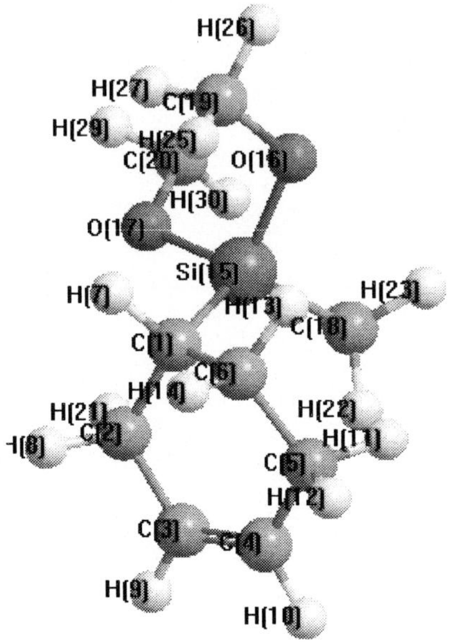

Figure 6. Final structure of the product obtained by modelling 1,2-addition of methyldimethoxysilane to cyclohexa-1,3-diene.

The 1,4-addition of modelling compound methyldimethoxysilane to cyclohexa-1,3-diene is also considered. The modelling reaction of 1,4-addition proceeds with formation of compound III according to the scheme 5.

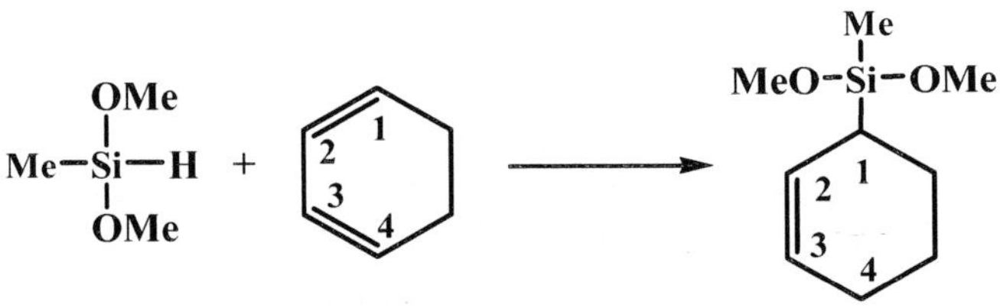

III

Scheme 5. 1,4-addition of modelling compound methyldimethoxysilane to cyclohexa-1,3-diene.

Figure 7 showed the starting position of compounds in modelling 1,4-hydrosilylation reaction of methyldimethoxysilane to cyclohexa-1,3-diene:

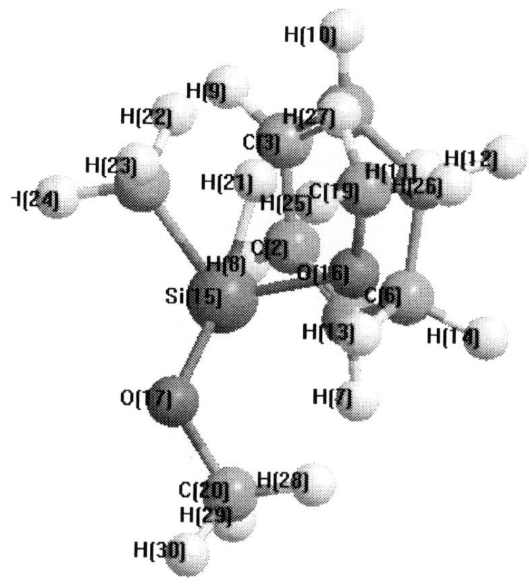

Figure 7. Starting position of compounds in modelling 1,4-hydrosilylation reaction of methyldimethoxysilane to cyclohexa-1,3-diene.

In Figure 8 the dependence of change of energy (ΔH) of the system on the distances (R_{C-Si}) between the silicon and the carbon atom is presented. Similarly, the above-mentioned distance between the double bonded carbon atom (C_1) and silicon was taken about 1.0 Å more than the bond, expected in a product. The change pace of distances (R_{C_1-Si}) between the silicon atom and double bonded the carbon atom (C_1) was about 0.05 Å.

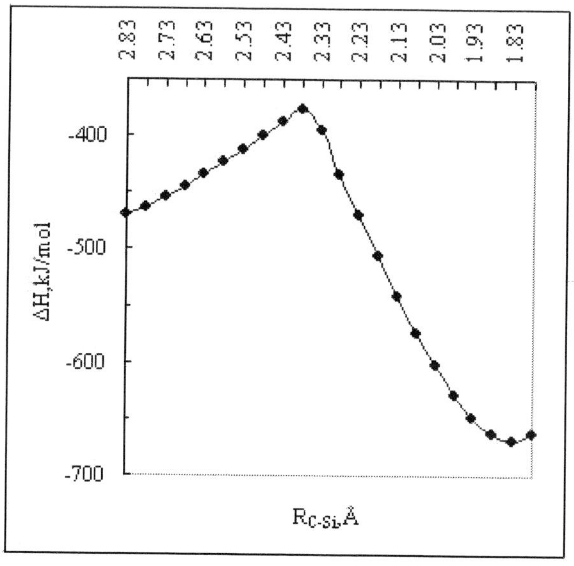

Figure 8. Dependence change of energy (ΔH) on distance (R_{C-Si}) between silicon and carbon atoms during modelling hydride addition of methyldimethoxysilane to cyclo-hexa-1,3-diene by 1,4-addition.

As it is seen in Figure 8, during approximation of the silicon to the carbon atom (C_1) up to the R_{C-Si}=2.38 Å distances the energy of system also rises. Thus the bond order between atoms of carbon C_1-C_2 and C_3-C_4 in a molecule of cyclohexa-1,3-diene decreases from 1.829 to 1.630 and from 1.840 to 1.627 accordingly. Thus the bond order of between atoms of carbon C_1-C_2 and C_3-C_4 in a molecule of cyclohexa-1,3-diene decreases from 1.829 up to 1.630 and from 1.840 up to 1.627 accordingly. In parallel the bond order between atoms of carbon C_2 and C_3 ($P_{C_2-C_3}$=1.063-1.205) is increased, and the process of formation of new bonds (P_{C_1-Si}=0.027-0.185 and P_{C_4-H}=0.014-0.170) is observed. From 2.33 Å distances the reduction of energy of the system is observed sharply, the double bonds C_1=C_2 and C_3=C_4 pass in ordinary C-C bonds ($P_{C_1-C_2}$=1.065 и $P_{C_3-C_4}$=1.046) and the bond order between the silicon and carbon atom C_1 reaches 0.779.

Thus hydrogen is completely eliminated from the silicon and (P_{Si-H}=0.011) and attaches to the carbon atom C_4 (P_{C_4-H}=0.878). The transition of an ordinary C_2-C_3 bond in a double bond also is observed, that is accompanied by sharp change of value of bond orders between atoms C_2 and C_3 ($P_{C_2=C_3}$=1.817). In Figure 9 one can see the structure of the final product obtained by modelling 1,4-addition of methyldimethoxysilane to cyclohexa-1,3-diene.

Table 2. Change of energy (ΔH) of the system depending on the change of distance (R_{C-Si}) between ≡C-Si≡ bonds,the charges values (q) on the atoms, dipole moments (μ) and bonds orders (Pij), during modelling hydride 1,2-addition of methyldimethoxysilane to cyclohexa-1,3-dienee.

$R_{1.15}$, Å	ΔH, kJ/mol	q_1	q_2	q_{15}	q_{21}	μ	P_{12}	$P_{1.15}$	$P_{15.21}$	$P_{2.21}$
2.81	-491.11	-.1862	-.1263	1.3613	-.2266	1.214	1.855	.013	.907	.001
2.76	-485.86	-.1776	-.1401	1.3738	-.2337	1.550	1.846	.020	.897	.003
2.71	-480.11	-.1860	-.1351	1.3746	-.2327	1.457	1.843	.023	.895	.004
2.66	-472.91	-.1966	-.1282	1.3755	-.2329	1.376	1.840	.027	.893	.005
2.61	-463.94	-.2099	-.1192	1.3770	-.2335	1.296	1.835	.031	.890	.006
2.56	-452.89	-.2261	-.1079	1.3793	-.2346	1.222	1.829	.035	.886	.008
2.51	-439.50	-.2457	-.0939	1.3829	-.2365	1.161	1.821	.041	.881	.011
2.46	-423.54	-.2702	-.0762	1.3888	-.2398	1.116	1.810	.047	.873	.015
2.41	-404.91	-.3011	-.0536	1.3984	-.2453	1.108	1.794	.056	.860	.020
2.36	-383.71	-.3438	-.0221	1.4143	-.2545	1.143	1.766	.068	.839	.030
2.31	-360.45	-.3979	.0187	1.4440	-.2714	1.302	1.723	.084	.802	.045
2.26	-335.94	-.4660	.0708	1.4943	-.2962	1.474	1.648	.106	.739	.073
2.21	-464.87	-.3891	-.0820	1.3615	-.0006	1.413	.996	.794	.011	.912
2.16	-498.63	-.3997	-.0869	1.3734	.0125	1.471	.993	.799	.009	.921
2.11	-530.82	-.4119	-.0907	1.3865	.0242	1.516	.991	.802	.008	.928
2.06	-560.93	-.4258	-.0938	1.4015	.0347	1.567	.990	.805	.006	.934
2.01	-594.18	-.4030	-.1208	1.3951	.0858	.472	.987	.847	.006	.939
1.96	-618.28	-.4236	-.1180	1.4164	.0879	.480	.985	.845	.005	.943
1.91	-637.40	-.4456	-.1152	1.4387	.0897	.504	.984	.842	.004	.947
1.86	-649.96	-.4688	-.1124	1.4621	.0917	.557	.983	.838	.003	.951
1.81	-653.95	-.4936	-.1087	1.4860	.0930	.647	.982	.835	.003	.954
1.76	-646.89	-.5198	-.1039	1.5104	.0936	.771	.980	.830	.002	.957

Table 3. Change of energy (ΔH) of the system depending on the change of distance (R_{C-Si}) between ≡C-Si≡ bonds, the charges values (q) on the atoms, dipole moments (μ) and bonds orders (Pij), during modelling hydride 1,4-addition of methyldimethoxysilane to cyclohexa-1,3-diene.

$R_{1.15}$, Å	ΔH, kJ/mol	q_1	q_4	q_{15}	q_{21}	μ	P_{12}	P_{23}	P_{34}	$P_{1.15}$	$P_{4.21}$
2.83	-469.79	-.1726	-.1349	1.3299	-.1804	1.760	1.829	1.063	1.840	.027	.014
2.78	-463.01	-.1758	-.1321	1.3233	-.1783	1.749	1.823	1.066	1.835	.032	.017
2.73	-454.27	-.1806	-.1274	1.3196	-.1756	1.648	1.815	1.070	1.827	.038	.022
2.68	-444.82	-.1862	-.1218	1.3157	-.1728	1.543	1.804	1.076	1.818	.046	.028
2.63	-434.63	-.1928	-.1151	1.3118	-.1699	1.433	1.791	1.084	1.805	.055	.037
2.58	-423.72	-.2002	-.1070	1.3079	-.1669	1.316	1.776	1.094	1.788	.068	.048
2.53	-412.15	-.2092	-.0974	1.3045	-.1636	1.206	1.755	1.107	1.767	.084	.064
2.48	-400.10	-.2191	-.0863	1.3017	-.1598	1.147	1.728	1.127	1.737	.105	.086
2.43	-387.98	-.2306	-.0730	1.3001	-.1552	1.129	1.689	1.156	1.694	.136	.118
2.38	-376.64	-.2443	-.0565	1.3007	-.1489	1.179	1.630	1.205	1.627	.185	.170
2.33	-394.47	-.2928	-.1072	1.2904	.0107	.755	1.065	1.817	1.046	.779	.878
2.28	-433.89	-.3027	-.1185	1.2977	.0327	.701	1.041	1.851	1.024	.809	.914
2.23	-470.50	-.3155	-.1215	1.3152	.0420	.672	1.034	1.861	1.019	.814	.921
2.18	-505.96	-.3295	-.1244	1.3339	.0507	.651	1.027	1.871	1.014	.819	.927
2.13	-541.80	-.3433	-.1265	1.3533	.0572	.692	1.015	1.887	1.010	.829	.933
2.08	-573.63	-.3598	-.1284	1.3733	.0639	.664	1.011	1.893	1.007	.830	.938
2.03	-602.61	-.3777	-.1299	1.3939	.0700	.633	1.008	1.898	1.004	.831	.942
1.98	-627.88	-.3968	-.1309	1.4154	.0752	.619	1.005	1.903	1.002	.830	.945
1.93	-648.31	-.4173	-.1312	1.4377	.0794	.626	1.002	1.908	1.000	.828	.949
1.88	-662.43	-.4391	-.1308	1.4605	.0827	.658	1.000	1.912	.998	.826	.952
1.83	-668.33	-.4621	-.1297	1.4839	.0851	.722	.997	1.915	.997	.823	.955
1.76	-662.17	-.4966	-.1277	1.5177	.0866	.872	.994	1.919	.996	.818	.957

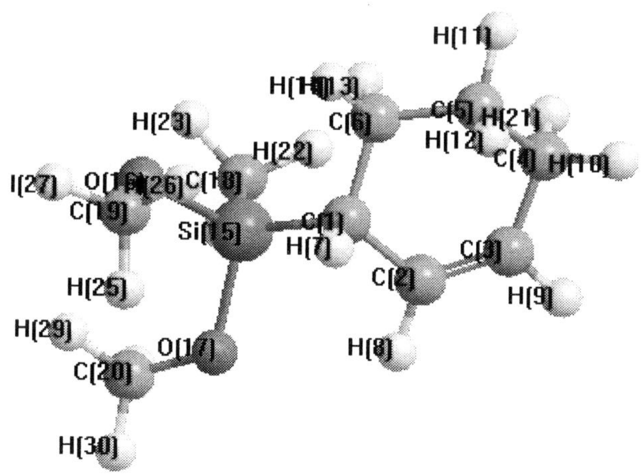

Figure 9. Final structure of the product obtained by modelling 1,4-addition of methyldimethoxysilane to cyclohexa-1,3-diene.

With consecutive reduction of distances between the silicon and carbon atoms the energy of the system monotonously decreases and with strengthening formed new bonds. The heat of formation of a product of modelling hydride 1,4-addition (ΔH=-656.87 kJ/mol) is in a good correlation with the calculated heat of formation (ΔH=-668.33 kJ/mol, see table 3) of compound III obtained by modelling hydride 1,4-addition of methyldimethoxysilane to cyclohexa-1,3-diene. Activation energy of 1,4-addition of the modelling reaction is equal to $E_{act} = 93.15$ kJ/mol.

Thus in comparison of the Figures 5 and 8, and also the energy of activation (the character of curve of change of energy and the value of bond orders) it can be concluded, that the course of the modelling reaction of a hydride addition of methyldimethoxysilane to cyclohexa-1,3-diene energetically is more favourable by 1,4-addition, which is in agreement with NMR spectral data.

In Figure 10 the dependence of reverse concentration on the time is presented, from which it is evident that at the beginning stages the hydrosilylation reaction is the reaction of the second order. From Figure 10 the reaction rate constants at various temperatures have been calculated: $k_{60^0C} \approx 1.4402 \times 10^{-2}$, $k_{70^0C} \approx 2.330 \times 10^{-2}$ and $k_{80^0C} \approx 3.2 \times 10^{-2}$ l/mol·s.

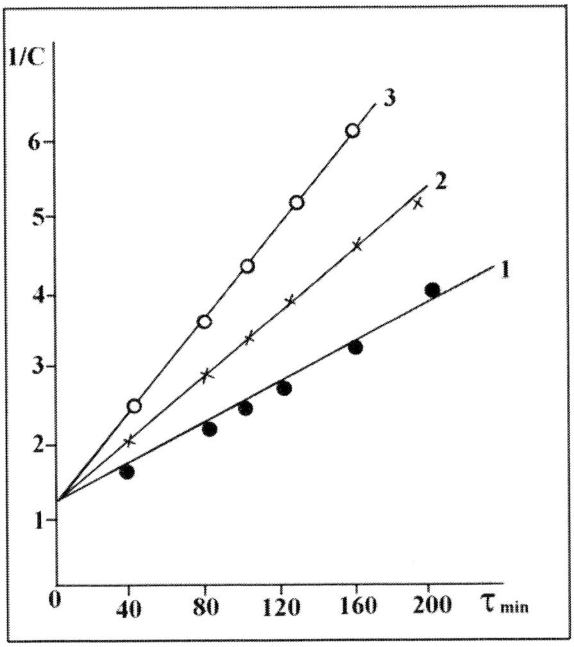

Figure 10. Dependence of reverse concentration on the time, where curve 1 is at 60^0C, curve 2 is at 70^0C and curve 3 is at 80^0C.

Figure 11 shows the dependence of the logarithm of the reaction rate constants on the reverse temperature. From these data, the activation energy of the hydrosilylation reaction was calculated to be equal $E_{activ} \approx 35.0$ kJ/mol. This value of activation energy is close to the activation energy of hydrosilylation of tricyclodecadiene with α,ω-bis(trimethylsiloxy)methylhydridesiloxanes ($E_{activ} \approx 32.3$ kJ/mol).

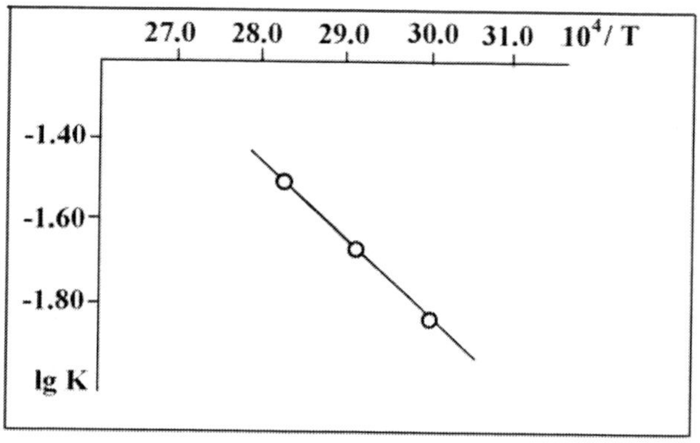

Figure 11. Dependence of reverse temperature on the logarithm of the rate constants of hydride addition of α,ω-bis(trimethylsiloxy)methylhydridesilo-xane to cyclohexa-1,3-diene.

The synthesized oligomers were studied by GPC method. Figure 12 shows molecular weight distribution of oligomer I, which showed that it has a polymodal character of distribution. The average weight molecular mass of oligomer is equal to $48x10^3$. In case of full hydrosilylation the value of average weight molecular mass does not exceed $4.3x10^3$. Increased molecular mass of obtained oligomers once again proved that the reaction proceeds by macromolecular interchain hydride addition with receiving of the branching oligomers. So in synthesized oligomers the elementary links have a different structure. With the rise of reaction temperature the intra- and interchain hydride addition increases.

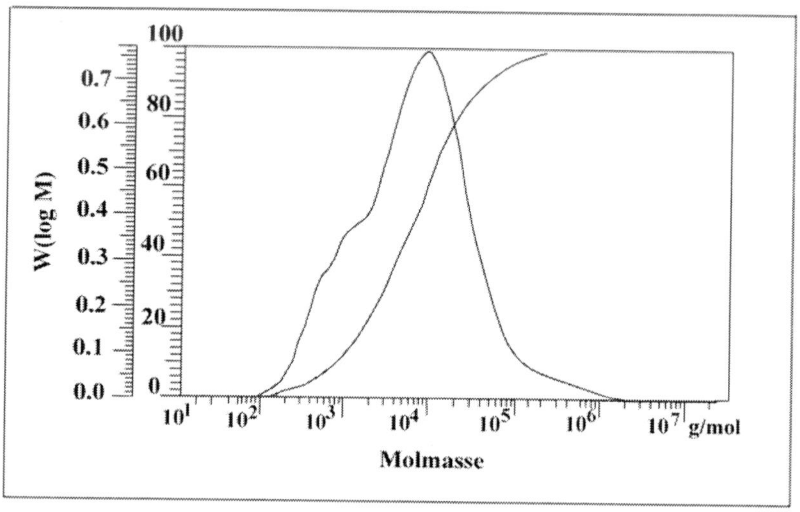

Figure 12. Gel permeation chromatographic curves of oligomer I.

Wide angle X-ray analysis showed that synthesized oligomers are amorphous one phase systems and value of interchain distance is equal to d_1=8.43 Å. As it is seen from table 1, the glass transition temperature of oligomers, depending on amount of cyclohexenyl groups in the side chain, changes in the range $-14^0C \div -10^0C$. Increasing glass transition temperature compared with linear polydime-thylsiloxane may be explained with the insertion of cyclohexenyl fragments in the chain and with the obtaining of partially branching systems.

Thermogravimetric investigations of oligomers have been carried out and 5% mass losses are observed at 230^{0}C. With the rise of temperature the mass losses increases and above 750^{0}C the changes of mass losses do not occur. With the insertion of cyclohexenyl fragments in the side chain the thermal-oxidative stability of oligomers increases.

By practical standpoint these above mentioned oligomers with unsaturated fragments in the chain are very interesting systems, because they contain unsaturated bonds, at the expense of the existing platinum hydrochloric acid in the system, gives possibility to change the properties of obtained siliconorganic elastomers in a wide range.

REFERENCES

[1] O.V. Mukbaniani, L.M. Khananashvili. *Intern. Journ. of Polymeric Materials*, 1994, *27*, 31.

[2] J.N. Aneli, Kh.E. Koberidze, O.V. Mukbaniani, M.G. Karchkhadze, L.M. Khananashvili. *Russian Polymer News,* 1998, *3,* 7.

[3] N.M. Galashina *Vysokomol. Soed.*, 1994, 36A, 640 (in Russian),

[4] J.N. Aneli, O.V. Mukbaniani, M.G. Karchkhadze, Kh.E. Koberidze, R.Sh. Tkeshelashvili, L.M. Khananashvili. Georgian Patent, 1996, *№1348101-96.*

[5] Mukbaniani O.V.; Khananashvili L.M.; Esartia I.G.; Khaduri S.D. International *Journal of Po-lymeric Materials,* 1994, 24, 211.

[6] O. Mukbaniani, L. Khananashvili and G. Zaikov. *Aging of Polymers, Polymer Blends and Po-lymer Composites*, Nova Science Publishers, Inc., New York, 2002, 1, 99.

[7] A.I. Nogaideli, L.M. Khananashvili, L.I. Nakaidze, T.V. Chogovadze, V.S. Tskhovrebashvili, R.V. Volkova, G.A. Nogaideli. *Doklady Akademye Nauk GSSR* 1978, 92, 341(in Russian).

[8] O. Mukbaniani, U. Scherf, M. Karchkhadze, Kh. Koberidze. Intern. *J. of Polym. Materials,* 2000, 48, 177.

[9] M.I.S. Dewar, E.G. Zeobish, E.F. Healy, J.S. Stewart. *J. American Chemical Society*, 1985, 107, 3902.

In: Modification Reactions of Oligomethylhydridesiloxanes ISBN: 1-60021-365-0
Editors: O.Mukbaniani, et al. pp. 179-200

Chapter 12

SYNTHESIS OF POLYSILOXANES WITH PENDANT UNSATURATED CYCLIC FRAGMENTS

O. Mukbaniani, T. Tatrishvili, G. Titvinidze, and S. Patsatsia
Iv. Javakhishvili Tbilisi State University, I. Chavchavadze Ave. 1, 0128 Tbilisi, Georgia

ABSTRACT

The synthesis of polysiloxanes with pendant unsaturated cyclic fragments have been successfully performed by hydrosilylation reactions of polymethylhydrosiloxane with 4-vinylcyclo-1-hexene in the presence of platinum hydrochloric acid (0.1 M solution in THF), Karstedt's catalyst ($Pt_2[(VinSiMe_2)_2O]_3$) and platinum on the carbon (5%). Reactions were carried out at various temperatures with 1:35, 1:70 and 1:105 ratios of initial compounds. It was shown, that not all active ≡Si-H groups participate in the hydrosilylation reaction. Some kinetic parameters of reactions were studied. The synthesized oligomers were characterized by FTIR, ^{1}H, ^{13}C, H,H-COSY and C,H-correlation NMR spectroscopy. Calculations using the quantum-chemical semi empirical AM1 method for modelling reactions between methyldimethoxysilane $[Me(MeO)_2SiH]$ and 4-vinylcyclo-1-hexene were performed to evaluate possible reaction paths. Synthesized oligomers were characterized by gel-permeation chromatography, differential scanning calorimetry, thermogravimetric and wide-angle X-ray diffractometry analyses.

Key words: hydrosilylation, methylhydridesiloxane oligomers, cross-linking, kinetics.

1. INTRODUCTION

Polysiloxanes, widely known as silicones, are of particular interest due to their extremely low glass transition temperatures and flexibility, their hydrophobic surface properties, good thermal stability and excellent flame retardant [1–4].

In the past years, catalytic grafting of polymethylhydrosiloxane has emerged as an alternative route for the synthesis of new silicon-organic polymers with different properties. Low cost, easy availability and the presence of catalytically transformable ≡Si-H bonds in polymethylhydrosiloxane (PMHS) make it very attractive and interesting for macromolecular grafting [5,6]. Organic functionalization of the polymer chain can lead to obtaining new polymers and oligomers with various physical and chemical properties. For example, appropriative substitution on the polysiloxanes backbone can lead to various materials such as cross-linking agents [7,8], liquid crystals [9,10], fluorescent and electro luminiscent polymers [11,12] and non-linear optical materials [13].

Hydrosilylation is a general term used, for the addition reaction of organic and inorganic silicon hydrides to molecules and macromolecules containing multiple bonds. The universal character of the process is reflected in the wide spectrum of usable unsaturated moieties [14]. Hydrosilylation is the most widely used reaction for grafting of polymethylhydrosiloxane. Chemical functionalization of polysiloxane chains involves preliminary introduction of side chains with functional groups. Introduced functional groups in side chains have the ability to undergo further reactions. The majority of functionalized siloxanes were designed to undergo thermally induced cross-linking, and to cure by hydrolytic processes or photochemically induced polymerization [4,15].

A number of papers currently deal with mechanistic aspects and directions of the hydrosilylation processes and also with the search for new, improved or less expensive convenient catalysts.

In this work, grafting of PMHS via hydrosilylation reactions were investigated in the presence of several platinum catalysts. The rates and depths of reactions were evaluated in the presence of different catalysts.

2. EXPERIMENTAL

2.1. Materials

Polymethylhydrosiloxane (Aldrich), platinum hydrochloric acid (Aldrich), Karstedt's catalyst ($Pt_2[(VinSiMe_2)_2O]_3$) or platinum(0)-1,3-divinyl-1,1,3,3-tetramethyldisiloxane complex (solution in PDMS), platinum on the carbon and 4-vinylcyclo-1-hexene (Aldrich) were used as received. Toluene and THF was dried over and distilled from sodium under an atmosphere of dry nitrogen. 0.1 M solution of platinum hydrochloric acid in THF was prepared and kept under nitrogen at low temperature.

2.2. Synthesis

2.2.1. Catalytic Hydrosilylation of PMHS with 4-Vinylcyclo-1-Hexene in the Presence of H_2ptcl_6.

Polymethylhydrosiloxane (0.9850 g, 0.43 mmol), 4-vinylcyclo-1-hexene (1.6595g, 0.0153 mol) and 5.5 ml dry toluene were placed into the three-necked flask equipped with a magnetic stirrer, reflux condenser and thermocouple. Then the catalyst 0.1 M solution of platinum hydrochloric acid in tetrahydrofuran ($5\div9\text{x}10^{-5}$ g per 1.0 g of starting substance) was introduced. After the reaction was completed, the solvent was partially eliminated, the reaction product was precipitated from toluene solution by n-hexane, and 2.4593 g (93 %) oligomer was obtained. The hydrosilylation reactions in the presence of other catalysts were carried out according to the above-mentioned method.

2.3. Determination of ≡Si-H content

We calculate the content of active ≡Si-H groups in oligomers on the basis of the following reaction:

$$2\equiv Si-H \xrightarrow[\text{solution}]{\text{KOH alcohol}} \equiv Si-O-Si\equiv + H_2\uparrow$$

We place in a two-necked flask an exactly weighed amount of ≡Si-H containing an oligomer, which should correspond to the evolution of a certain volume of hydrogen gas. We fit a pressure-equalized dropping funnel with 0.5 M KOH alcohol solution onto the flask. We connect the flask with a gas buret, which we connect with a levelling bottle at the bottom part, by means of a silicon tube. We allow the system to stand for 10 min. The water level in the levelling bottle which adjusts to the same level as that in the gas buret (V_1, ml) is recorded. Next we add the 0.5 M KOH alcohol solution slowly. As hydrogen gas evolves, the water level goes downward. Thus, we need to keep both water levels the same as each other so that we could equalize the pressure in the system to atmosphere. After the evolution of hydrogen gas completes (around 3-5 min) we record another 5 min later the value of the water level in the buret (V_2, ml). The difference between V_2 and V_1 gives the volume of eliminated hydrogen gas (V, ml). The Si-H content of an oligomer is given by the following:

$$\equiv Si\text{-}H \text{ content} = 1000\ PV/WRT\ (\text{mol}/100\text{g})$$

Where p – atmospheric pressure at measurement (atm), V – gas volume captured (ml), W – sample weight (g), R = 0.0082 (gas constant) ($l \cdot atm \cdot mol^{-1} \cdot K^{-1}$), and T – temperature (K at measurement) [16].

2.4. Characterization

FTIR spectra were obtained on a Nicolet Nexus 470 machine with MCTB detector. ^{1}H, H,H COSY, ^{13}C NMR and H,C-correlation NMR spectra were recorded on a 500 MHz Bruker NMR spectrometer, using $CDCl_3$ as the solvent and an internal standard. Thermogravimetric analysis (TGA) was performed on a Perkin-Elmer TGA-7 machine at the heating rate of 10 K/min. Differential scanning calorimetric investigation (DSC) was performed on a Perkin Elmer DSC-7 apparatus. Thermal transitions including glass transition temperatures T_g were taken at the maxima of the peaks. The heating and cooling scanning rates were 10 K/min.

Gel-permeation chromatography investigation was carried out with the use of a Waters Model 6000A chromatograph with an R 401 differential refractometer detector. The column set comprised 10^3 and 10^4 Å Ultrastyragel columns. Sample concentration was approximately 3% by weight in toluene and typical injection volume for the siloxane was 5 μL. standardization of the GPC was accomplished by the use of styrene or polydimethylsiloxane standards with the known molecular weight.

Wide-angle X-ray diffractograms were taken on a DRON-2 (Burevestnik, Saint Petersburg, Russia) instrument. A-CuK_α was measured without a filter; the angular velocity of the motor was $\omega \approx 2$ deg/min.

3. Results and Discussion

For synthesis of methylsiloxane oligomers with unsaturated cyclic fragments in the side chain we carried out hydrosilylation reactions of 4-vinylcyclo-1-hexene with PMHS in the presence of platinum hydrochloric acid (0.1 M in THF), Karstedt's catalyst ($Pt_2[(VinSiMe_2)_2O]_3$) and platinum on the carbon. Hydrosilylation reactions were carried out with various molar ratios of initial compounds (1:35, 1:70 and 1:105). From the literature [17] it is known that not all active ≡Si-H groups participate in hydrosilylation reactions in case of stoichiometric relationships in view of ≡Si-H bonds. We have taken 4-vinylcyclo-1-hexene with excess for achievement of full hydrosilylation of polymethylhydrosiloxane.

Preliminary heating of initial compounds in the temperature range of $40 \div 100^0$C in the presence of catalysts showed that in these conditions there is no polymerization of 4-vinylcyclo-1-hexene, nor scission of siloxane backbones, nor elimination of methane from methylhydridesiloxane. No changes in the NMR and FTIR spectra of initial compounds were found.

Without solvent, reactions proceed vigorously (especially in case of platinum hydrochloric acid and Karstedt's catalyst) and in low stages of ≡Si-H bonds conversion (~30%) gelation takes place. For prevention of gelation and for investigation of kinetic parameters we have carried out reactions in the toluene solution.

As it's seen 4-vinylcyclo-1-hexene involves two active centres for hydride addition of hydrosilane. Thus hydrosilylation reaction of 4-vinylcyclo-1-hexene with polymethylhydridesiloxane can proceed according to four directions: anti-Markovnikov (A) and Markovnikov (B) addition to vinyl group and hydrosilylation to unsaturated bond in the cyclic fragment in the para (C) or meta (D) position:

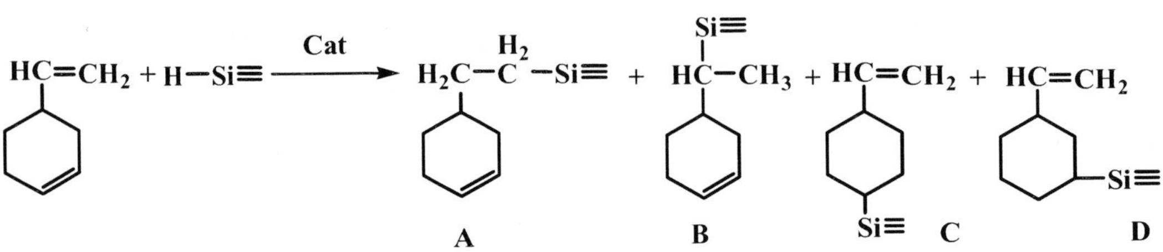

In general, we can represent hydrosilylation of α,ω–bis(trimethylsiloxy)methylhydrosiloxane to 4-vinylcyclo-1-hexene by the following scheme:

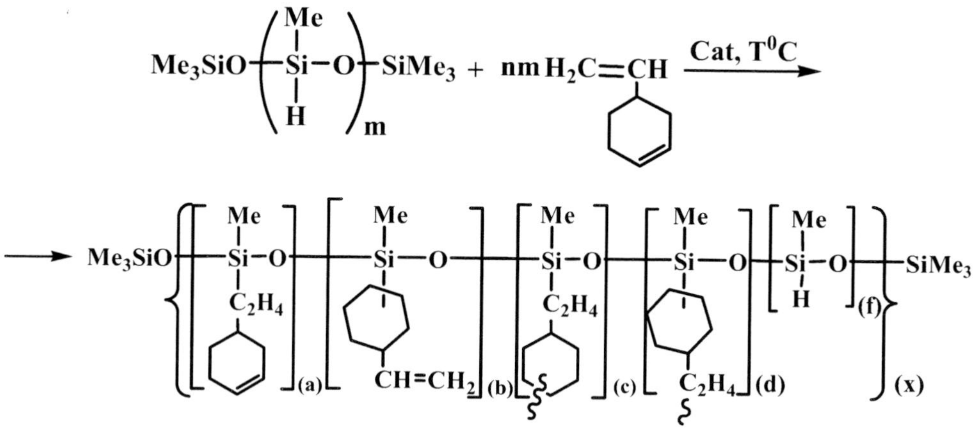

Scheme 1. Hydrosilylation reaction of α,ω–bis(trimethylsiloxy)methylhydrosiloxane to 4-vinylcyclo-1-hexene.

Where: [(a)+(b)+(c)+(d)+(f)](x)=m≈35; Cat - $H_2PtCl_6{\cdot}6H_2O$. 40^0C (I^1), 50^0C(I^2) and 60^0C (I)(ratio 1:35). 40^0C (II^1), 50^0C (II^2), 60^0C (II) (ratio 1:70), 60^0C (III) (ratio 1:105). 80^0C (IV) (ratio 1:140). Cat-Pt/C - 80^0C (V^1), 90^0C(V^2), 100^0C (V)(ratio 1:70). Karstedt's catalyst-70^0C(VI^1), 80^0C(VI^2), 90^0C (VI) (ratio 1:70).

We have seen in Figure 1 that hydrosilylation proceeds rapidly during the first hour and then slows down. The synthesized oligomers are vitreous liquid products, which are well soluble in ordinary aromatic type organic solvents with the specific viscosity $\eta_{sp} \approx 0.09 \div 0.1$. Structures and compositions of the oligomers were established by elementary and functional analyses, FTIR spectra, ^1H, ^{13}C, H,H-COSY and C,H-correlation NMR spectral data. Some resulting physical-chemical properties of the synthesized oligomers are presented in table 1.

We have studied some kinetic parameters for hydrosilylation reactions of PMHS with 4-vinylcyclo-1-hexene in the presence of H_2PtCl_6 with 1:70 ratios of initial compounds (1:2 stoichiometric relationships between initial compounds in view of ≡Si-H bonds). From Figure 1 one can observe dependence of changes of active ≡Si-H groups on the time.

From Figure 1 it is evident that at 40°C the hydrosilylation reaction proceeds with ~75% conversion of active ≡Si-H groups, whereas at 60°C the hydrosilylation reaction proceeds with ~91% conversion. Consequently, with the rise of temperature, the depth of the hydrosilylation reaction increases. Figure 2 shows dependence of reverse concentration on time during the hydrosilylation reaction of polymethylhydridesiloxane to 4-vinylcyclo-1-hexene. One can see that at the initial stages the hydrosilylation reaction is of second order. From the literature [17-19] it is known that hydrosilylation reactions at initial stages are usually of a second order in case of stoichiometric relationships between initial compounds in view of ≡Si-H bonds. As we can see from experimental results increasing concentration of one initial compound does not change the reaction order.

Table 1. Some physical-chemical properties of synthesized oligomers.

№	Yield %	Catalyst	Reaction tempera-ture / ^{0}C	Ratio of initial compounds	% conversion ofactive ≡Si-H groups	η_{sp} *	Å, d_1	T_g, ^{0}C
I	85	H_2PtCl_6	60	1:35	89	0.1	9.83	-57, -34
I^1	82	H_2PtCl_6	40	1:35	80	0.09	-	-
I^2	83	H_2PtCl_6	50	1:35	82	0.09	-	-
II	88	H_2PtCl_6	60	1:70	91	0.1	-	-
II^1	78	H_2PtCl_6	40	1:70	75	0.09	-	-55, -30
II^2	82	H_2PtCl_6	50	1:70	83	0.09	9.82	-
III	90	H_2PtCl_6	60	1:105	92	0.1	-	-
IV	91	H_2PtCl_6	80	1:140	99	0.1	-	-
V	94	Pt/C	100	1:70	85	0.09	-	-
V^1	92	Pt/C	80	1:70	66	0.09	-	-
V^2	92	Pt/C	90	1:70	76	0.09	-	-
VI	91	Karstedt's	90	1:70	87	0.09	-	-58, -28
VI^1	88	Karstedt's	70	1:70	80	0.09	9.82	-
VI^2	88	Karstedt's	80	1:70	82	0.09	-	-

*In 1% solution of toluene, at 25^{0}C.

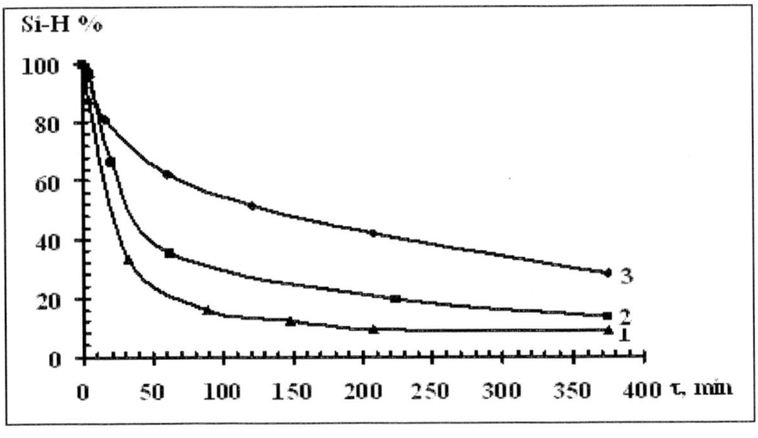

Figure 1. Dependence of changes of concentration of active ≡Si-H groups on the time, upon hydrosilylation reaction of α,ω-bis(trimethylsiloxy)methylhydrosiloxane with 4-vinylcyclo-1-hexene, where curve 1 represents the values obtained at 60°C, curve 2 represents the values obtained at 50°C and curve 3 represents the values obtained at 40°C (H_2PtCl_6).

The reaction rate constants of hydrosilylation reactions of methylhydridesiloxane to 4-vinylcyclo-1-hexene at various temperatures were determined, which values are: $k_{60°C} \approx 1.6156$, $k_{50°C} \approx 0.6157$ and $k_{40°C} \approx 0.2534$ l/mol·s. The reaction temperature coefficient is equal to γ=2,5.

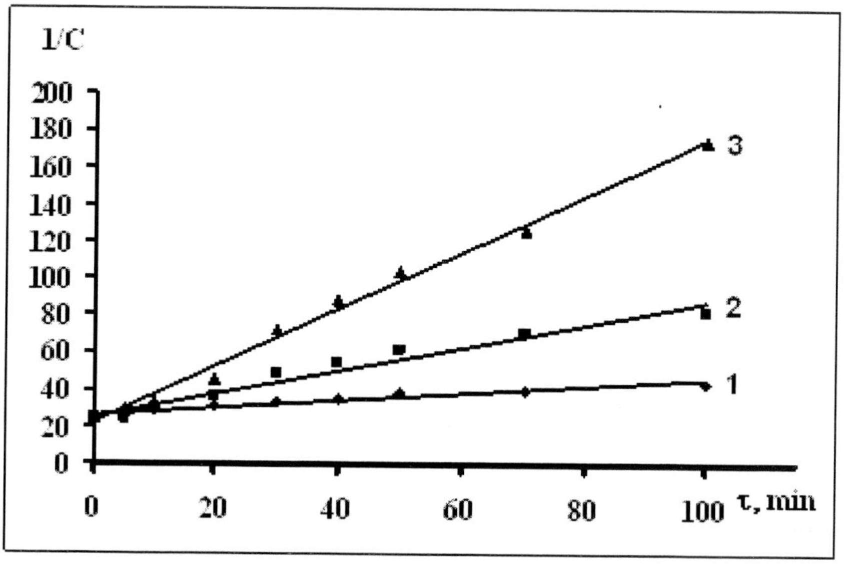

Figure 2. Dependence of inverse concentration on the time during the hydrosilylation reaction of methylhydridesiloxane to 4-vinylcyclo-1-hexene. Where, curve 1 is at 60^0C, curve – 2 at 50^0C and curve – 3 at 40^0C.

From the dependence of the hydrosilylation reaction rate constants' logarithm on the reverse temperature (Figure 4), the activation energy of the hydrosilylation reaction was calculated, which is equal to: E_{act}≈16,1 kJ/mol.

We have also studied hydrosilylation reactions of PMHS with 4-vinylcyclo-1-hexene in the presence of platinum hydrochloric acid (0.1 M in THF) with 1:105 ratios of initial compounds (1:3 stoichiometric relationships between initial compounds in view of ≡Si-H bonds). In case of excess of 4-vinylcyclo-1-hexene 3-times reaction proceeds with ~92% conversion of active ≡Si-H bonds, these results are close to the results in case of excess of 4-vinylcyclo-1-hexene 2-times. In the case of 1:140 ratio (1:4 stoichiometric relationship between initial compounds in view of ≡Si-H bonds) results 99% conversion of active ≡Si-H bonds.

Hydrosilylation reactions of PMHS with 4-vinylcyclo-1-hexene in the presence of platinum on the carbon at temperatures which was chosen for H_2PtCl_6 proceed with low rate and depth (~50%). Because of this for Pt/C we have chosen temperature range 80-100°C. At 80°C conversion of active ≡Si-H bonds is equal to 66% (oligomer V^1) and at 100°C – 87% (oligomer V).

Hydrosilylation reactions of PMHS with 4-vinylcyclo-1-hexene in the presence of Karstedt's catalyst at 60^0C proceed with a high degree of active ≡Si-H bonds conversion - 85% (oligomer VI), but the activity of the catalyst is less than in case of H_2PtCl_6.

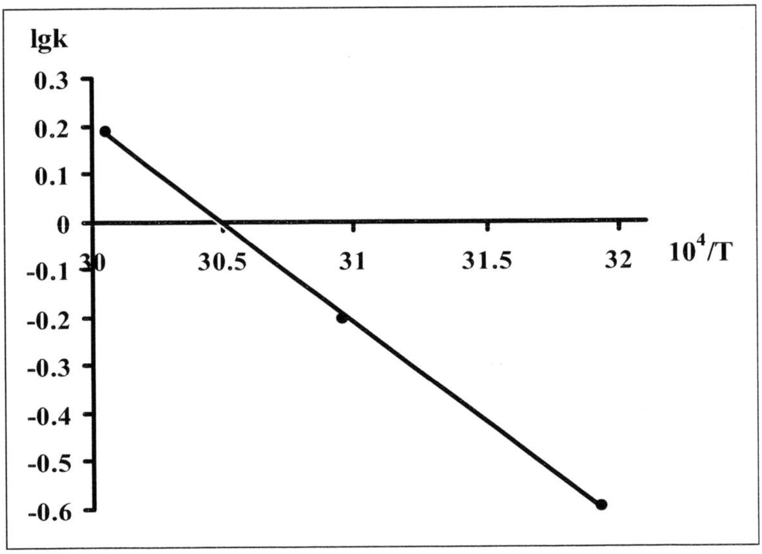

Figure 3. Dependence of reaction rate constants' logarithm on the reverse temperaure upon hydrosilylation of α,ω-bis(trimethylsiloxy)methylhydridesiloxane to 4-vinylcyclo-1-hexene.

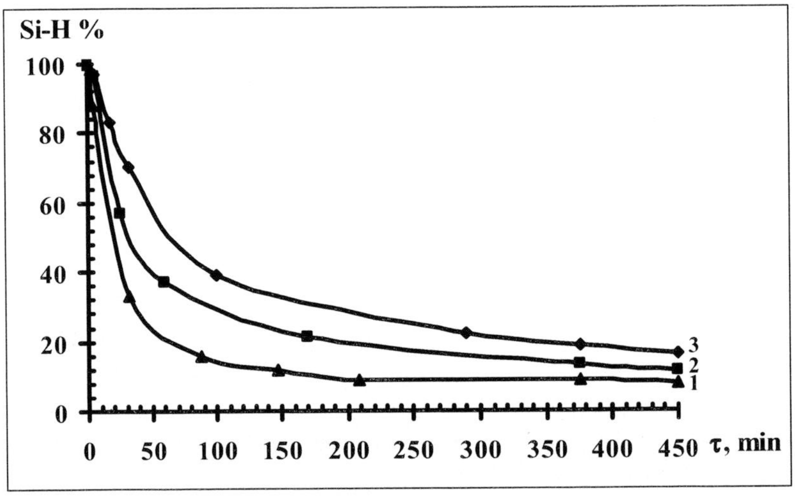

Figure 4. Dependence of changes of concentration of active ≡Si-H groups on the time, upon hydrosilylation reaction of α,ω-bis(trimethylsiloxy)methylhydridesiloxane with 4-vinylcyclo-1-hexene, where curve 1 represents the values obtained at 60°C with 1:70 ratio in the presence of H2PtCl6, curve 2 represents the values obtained at same conditions in the presence of Karstedt's catalyst and curve 3 represents the values in the presence of Pt/C at 100°C.

As we mentioned above during the hydrosilylation reactions, the decreases of active ≡Si-H groups' concentrations with time were observed. Not all active ≡Si-H groups participate in hydrosilylation reactions in spite of the excess of 4-vinylcyclo-1-hexene. In Figure 4 we show changes of concentrations of active ≡Si-H groups with the time during hydrosilylation reactions in the presence of different catalysts at optimal conditions.

Comparisons were done for optimal conditions of reactions, for H_2PtCl_6 and Karstedt's catalysts it was 60°C, but for Pt/C it was 100°C, here we want to mention that in the presence of Pt/C at 60°C reaction does not proceed. Figure 4 evidently shows the influence of the catalysts on the reaction rate and depth. From Figure 4 we can conclude that the activity of catalysts for hydrosilylation reactions of PMHS with 4-vinylcyclo-1-hexene decreases in the next rank: H_2PtCl_6>Karstedt's catalyst>Pt/C.

Figure 5 shows FTIR spectra of oligomer I. From the literature it is known that hydrosilylation of 4-vinylcyclo-1-hexene occurs only with a vinyl fragment [20-22]. In contrast to literature data we have established from FTIR spectra that hydrosilylation of PMHS with 4-vinylcyclo-1-hexene proceeds also with participation of an unsaturated bond of cyclic fragment. We observe absorption bands characteristic for asymmetric valence oscillation of linear ≡Si-O-Si≡ bonds at 1025 cm^{-1}.

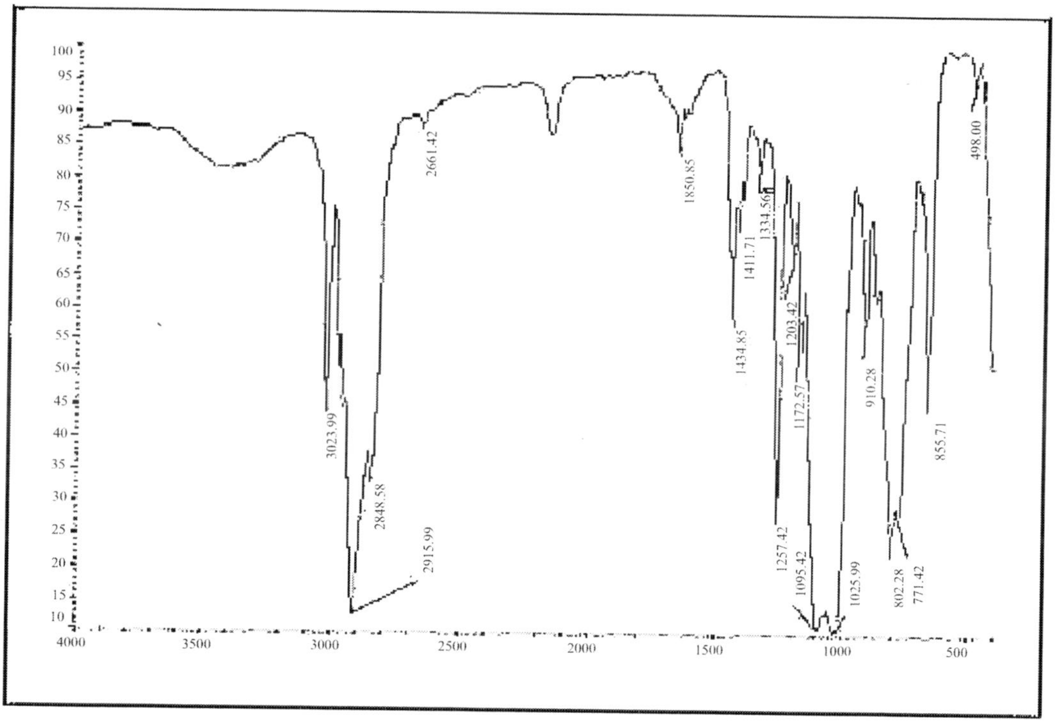

Figure 5. FTIR spectrum of oligomer I.

In the 840 and 1259 cm^{-1} regions we observe the absorption bands characteristic for $-OSiMe_3$ and ≡Si-Me accordingly. An absorption band in the 1650 cm^{-1} region characteristic for -CH=CH- bonds of cyclic fragment proves that the reaction proceeds with a substituted vinyl group. In addition, we observe an absorption band in the region 2165 cm^{-1} characteristic for un-reacted ≡Si-H bonds. In the spectra one can observe a signal in the 2846-2915 cm^{-1} region characteristic for the cyclohexane fragment, which proves that reaction proceeds with participation of an unsaturated bond of the cyclic fragment.

In Figure 6 we show ^{1}H and H,H-COSY NMR spectrum of oligomer I, where one can observe singlet signals for protons in ≡Si-Me and -SiMe$_3$ groups with chemical shifts δ≈0.01 ppm and δ≈0.03 ppm. In spectra one can observe signals δ≈0.45 and δ≈1.2 ppm characteristic for methylene protons accordingly in ≡Si-CH$_2$- and C$_6$H$_9$-CH$_2$- fragments. This shows that reaction proceeds by participation of vinyl group. In the spectra one can observe signals with low intensities with chemical shifts δ≈0.8 and δ≈1.35 ppm accordingly characteristic for methyl protons and methine protons in =CH-CH$_3$ fragment. These two signals with low intensities show that reaction proceeds according to the Markovnikov rule with a low degree. Signals with chemical shifts δ≈1.5, 1,7 and 2.0 ppm correspond to methylene protons of cyclohexene. A signal with chemical shift δ≈5.65 ppm corresponds to -CH=CH- methine protons of cyclic fragment. In the spectra one can observe signals with low intensities with chemical shifts δ≈5.0 and 5.8 ppm characteristic for methylene and methine protons of the vinyl group, this proves once again the proceeding of hydrosilylation according to the Markovnikov rule with low degree.

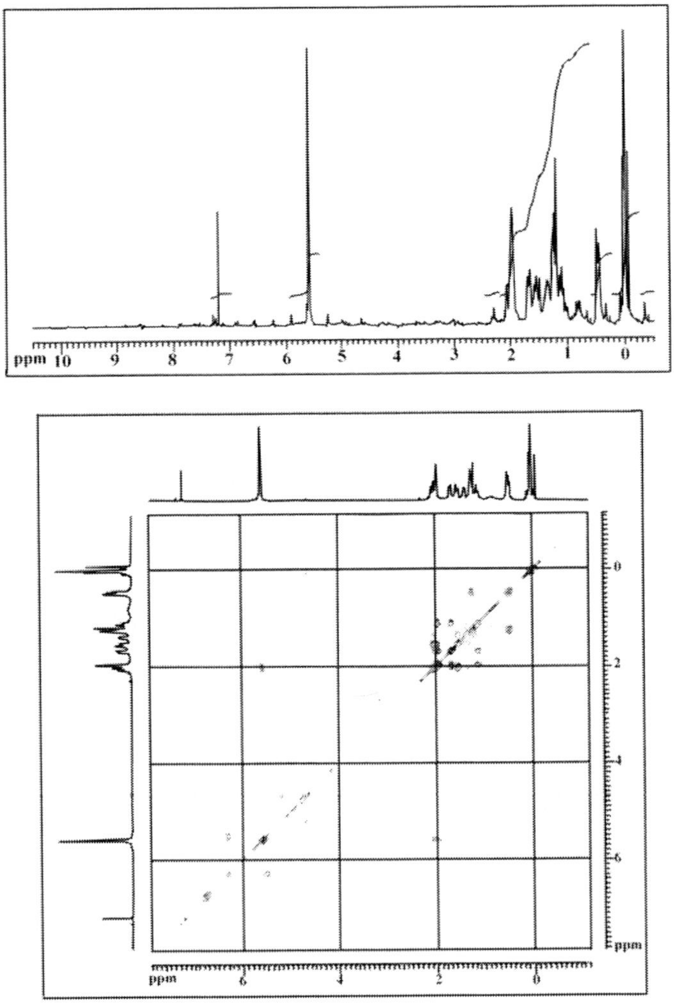

Figure 6. ^{1}H and H,H-COSY NMR spectra of oligomer I.

A signal with chemical shift δ≈4.1 ppm characteristic for ≡Si-H bond shows that full hydrosilylation does not proceed and that various linked oligomers are obtained.

^{13}C and C,H-correlation NMR spectra of oligomer I (Figure 7) is in full correlation with ^{1}H and H,H-COSY NMR spectra of oligomer I and proves that the reaction proceeds according to anti-Markovnikov and Markovnikov rule. In the ^{13}C NMR spectra one can observe signals δ≈0.01 and δ≈0.29 ppm characteristic for carbon atoms in ≡Si-Me and $-SiMe_3$ fragments, signal δ≈14.98 ppm for $≡Si-CH_2-$ fragment, signal δ≈25.68 ppm for methine carbon in $=CH-CH_3$, also signals δ≈126.91 and δ≈127.32 ppm characteristic for methine carbons of cyclic -CH=CH- fragment.

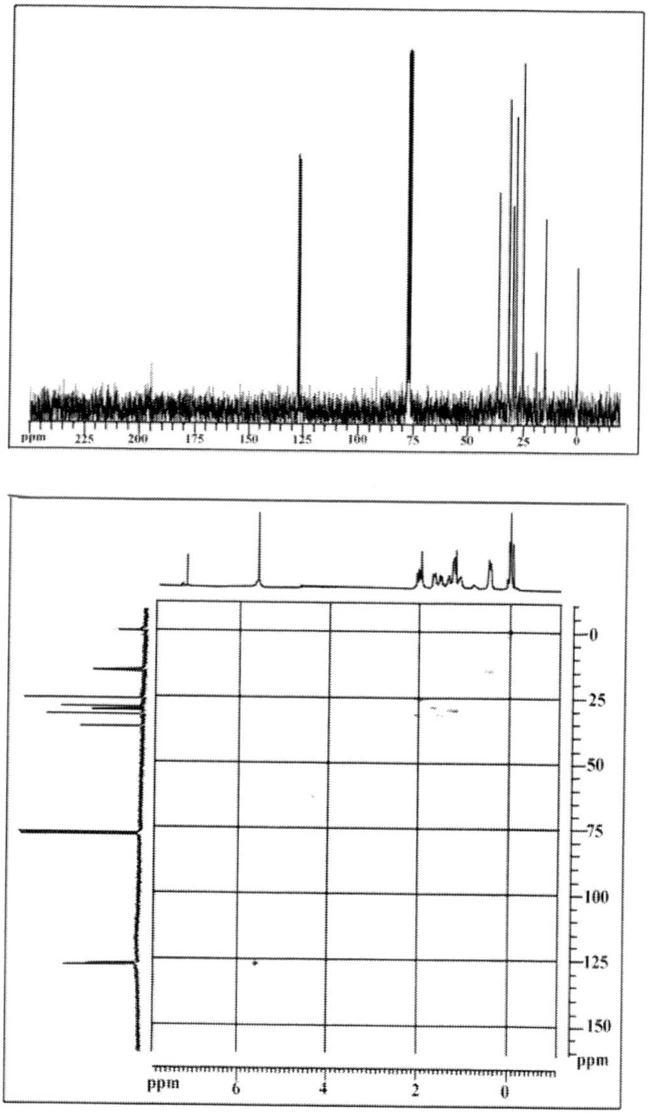

Figure 7. ^{13}C and C,H-correlation NMR spectra of oligomer I.

In the spectra also one can observe signals δ≈28.94, 30.01, 30.10, 31.98 and 36.68 ppm for methine and methylene carbon atoms of a cyclic fragment. C,H-correlation NMR spectra of oligomer I is in full correlation with ^{1}H NMR and ^{13}C NMR spectrum data.

Determination of the ratio of the fragments obtained from the anti-Markovnikov and Markovnikov rules, also from hydride addition towards a double bond of the cyclic fragment, is not available, because the spectra have a complicated structure.

For full characterization of the hydrosilylation reaction of methylhydridesiloxane to 4-vinyl-1-cycloxene we have carried out calculations using a semiempirical AM1 method for modelling reactions between methyldimethoxysilane [Me(MeO)$_2$SiH] and 4-vinylcyclo-1-hexene for full characterization of hydride addition of polymethylhydridesiloxane to 4-vinylcyclo-1-hexene and for ascertainment of suitability of calculations for prediction of reactions' directions. We have used the software Chem3D Ultra 9.0 from Cambridge Soft.

By quantum-chemical half empiric AM1 method we have calculated for all initial, intermediate and final products, in the modelling reaction of hydrosilylation of methyldimethoxysilane with 4-vinylcyclo 1-hexene, the heats of formations (ΔH_f), changes of energy (ΔH) of the system depending on the change of distance ($R_{C\text{-}Si}$) between ≡C-Si≡ bonds, also the charge values (q) on the atoms, dipole moments (μ) and bonds orders (Pij). The obtained results are presented on the diagram 1.

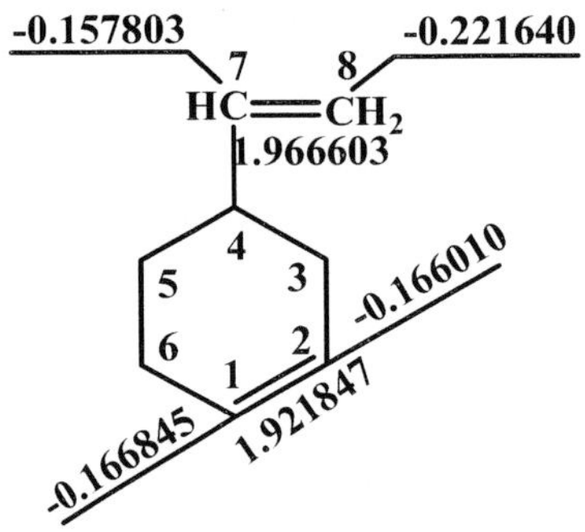

Diagram 1. The charges on the atoms and bond orders, in 4-vinyl-1-cycloxene molecule.

As it's seen from the diagram, the higher values of the charge observed are on the carbon atoms in 1,2 and 7 position. This indicates that these carbon atoms are reactionable centers for electrophilic addition. Also, bond orders in 7-8 position are characterized with higher values.

We have considered the hydrosilylation of [Me(MeO)$_2$SiH] with 4-vinylcyclo-1-hexene in view of the anti-Markovnikov and Markovnikov rules with a substituted vinyl group and also hydrosilylation with a double bond owing to the cyclic fragment.

First we have considered the hydrosilylation reaction of modelling compound methyldimethoxysilane [Me(MeO)$_2$SiH] with 4-vinylcyclo-1-hexene according to the anti-Markovnikov rule. The model reaction proceeds according to the following scheme 2 with formation of compound VII.

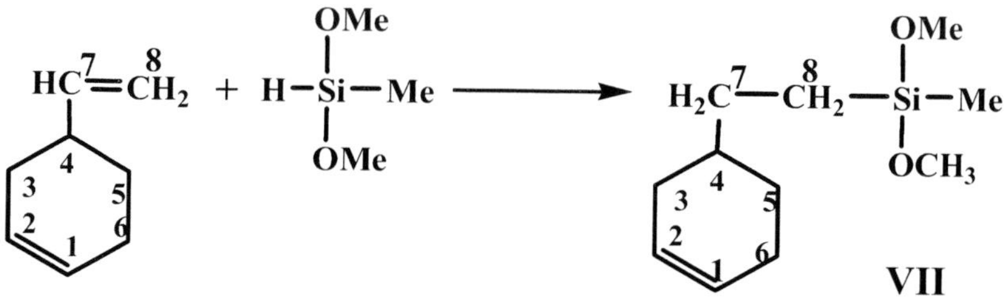

VII

Scheme 2. Modelling hydrosilylation reaction of methyldimethoxysilane to 4-vinylcyclo-1-hexene according to the anti-Markovnikov rule.

The distance between the double bonded carbon atom C_8 and silicon was assumed 1.0 Å longer than the bond length expected in the product. The change of distance $R_{C\text{-}Si}$ between the atom of silicon and the double bonded carbon atom C_8 was assumed = 0.05 Å. The initial position of reacting compounds is presented in Figure 8.

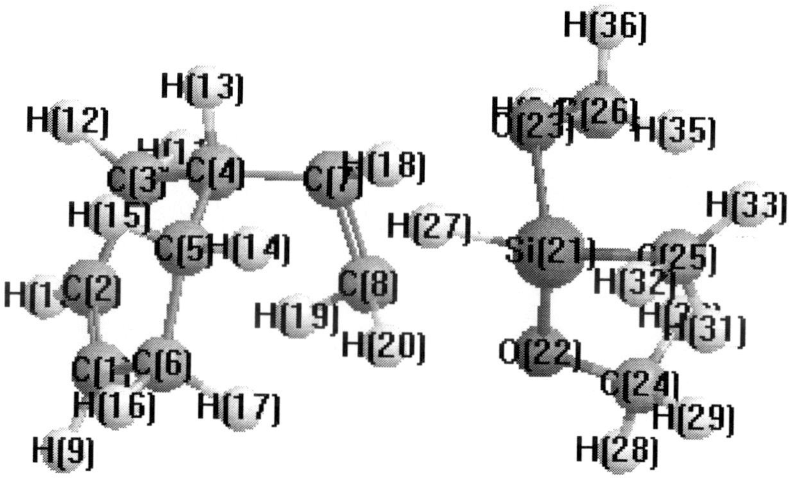

Figure 8. Starting position of compounds in modelling hydrosilylation reaction of methyldimethoxysilane and 4-vinylcyclo-1-hexene (anti-Markovnikov rule).

The change of energy (ΔH) calculated from AM1 as a function of the distance $R_{C\text{-}Si}$ is presented in Figure 9.

As it is seen from Figure 9, at approximation of silicon to atom of carbon (C_8) up to $R_{C\text{-}Si}$=2.25 Å distances the energy of the system rises. At the same time the bond orders between carbon C_7 and C_8 atoms in a molecule of 4-vinylcyclo-1-hexene decreases from 1.9459 up to 1.6859. The bond order between hydrogen and silicon atoms also decreases ($R_{Si\text{-}H}$=0.9191-0.7351) and in the modelling compound methyldimethoxysilane the process of formation of new bonds ($P_{C\text{-}Si}$ =0.0172-0.121887 and $P_{C_7\text{-}H}$=0.0008-0.091137) is observed. In case of distances up to 2.20 Å between the silicon atom and carbon (C_8) atom the energy of system sharply decreases, the double C=C bond passes in ordinary C-C bond ($P_{C_7\text{-}C_8}$ =1.0126), and the bond order between silicon and carbon atom C_8 it reaches to 0.8298. Thus the hydrogen is completely eliminated from silicon ($P_{Si\text{-}H}$= 0.0111) and attaches to carbon atom C_7 ($P_{C_7\text{-}H}$= 0.934).

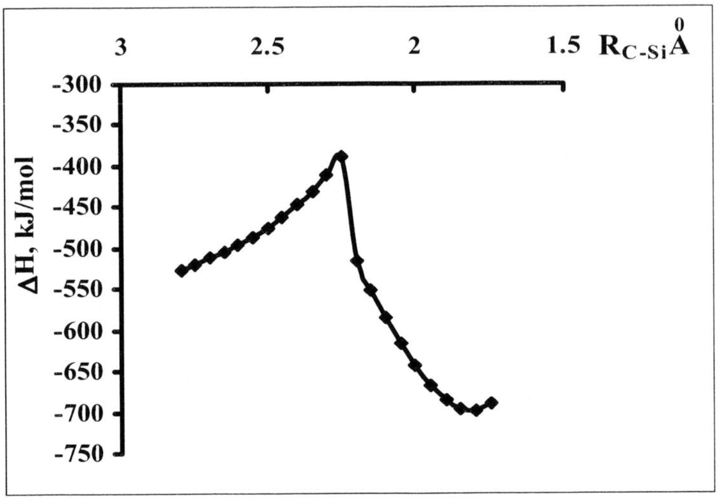

Figure 9. The change of energy (ΔH) as a function of the distance $R_{C\text{-}Si}$ between silicon and carbon atoms from modelling reaction of methyldimethoxysilane with 4-vinylcyclo-1-hexene, according to anti-Markovnikov rule.

With consecutive reduction of distances between the silicon atom and carbon (C_8) atom the energy of the system monotonously decreases and with strengthening formed new bonds. In Figure 10 the structure of modelling product obtained by hydrosilylation of methyldimethoxysilane to 4-vinylcyclo-1-hexene according to the anti-Markovnikov rule is presented.

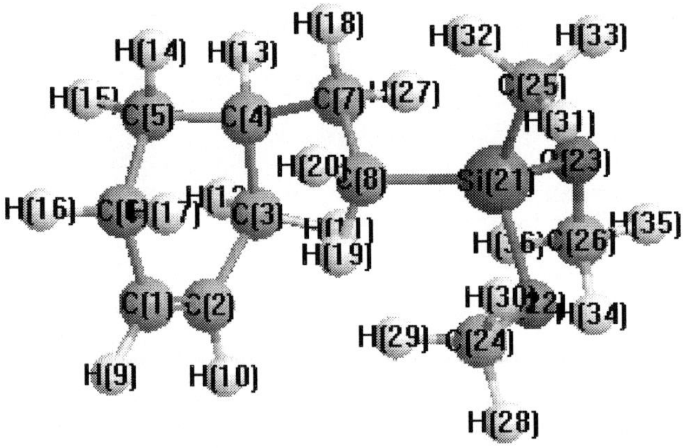

VII

Figure 10. The structure of obtained product VII in modelling reaction of hydrosilylation of methyldimethoxysilane to 4-vinylcyclo-1-hexene according to the anti-Markovnikov rule.

We have considered the hydrosilylation reaction of modelling compound methyldimethoxysilane [$Me(MeO)_2SiH$] with 4-vinylcyclo-1-hexene according to the Markovnikov rule. The model reaction proceeds according to the following scheme with formation of compound VIII.

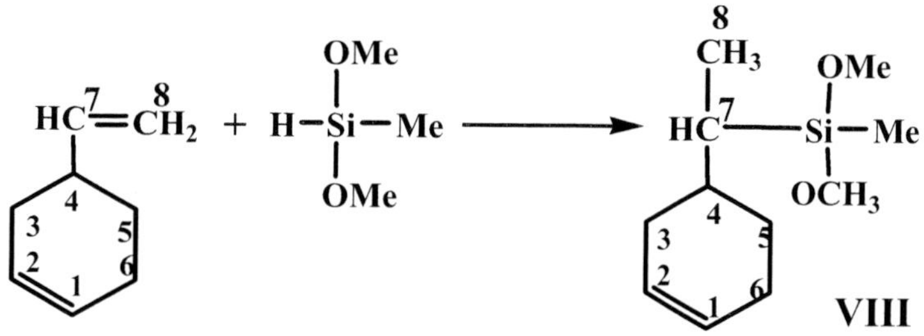

Scheme 2. Modelling hydrosilylation reaction of methyldimethoxysilane to 4-vinylcyclo-1-hexene according to the Markovnikov rule.

In Figure 11 the starting position of compounds in model hydrosilylation reaction of methyldimethoxysilane and 4-vinylcyclo-1-hexene is presented

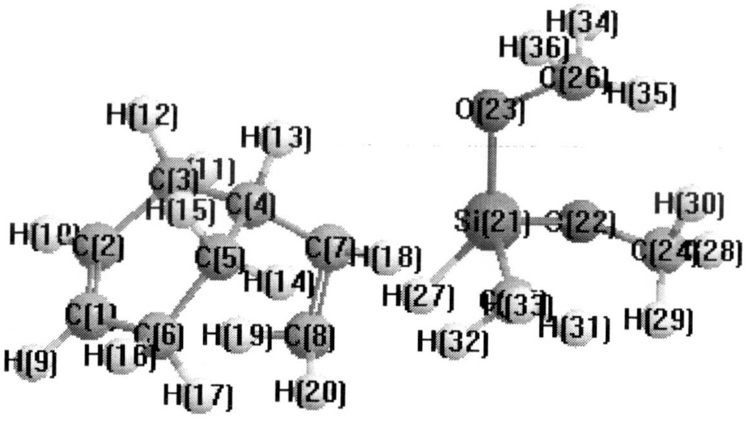

Figure 11. Starting position of compounds in model hydrosilylation reaction of methyldimethoxysilane and 4-vinylcyclo-1-hexene (Markovnikov rule).

Here also the distance between the double bonded carbon atom C_7 and silicon was assumed 1.0 Å longer than the bond length expected in the product. The change of distance $R_{C\text{-}Si}$ between the atom of silicon and the double bonded carbon atom C_7 was assumed = 0.05 Å. The change of energy (ΔH) calculated from AM1 as a function of the distance $R_{C\text{-}Si}$ is presented in Figure 12.

As it is seen from Figure 12, at approximation of the silicon to atom of carbon (C_7) up to $R_{C\text{-}Si}$=2,25 Å distances the energy of the system rises. At the same time the bond orders between carbon C_7 and C_8 atoms in a molecule of 4-vinylcyclo-1-hexene decreases from 1.9518 up to 1.7158.

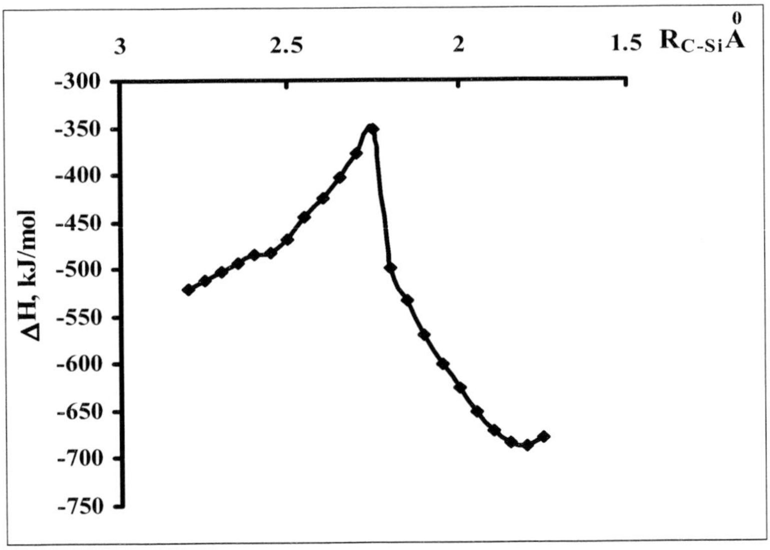

Figure 12. The change of energy (ΔH) as a function of the distance $R_{C\text{-}Si}$ between silicon and carbon atoms from modelling reaction of methyldimethoxysilane with 4-vinylcyclo-1-hexene according to Markovnikov rule.

The bond order between hydrogen and silicon atoms also decreases ($R_{Si\text{-}H}$=0.9168-0.7379) and in the modelling compound methyldimethoxysilane the process of formation of new bonds ($P_{C\text{-}Si}$=0.0173-0.1029 and $P_{C_8\text{-}H}$=0.0004-0.077709) is observed. In case of distances up to 2.20 Å between the silicon atom and carbon (C_7) atom the energy of the system sharply decreases, the double C=C bond passes in ordinary C-C bond ($P_{C_7\text{-}C_8}$=1.0132), and as to the bond order between the silicon and carbon atom C_7 that it reaches to 0.8488. Thus the hydrogen is completely eliminated from silicon ($P_{Si\text{-}H}$ =0.0113) and attaches to the carbon atom C_8 ($P_{C_8\text{-}H}$=0.9526). In Figure 13 the structure of modelling product obtained by hydrosilylation of methyldimethoxysilane to 4-vinylcyclo-1-hexene according to the anti-Markovnikov rule is presented.

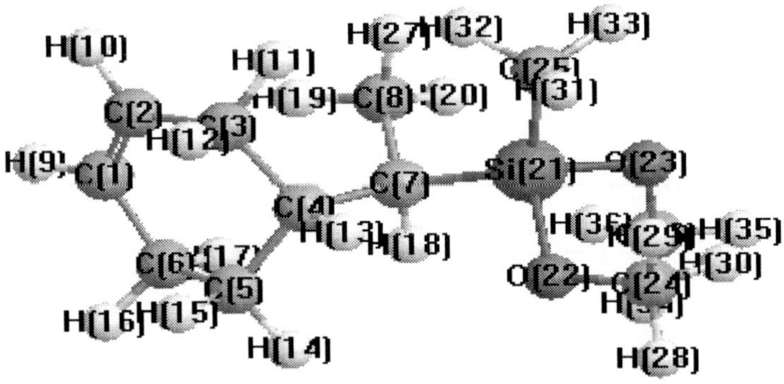

VIII

Figure 13. The structure of obtained product VIII in modelling reaction of hydrosilylation of methyldimethoxysilane to 4-vinylcyclo-1-hexene according to the Markovnikov rule.

We have also considered the hydrosilylation reaction of the modelling compound methyldimethoxysilane [Me(MeO)$_2$SiH] with 4-vinylcyclo-1-hexene towards –CH=CH– double bond in the cyclic fragment. Here we want to mention that from the literature this direction is considered not available. The model reaction proceeds according to the following scheme with formation of compound IX.

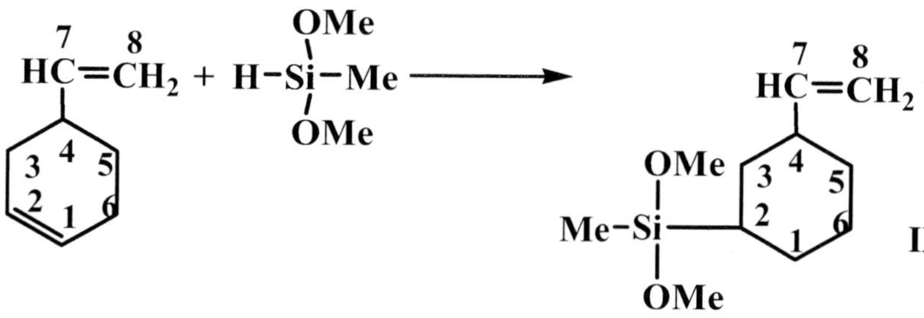

Scheme 3. Modelling hydrosilylation reaction of methyldimethoxysilane to 4-vinylyclo-1-hexene at 1-2 position.

The curve of the change of energy (ΔH) calculated from AM1 as a function of the distance $R_{C\text{-}Si}$ was done according to the above considered method and is presented in Figure 14.

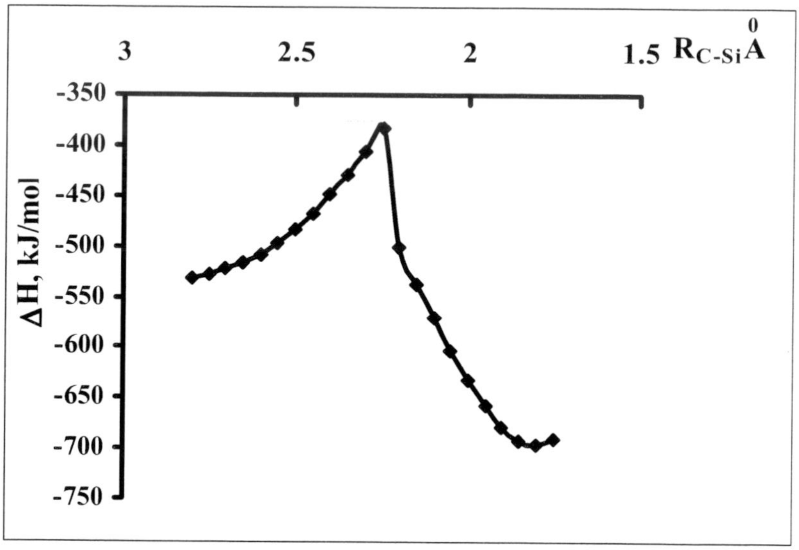

Figure 14. The change of energy (ΔH) as a function of the distance $R_{C\text{-}Si}$ between silicon and carbon atoms from modelling reaction of methyldimethoxysilane with 4-vinylcyclo-1-hexene towards double bond in the cyclic fragment.

In Figure 15 the structure of modelling product obtained by hydrosilylation of methyldimethoxysilane to 4-vinylcyclo-1-hexene towards double bond in the cyclic fragment is presented.

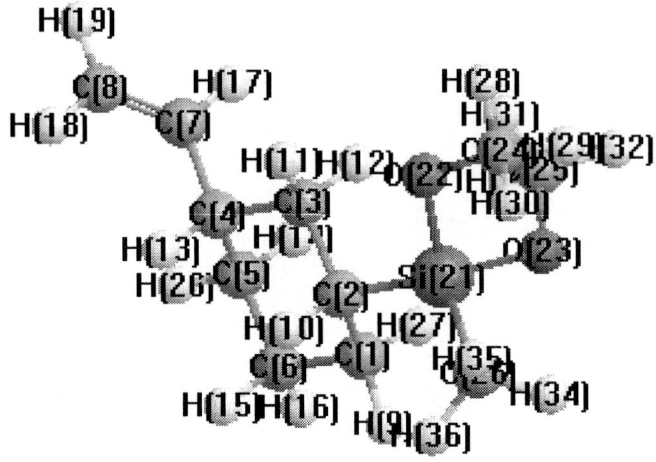

Figure 15. The structure of obtained product IX in modelling reaction of hydrosilylation of methyldimethoxysilane to 4-vinylcyclo-1-hexene (1-2 addition).

It must be denoted that activation energies and heats of formation in the modelling reaction of hydrosilylation of methyldimethoxysilane to 4-vinylcyclo-1-hexene during 2-1 addition approximately is the same.

With the help of the modelling reactions we have calculated activation energies E_{act} and the heats of formation (ΔH) of the hydrosilylation products. One can observe that ΔH monotonously decreases with decreasing distance while strengthening of newly formed bonds takes place. Activation energies for the modelling reactions were calculated from the difference of the maxima and the initial points from Figure 9, 12 and 14. In the following table are given activation energies and heats of formations for all three directions.

Table 2. Heats of formations and activation energies for modelling hydrosilylation reaction of methyldimethoxysilane with 4-vinylcyclo-1-hexene.

Reaction Direction	Heat of formation ΔH kJ/mol	Activation energy E_{act} kJ/mol
anti-Markovnikov rule	-172.0	136.7
Markovnikov rule	-166.6	142.7
Hydrosilylation towards the double bond of cyclic fragment (1-2 addition)	-164.2	149.0

Comparison of activation energies and heats of formations shows that the reaction is able to occur according to all three directions. Energetically the most favourable direction for the modelling hydrosilylation reaction of methyldimethoxysilane with 4-vinylcyclo-1-hexene is according to the anti-Markovnikov rule, after according to the Markovnikov rule and the less favourable is hydrosilylation towards the double bond in the cyclic fragment. The results of theoretical calculations are in good agreement with NMR spectral data. Thus this method proves validity of the method for prediction of the hydrosilylation reaction direction as a rough approximation.

The purpose of the quantum-mechanical calculations has been the development of the method for the prediction of hydrosilylation reactions' directions. Chemists can us this method as an aid to experiment, in particular in studies of reaction mechanisms. A major problem in studying reactions by any current theoretical model is the lack of experimental data; because of this we have performed calculations parallel with experiment.

The synthesized oligomers were studied by GPC method. Figure 16 shows the molecular weight distribution of oligomer III. From this figure it is evident that oligomer III has bimodal molecular weight distribution. Its molecular weight is equal to: $\bar{M}_n \approx 1,36x10^4$, $\bar{M}_\omega = 2,51x10^4$, D=1.87.

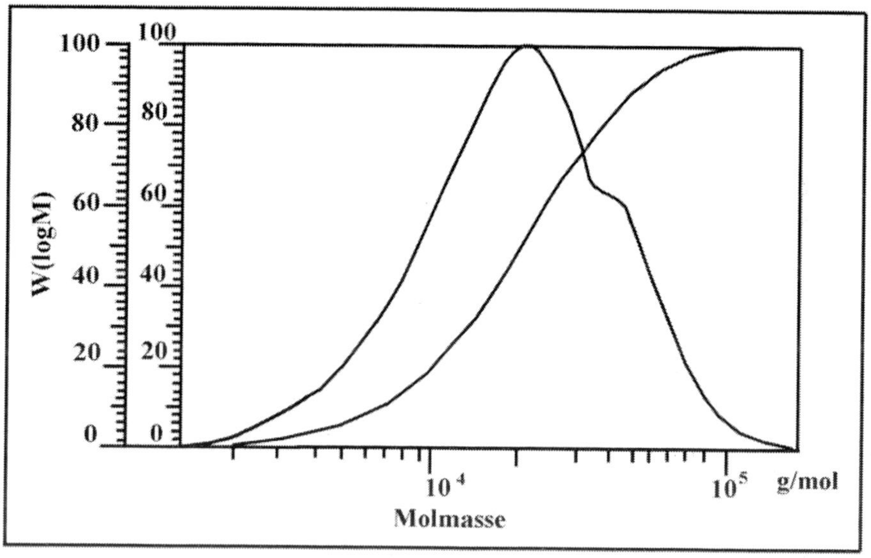

Figure 16. Gel-permeation chromatographic curves of oligomer III.

As is obvious, the average molecular weights of synthesized oligomers exceed several times the theoretical values of molecular weights calculated in case of full hydrosilylation. It indicates that during the inter-molecular hydrosilylation reaction branching process also takes place.

It must be denoted that oligomer III was additionally heated for 4-5 hours at 80^0C temperature and at this time the value inter-molecular hydrosilylation takes place and in organic solvents practical insoluble polymers and obtained Thus synthesized oligomers are thermo reactive systems where cross-linking reactions proceeds with participation of the catalyst remainder in the system.

Synthesized oligomers were investigated by DSC analyses. We show in Figure 17 DSC curve for oligomer I. There is only a single endothermic peak which corresponds to the glass transition and is equal to $T_g \approx -57^0$C. Glass transition temperature of the oligomer I is close to the value of glass transition temperatures of the products of hydrosilylation reactions of PMHS with styrene (-50^0C), α-styrene (-56^0C) [19].

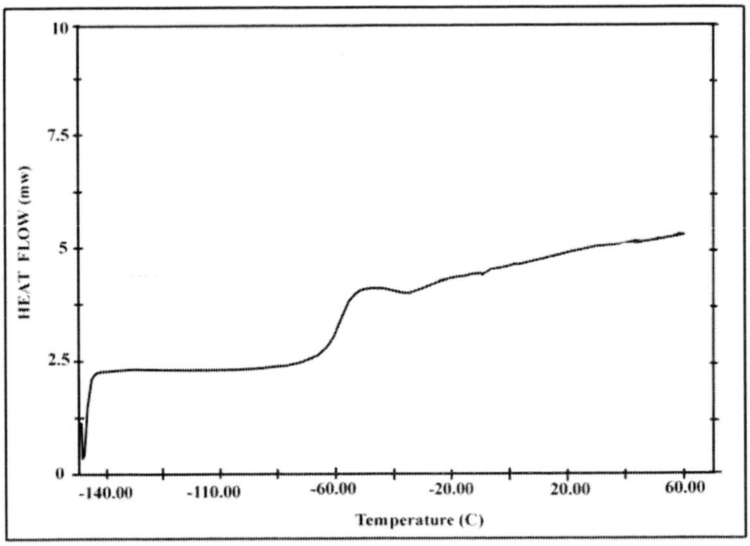

Figure 17. DSC curve of oligomer I.

We have performed wide-angle X-ray scattering analysis for our oligomers. Figure 18 shows that the oligomers I is a one-phase amorphous system. Diffraction patterns display two maxima. The main one on the left side corresponds to the maximum of the interchain distance d_1 while the second maximum corresponds to d_2, which characterizes both intra-molecular and interchain interactions [23].

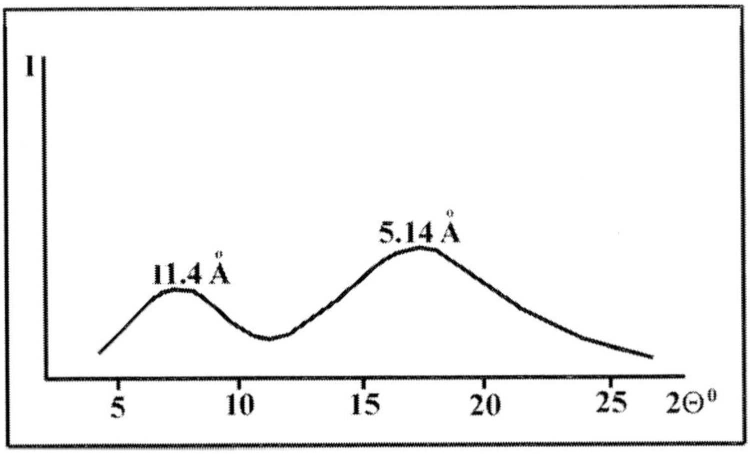

Figure 18. X-ray patterns for oligomer I.

4. Conclusion

Hydrosilylation reactions of polymethylhydridesiloxane with 4-vinylcyclo-1-hexene in the presence of platinum hydrochloric acid (0.1 M in THF), Karstedt's catalyst ($Pt_2[(VinSiMe_2)_2O]_3$) and platinum on the carbon were investigated. Synthesized oligomers were characterized by FTIR, ^{1}H, ^{13}C, H,H-COSY and C,H-correlation NMR spectroscopy. It was shown that the hydrosilylation reaction proceeds as according to the anti-Markovnikov and Markovnikov rules, also according to the hydrosilylation of a double bond of the cyclic fragment. Synthesized oligomers were characterized by gel-permeation chromatography, differential scanning calorimetry, thermogravimetric and wide-angle X-ray diffractometry analyses. Calculations using the quantum-chemical semi empirical AM1 method for modelling reactions between methyldimethoxysilane [$Me(MeO)_2SiH$] and 4-vinylcyclo-1-hexene were performed. The results of theoretical calculations are in good agreement with NMR spectral data and show that hydride addition is able to occur according to all three directions.

References

[1] I. Yilgor, Jr. W.P. Steckle, E. Yilgor, R.G. Freelin, J.S. Riffle. *J. Polym. Sci. Part A: Polym. Chem.*, 1989, 27, 3673.

[2] S. Garin, L. Lecamp, B. Youssef, C. Bunel. *Eur. Polym. J.* 1999, 35, 473.

[3] S.S. Hou, Y.P. Chung, C.K. Chan, C.P. *Kup. Polymer* 2000, 41, 3263.

[4] H.R. Kricheldorf, (Ed.), *Silicon in Polymer Synthesis,* Springer, Berlin, 1996.

[5] B.P.S. Chauhan, J.S. Rathore, N. Glloxhani. *Appl. Organometal. Chem.* 2005, 19, 542.

[6] B.P.S. Chauhan, P. Boudjouk. *Tetrahedron Lett.* 2000, 41,1127.

[7] X. Coquert, A. El Achari, A. Hajaiej, A. Lablache-Combier, C. Loucheux, L. Randrianarisoa. *Macromol. Chem.* 1991, 192, 1517.

[8] G. Titvinidze, T. Tatrishvili, O. Mukbaniani, *Georgian Chemical Journal,* 2005, 5, 249 (in Russian).

[9] D.-s. Yao, B.-y. Zhang, Q.-j. Sun, L.-f. Zhang. *J. Appl. Polym. Sci.* 2005, 95, 946.

[10] L.-C. Chien, L.G. Cada. *Macromolecules* 1994, 27, 3721.

[11] K.D. Belfield, C. Chinna, O. Najjar. *Macromolecules* 1998, 31, 2918.

[12] I. Toulokhonova, B. Bjerke-Kroll, R. West. *J. Organomet. Chem* 2003, 686, 101.

[13] J. Hua, Z. Li, K. Long, J. Qin, Sh. Li, Ch. Ye, Z. Lu. *J. Polym. Sci. Part A: Polym. Chem.* 2005, 43, 1317.

[14] T. Ganicz, U. Mizerska, M. Moszner, M. O'Brien, R. Perry, W.A. Sta´nczyk. *Appl. Catal. A: General* 2004, 259, 49.

[15] B. Marciniec, J. Gulinski, L. Kopylova, H. Maciejewski, M. Grundwald-Wyspianska, M. Lewandowski. *Appl. Oranomet. Chem.*, 1997, 11, 843.

[16] T. Iwahara, M. Kusakabe, M. Chiba, K. Yonezawa. *J. Polym. Sci. A* 1993, 31, 2617.

[17] G. Titvinidze, "*Hydride addition reactions of α,ω-bis(trimethylsiloxy)methylhydrosiloxane with some unsaturated bonds containing compounds*"; PhD thesis, Ivane Javakhishvili University, Tbilisi, 2003.

[18] O. Mukbaniani, T. Tatrishvili, G. Titvinidze, N. Mukbaniani, L. Lezhava, N. Gogesashvili. *J. Appl. Polym. Sci.*, 2006, 100, 2511.
[19] G. Titvinidze, T. Tatrishvili, N. Mukbaniani, O. Mukbaniani. Proceedings of Georgian Academy of Sciences, 2004, 30, 53.
[20] V.P. Yuriev, I.M. Salimgareva. Hydrosilylation Reactions of Olefins; *Nauka:* Moscow, 1982; p. 76 (in Russian).
[21] V.P. Yuriev, I.M. Salimgareva, O.Zh. Zhebarov, V.V. Kaverin, S.S. Rafikov. *Doklady Akademye Nauk USSR* 1976, 229, 892 (in Russian).
[22] V.P.; Yuriev, I.M. Salimgareva, V.V. Kaverin, G.A. Tolstikov. *Zhurn. Obshch. Kimii* 1977, 47, 355 (in Russian).
[23] O.V. Mukbaniani, G.E. Zaikov. *Cyclolinear Organosilicon Copolymers: Synthesis, Properties, Application; VSP,* Utrecht–Boston, 2003.

In: Modification Reactions of Oligomethylhydridesiloxanes ISBN: 1-60021-365-0
Editors: O.Mukbaniani, et al. pp. 201-217

Chapter 13

SYNTHESIS AND CHARACTERIZATION OF POLYSILOXANES WITH PENDANT BICYCLIC FRAGMENTS

O. Mukbaniani, T. Tatrishvili, G. Titvinidze, and S. Patsatsia
Iv. Javakhishvili Tbilisi State University, I. Chavchavadze Ave. 1,
0128 Tbilisi, Georgia

ABSTRACT

The hydrosilylation reactions of 5-vinyl-2-norbornene with α,ω–bis(trimethylsiloxy) methylhydrosiloxane catalyzed by platinum hydrochloric acid (0.1 M in THF), Karstedt's catalyst ($Pt_2[(VinSiMe_2)_2O]_3$) and platinum on the carbon have been studied. Dependence of reaction rate and active ≡Si-H bonds' conversion depth on the catalyst was investigated. Even in the excess of the 5-vinyl-2-norbornene, not all active ≡Si-H groups participate in the hydrosilylation reaction. The reaction order, activation energies and rate constants have been determined for hydrosilylation reactions in the presence of H_2PtCl_6. The synthesized oligomers were characterized by FTIR, ^1H, ^{13}C, ^1H-^1H COSY and C,H-correlation NMR spectroscopy. Synthesized polysiloxanes were characterized by wide-angle X-ray diffractometry. In addition, calculations using the quantum-chemical semi-empirical AM1 method for modelling reaction between methyldimethoxysilane [$Me(MeO)_2SiH$] and 5-vinyl-2-norbornene were performed to evaluate possible reaction paths.

Key words: polysiloxanes, AM1 theory, hydrosilylation, comb-type oligomers, kinetics.

1. INTRODUCTION

Hybrid organic-inorganic materials, where molecular organic and inorganic fragments are combined, have been considered potentially attractive for the purpose of developing new materials with a broad spectrum of interesting properties. In comparison with organic and inorganic constituents and polymers separately, hybrid organic-inorganic materials have a lot of advantages [1-5].

Polysiloxanes, usually referred to as silicones, find numerous applications in different fields of chemistry and engineering. All of these applications rely on the unique physical and chemical properties of silicones in the bulk form and at interfaces. Widely applied and recently explored hydrosilylation of polymeric systems [6,7] and especially hydrosilylation curing of polyvinylsiloxanes by polyhydrosiloxanes have attracted great interest due to the practical outcome and recent development in silicon-containing polymeric systems. From the above mentioned it is evident that synthesis and investigation of new polysiloxanes with double bond containing fragments attracts much attention due to their wide applications.

The hydrosilylation process is the most widely used method for preparation of organofunctional polymethylsiloxanes starting from poly(methylhydro)siloxanes [8, 9].

Transition metal catalyzed hydrosilylation of organic substrates possessing an unsaturated carbon–carbon bond is an effective, still attractive and most important commercial way of synthesis of carbofunctional silanes, siloxanes and oligo(poly)siloxanes [6,7,10]. Since Speier's first discovery that H_2PtCl_6 is an efficient hydrosilylation catalyst [11] the effect of different catalysts on the hydrosilylation process has been extensively studied [12].

In this work, we have studied modification of PMHS via hydrosilylation reactions in the presence of platinum catalysts, which are reported to be among the most active catalysts for hydrosilylation reactions [12]. We have investigated the dependence of reaction rate and active ≡Si-H bonds' conversion depth on the catalyst.

2. EXPERIMENTAL

2.1. Materials

Polymethylhydridesiloxane (Aldrich), platinum hydrochloric acid (Aldrich), Karstedt's catalyst ($Pt_2[(VinSiMe_2)_2O]_3$) or platinum(0)-1,3-divinyl-1,1,3,3-tetramethyldisiloxane complex (solution in PDMS), platinum on the carbon and 5-vinyl-2-norbornene (Aldrich) were used as received. Toluene was dried over and distilled from sodium under an atmosphere of dry nitrogen. Tetrahydrofuran (THF) was dried over and distilled from a K–Na alloy under an atmosphere of dry nitrogen. 0.1 M solution of platinum hydrochloric acid in THF was prepared and kept under nitrogen at low temperature.

2.2. Hydrosilylation Reaction of PMHS With 5-Vinyl-2-Norbornene in the Presence of H_2ptcl_6

Polymethylhydridesiloxane (0.8769 g, 0.3876 mmol), 5-vinyl-2-norbornene (1.6279 g, 0.0135 mol) and 5.5 ml dry toluene were placed into the three-necked flask equipped with a magnetic stirrer, reflux condenser and thermocouple (40^0C). Then the catalyst 0.1 M solution of platinum hydrochloric acid in tetrahydrofuran ($5\div9\text{x}10^{-5}$ g per 1.0 g of starting substance) was introduced. After the reaction completed, the solvent was partially eliminated, the reaction product was precipitated from the toluene solution by n-hexane, and 1.9 g (86 %) oligomer was obtained. The hydrosilylation reactions in the presence of other catalysts were carried out according to the above-mentioned method.

2.3. Determination of ≡Si-H Content

We calculate the content of active ≡Si-H groups in oligomers on the basis of the following reaction:

$$2\equiv\text{Si}-\text{H}\xrightarrow[\text{solution}]{\text{KOH alcohol}}\equiv\text{Si}-\text{O}-\text{Si}\equiv + H_2\uparrow$$

We place in a two-necked flask an exactly weighed amount of ≡Si-H containing oligomer, which should correspond to the evolution of a certain volume of hydrogen gas. We fit a pressure-equalized dropping funnel with 0.5 M KOH alcohol solution onto the flask. We connect the flask with a gas buret, which we connect with a levelling bottle at the bottom part, by means of a silicon tube. We allow the system to stand for 10 min. The water level in the levelling bottle adjusts to the same level as that in the gas buret (V_1, ml) and is recorded. After, we add the 0.5 M KOH alcohol solution slowly. As the hydrogen gas evolves, the water level goes downward. Thus, we need to keep both water levels the same as each other so that we could equalize the pressure in the system to the atmosphere. After the evolution of hydrogen gas completes (around 3-5 min) we record another 5 min later the value of the water level in the buret (V_2, ml). The difference between V_2 and V_1 gives the volume of eliminated hydrogen gas (V, ml). The Si-H content of an oligomer is given by the following:

$$\equiv\text{Si-H content} = 1000\ PV/WRT\ (\text{mol}/100\text{g})$$

Where p – atmospheric pressure at measurement (atm), V – gas volume captured (ml), W – sample weight (g), R = 0.0082 (gas constant) ($\text{l}\cdot\text{atm}\cdot\text{mol}^{-1}\cdot\text{K}^{-1}$), and T – temperature (K at measurement) [13].

2.4. Characterization

FTIR spectra were recorded on a Nicolet Nexus 470 machine with a MCTB detector. ^{1}H, ^{1}H-^{1}H COZY, ^{13}C NMR and H,C-correlation NMR spectra were recorded on a 500 MHz Brukker NMR spectrometer, using $CDCl_3$ as the solvent and an internal standard. Gel-permeation chromatography investigation was carried out with the use of a Waters Model 6000A chromatograph with an R 401 differential refractometer detector. The column set comprised 10^3 and 10^4 Å Ultrastyragel columns. Sample concentration was approximately 3% by weight in toluene and typical injection volume for the siloxane was 5 μL. standardization of the GPC was accomplished by the use of styrene or polydimethylsiloxane standards with the known molecular weight.

Wide-angle X-ray diffractograms were taken on a DRON-2 (Burevestnik, Saint Petersburg, Russia) instrument. A-CuK_α was measured without a filter; the angular velocity of the motor was $\omega \approx 2$ deg/min.

3. Results And Discussion

There is sufficient literature data on hydrosilylation of norbornene [14]. For example, the hydrosilylation of norbornene with $HSiCl_3$ takes place under mild conditions in the presence of an optically active 2-(diphenylphosphino)-2'-methoxy-1,1'-binaphthyl (MeO–MOP) complex of palladium to produce exo-2-(trichlorosilyl)norbornene with 99% quantitative yield [15,16].

5-vinyl-2-norbornene contains two active double bonds, which can participate in hydrosilylation reactions. Thus the hydrosilylation reaction can proceed into two main directions: with a substituted vinyl group and with a double bond in bicycle. With the vinyl group hydride addition can proceed according to anti-Markovnikov and Markovnikov rule.

For the purpose to synthesize new functional polysiloxanes with pendant bicyclic fragments including unsaturated bonds we have carried out hydrosilylation reactions of α,ω–bis(trimethylsiloxy)methylhydrosiloxane with 5-vinyl-2-norbornene in the presence of platinum hydrochloric acid (0.1 M in THF), Karstedt's catalyst ($Pt_2[(VinSiMe_2)_2O]_3$ in PDMS) and platinum on the carbon. For the purpose to utilize all active ≡Si-H bonds reactions were carried out with various molar ratios of initial compounds (1:35, 1:70 and 1:105). From the literature [17,18] it is known that not all active ≡Si-H groups participate in hydrosilylation reactions in case of stoichiometric relationships in view of ≡Si-H bonds.

Preliminary heating of initial compounds in the temperature range of $20 \div 60^0$C in the presence of catalysts showed that in these conditions polymerization of 5-vinyl-2-norbornene and destruction of the siloxane backbone were not observed. No changes in the NMR and FTIR spectra of initial compounds were found.

We have carried out the above mentioned reactions in the melt conditions, but reactions proceed vigorously (especially in case of platinum hydrochloric acid and Karstedt's catalyst) and in low stages of conversion of ≡Si-H bonds (~25%) gelation takes place via intermolecular hydrosilylation. For prevention of gelation and for investigation of kinetic parameters of the reaction we have carried out hydrosilylation in the absolute dry toluene.

In general, hydrosilylation of α,ω–bis(trimethylsiloxy)methylhydrosiloxane to 5-vinyl-2-norbornene proceeds by the following scheme 1. Where: [(a)+(b)+(c)+(d)+(f)](x)=m≈35; Cat - $H_2PtCl_6{\cdot}6H_2O$ - 20^{0}C (I^1), 30^{0}C(I^2) and 40^{0}C (I) (n=1). 20^{0}C (II^1), 30^{0}C (II^2), 40^{0}C (II) (n=2), 20^{0}C (III^1), 30^{0}C (III^2), 40^{0}C (III) (n=3). Cat-Pt/C - 40^{0}C (IV^1), 50^{0}C(IV^2), 60^{0}C (IV) (n=2). Karstedt's catalyst- 20^{0}C (V^1), 30^{0}C (V^2), 40^{0}C (V) (n=2).

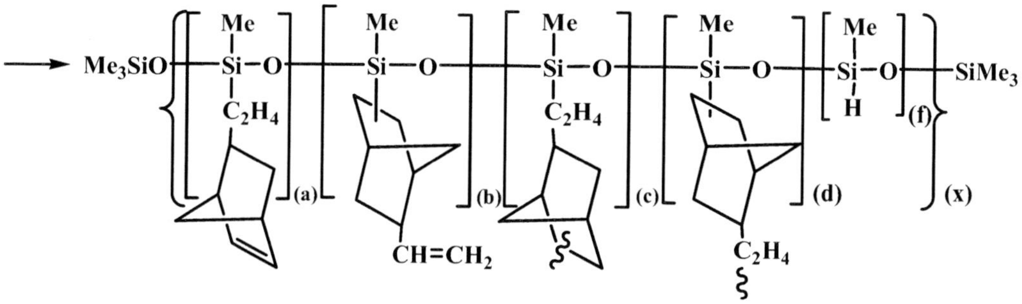

Scheme 1. Hydrosilylation reaction of α,ω–bis(trimethylsiloxy)methylhydrosiloxane to 5-vinyl-2-norbornene.

The synthesized oligomers are vitreous liquid products, which are well soluble in ordinary organic solvents with the specific viscosity $\eta_{sp} \approx 0.08 \div 0.1$. Structure and composition of the oligomers were established by elementary and functional analyses, FTIR spectra, ^{1}H, ^{13}C, H,H-COSY and C,H-correlation NMR spectral data. Synthesized polysiloxanes were characterized by wide-angle X-ray diffractometry. Some characteristic data for hydrosilylation reaction and synthesized oligomers are presented in table 1.

Table 1. Some characteristic data for hydrosilylation reaction and synthesized oligomers.

№	Yield %	Catalyst	Reaction tempera-ture / ^{0}C	n	% conversion of active ≡Si-H groups	η_{sp} *	d_1 Å	$\bar{M}_\omega x10^{-4}$
I	86	H_2PtCl_6	40	1	83	0.09	-	-
I^1	84	H_2PtCl_6	20	1	80	0.08	-	-
I^2	85	H_2PtCl_6	30	1	81	0.08	-	-
II	86	H_2PtCl_6	40	2	92	0.09	9.31	$2.97x10^3$
II^1	85	H_2PtCl_6	20	2	82	0.09	-	-
II^2	85	H_2PtCl_6	30	2	87	0.08	-	-
III	88	H_2PtCl_6	40	3	95	0.1	-	-
III^1	86	H_2PtCl_6	20	3	89	0.08	-	-
III^2	87	H_2PtCl_6	30	3	93	0.09	-	-
IV	90	Pt/C	60	2	72	0.1	9.27	-
IV^1	87	Pt/C	40	2	69	0.09	-	-
IV^2	89	Pt/C	50	2	72	0.1	-	-
V	88	Karstedt's	40	2	95	0.1	9.32	-
V^1	85	Karstedt's	20	2	88	0.09	-	-
V^2	86	Karstedt's	30	2	91	0.09	-	-

During the hydrosilylation reaction, a decrease of active ≡Si-H groups' concentration with the time was observed. The hydrosilylation reaction was performed in a dry toluene solution ($C \approx 5.7 \cdot 10^{-2}$ mol/l). As it's evident from Table 1 not all active ≡Si-H groups participate in the hydrosilylation reaction. In Figure 1 we show changes of concentration of active ≡Si-H groups with time (determined as described in Section 2.3) in case of hydrosilylation reactions of PMHS with 5-viny-2-norbornene in the presence of H_2PtCl_6 at 1:70 ratios of initial compounds (1:2 stoichiometric relationships between initial compounds in view of ≡Si-H bonds). From Figure 1 it is evident that at 20°C the hydrosilylation reaction proceeds with ~87% conversion of active ≡Si-H groups, whereas at 40°C the hydrosilylation reaction proceeds with ~92% conversion. Consequently, with the rise of temperature, the depth of the hydrosilylation reaction increases.

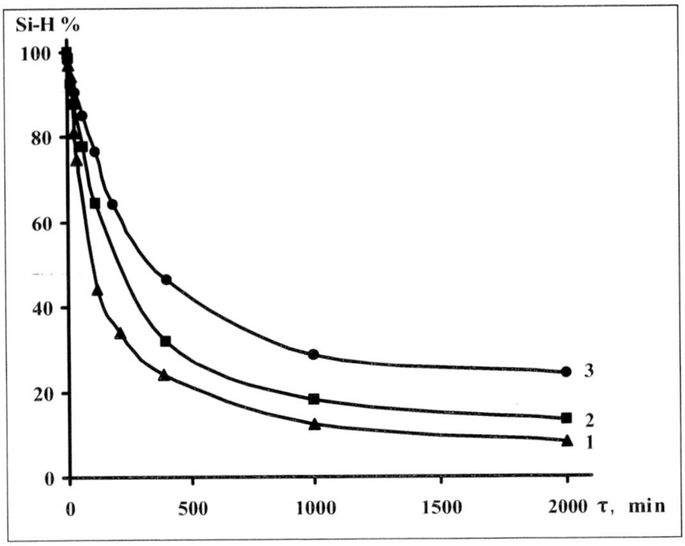

Figure 1. Dependence of changes of concentration of active ≡Si-H groups on the time, during hydrosilylation reactions of PMHS with 5-vinyl-2-norbornene, where curve 1 represents the values obtained at 40°C, curve 2 represents the values obtained at 30°C and curve 3 represents the values obtained at 20°C (Catalyst - H_2PtCl_6).

Figure 2 shows dependence of the reciprocal concentration of ≡Si-H groups on the time during the hydrosilylation reaction of α,ω–bis(trimethylsiloxy)methylhydrosiloxane with 5-vinyl-2-norbornene. One can see that at the initial stages the hydrosilylation reaction is of second order. From the literature [17-19] it is known that hydrosilylation reactions of PMHS with unsaturated compounds as usual at initial stages are of a second order in case of stoichiometric relationships between initial compounds in view of ≡Si-H bonds. As we can see from experimental results the increase of concentration of one initial compound does not change reaction order. The reaction rate constants of hydride addition reactions of polymethylhydridesiloxane to 5-vinyl-2-norbornene at various temperatures were determined: $k_{40°C} \approx 1.5206$, $k_{30°C} \approx 0.7011$ and $k_{20°C} \approx 0.3084$ l/mol·s. The reaction temperature coefficient is equal to γ=2,2.

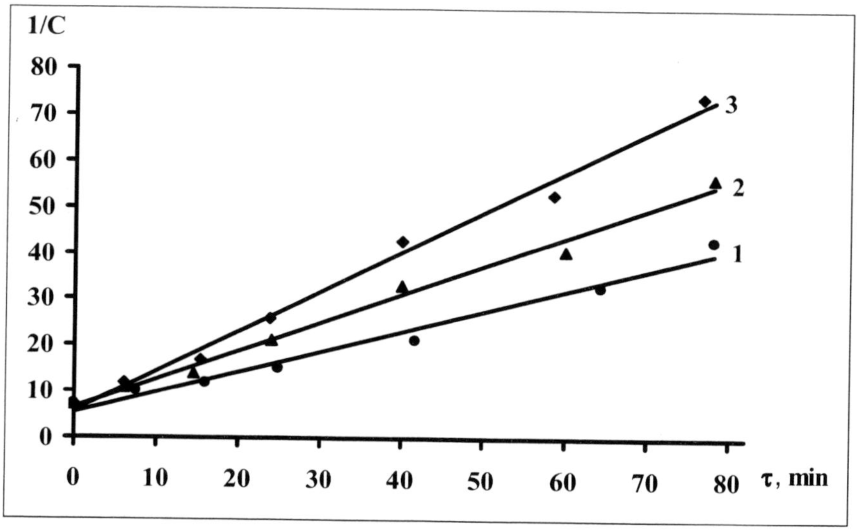

Figure 2. Dependence of reaction rate constants' logarithm on the reverse temperature, during hydrosilylation reaction of polymethylhydridesiloxane to 5-vinyl-2-norbornene.

We have calculated the activation energy for the reactions using logarithmic reaction rate constants as functions of the reciprocal temperature (Figure 3). The activation energy is equal to E_{act}≈20,3 kJ/mol.

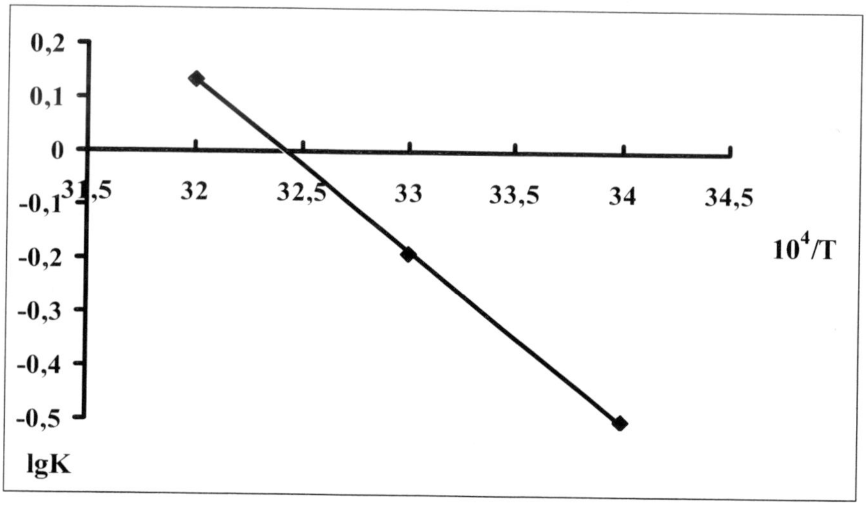

Figure 3. Dependence of reaction rate constants' logarithm on the reverse temperature during hydrosilylation reaction of polymethylhydridesiloxane to 5-vinyl-2-norbornene.

We have also investigated hydrosilylation reactions of polymethylhydridesiloxane with 5-vinyl-2-norbornene in the presence of platinum hydrochloric acid (0.1 M in THF) with 1:105 ratios of initial compounds (1:3 stoichiometric relationships between initial compounds in view of ≡Si-H bonds). In case of excess of 5-vinyl-2-norbornene 3-times reaction proceeds deeper than in the case of ratios 1:70 of initial components. From Table 1 we can compare results of ≡Si-H bonds' conversion, obtained at the same temperatures with different ratios of reacting compounds. From table 1 it is evident that with the increase of concentration of 5-vinyl-2-norbornene reaction proceeds with a higher conversion of ≡Si-H bond.

Hydrosilylation reactions of PMHS with 5-vinyl-2-norbornene in the presence of platinum on the carbon proceed with low rate and depth. Reactions were carried out at higher temperatures than in the case of platinum hydrochloric acid and Karstedt's catalyst, but the conversion of active ≡Si-H bonds was lower (table 1).

Hydrosilylation reaction of polymethylhydridesiloxane with 5-vinyl-2-nornornene in the presence of Karstedt's catalyst at 40^0C in the case of 1:70 ratio of initial compounds proceeds with a high degree of active ≡Si-H bonds conversion ~95% (oligomer V). Here we want to mention that in the presence of H_2PtCl_6 at 40^0C temperature in case of 1:105 ratio of initial compounds the reaction proceeds with the same degree of active ≡Si-H bonds conversion ~95% (oligomers III). It shows superiority of the Karstedt's catalyst in comparison with platinum hydrochloric acid.

During the hydride addition reactions, the changes of active ≡Si-H bonds' concentrations with time were observed. It was shown that not all active ≡Si-H groups participate in hydrosilylation reactions in spite of the excess of 5-vinyl-2-norbornene; it can be explained by steric effect of the pendant bicyclic fragments. In Figure 4 we show the changes of concentrations of active ≡Si-H groups with time during hydrosilylation reactions in the presence of different catalysts at 40^0C temperature. Figure 4 evidently shows the influence of the catalysts on the reaction rate and depth. From Figure 4 we can conclude that the activity of catalysts for hydrosilylation reactions of PMHS with 5-vinyl-2-nornornene decreases in the next rank: Karstedt's catalyst> H_2PtCl_6>Pt/C.

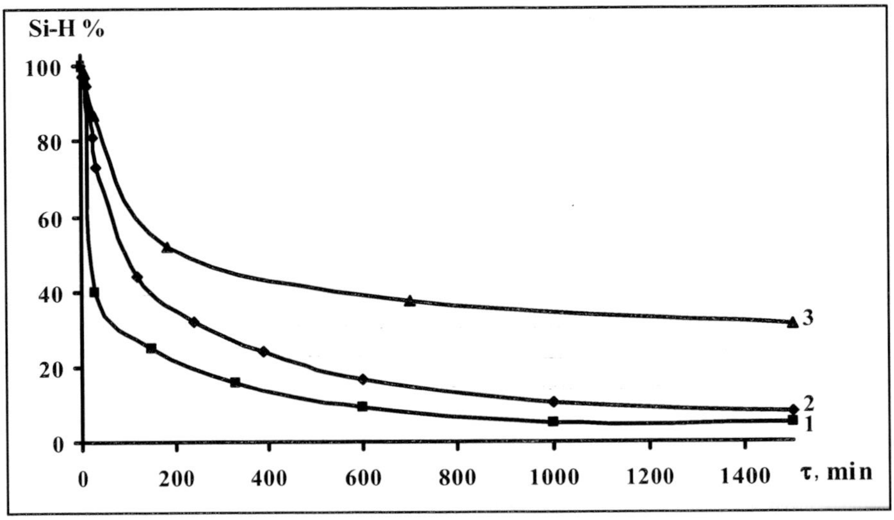

Figure 4. Dependence of changes of concentration of active ≡Si-H groups on the time, during hydrosilylation of PMHS with 5-vinyl-2-norbornene, where curve 1 represents the values obtained at 40°C with 1:70 ratio in the presence of Karstedt's catalyst, curve 2 and 3 accordingly represent the values obtained at same conditions in the presence of H_2PtCl_6 and Pt/C at 40°C.

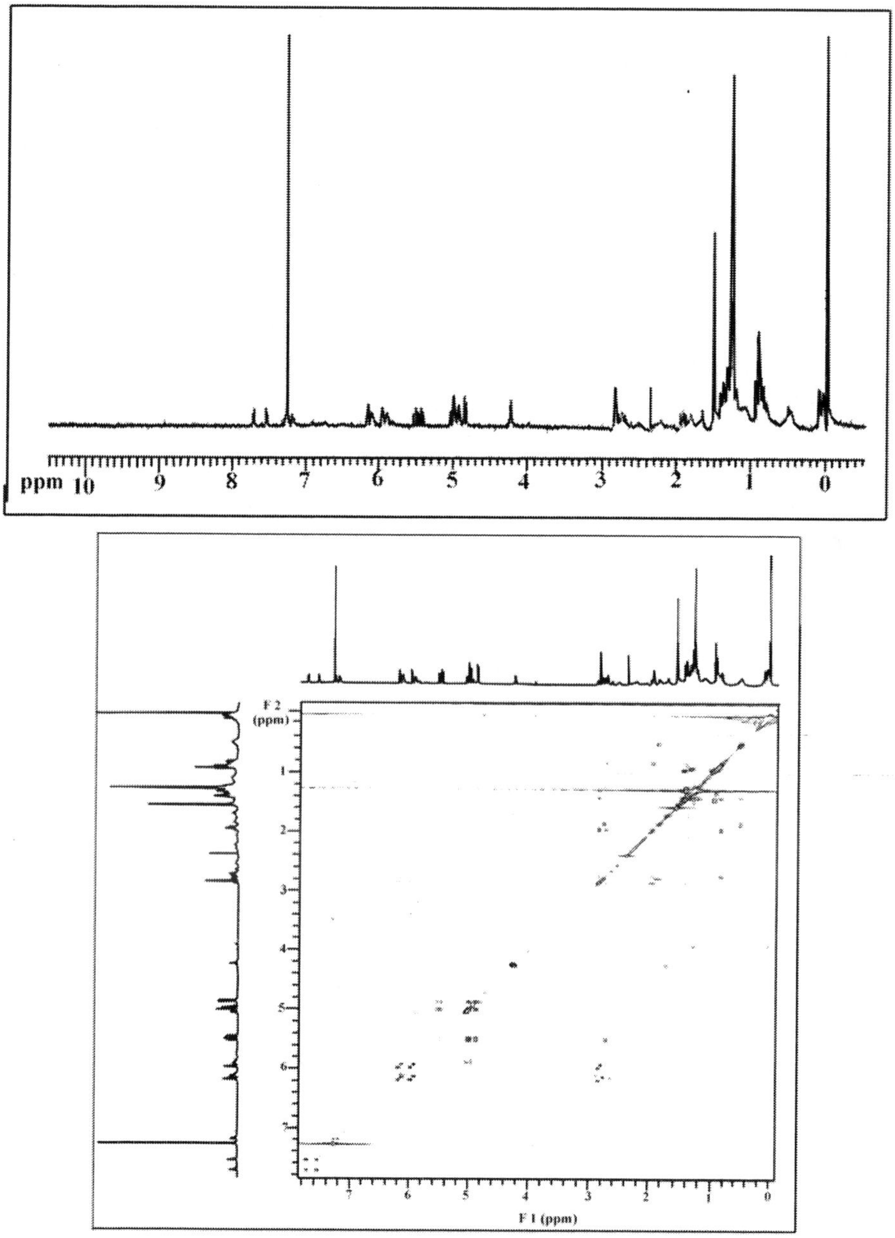

Figure 5. ${}^{1}H$ and H,H-COSY NMR spectra of oligomer II.

In Figure 5 ${}^{1}H$ and H,H-COSY NMR spectra of oligomer II (Figure 5) are shown, where one can observe signals for methyl protons of ≡Si-Me and $-SiMe_3$ fragments with the chemical shift δ=0,01 and δ=0,03 ppm. Signals with chemical shift δ≈0.45 and δ≈2.7 ppm are accordingly characteristic for methylene protons of $\equiv Si-CH_2-$ and $C_7H_9-CH_2-$ fragments have low intensities, which shows that reaction according to the anti-Markovnikov rule proceeds with a low degree. In the ${}^{1}H$ NMR spectra one can observe signals with chemical shift δ≈0.8 and δ≈1,3 ppm characteristic for methyl protons of $=CH-CH_3$ fragment and methine protons of $=CH-CH_3$ fragment accordingly.

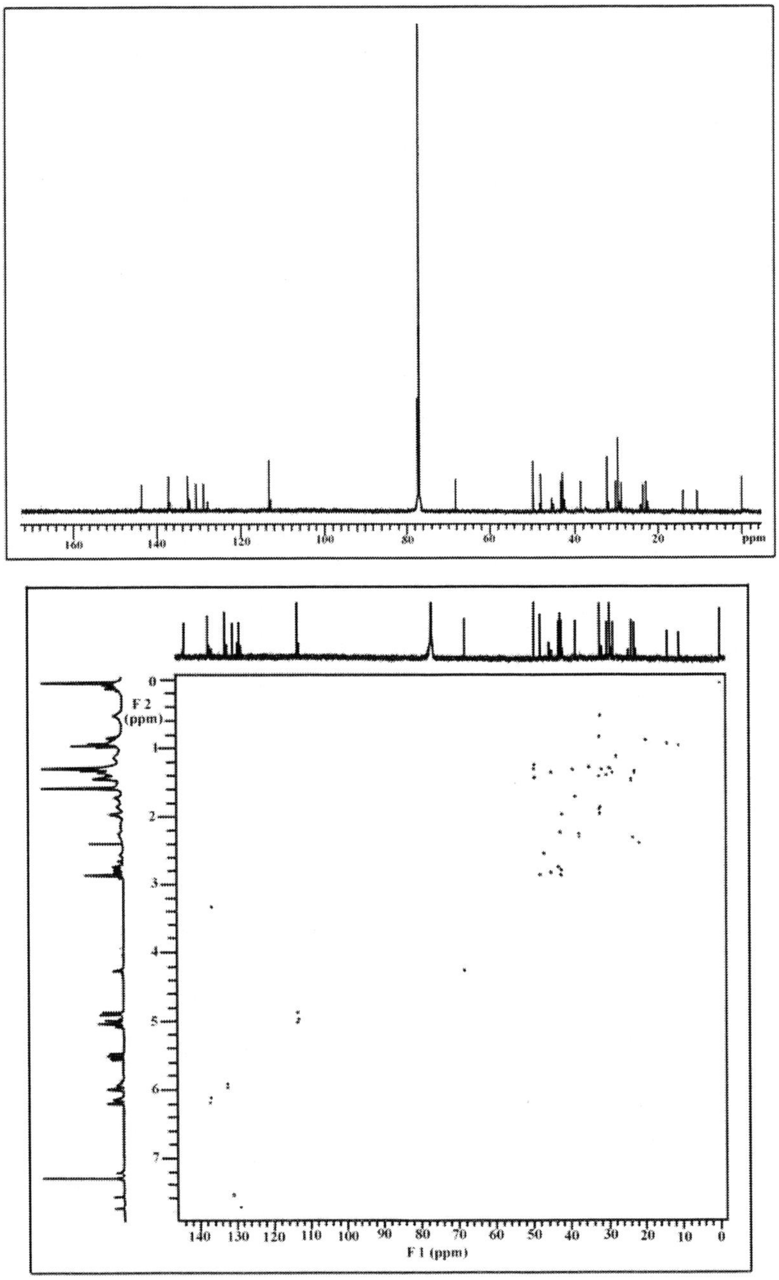

Figure 6. ^{13}C and C,H-correlation NMR spectra of oligomer II.

These two signals show that the reaction proceeds according to Markovnikov rule. From the comparison of intensities of signals characteristic for proceeding of hydrosilylation according to two directions, we can conclude that the favourable direction is according to the Markovnikov rule. Multiplet signals with a center of chemical shifts $\delta \approx 1.4$, 1.6, 1.95 and 2.4 ppm correspond to methylene and methine protons of norbornene fragment and partially overlapping of resonance signals occurs. In the spectra one can observe low intensity multiplet signals with a center of chemical shifts $\delta \approx 4.85 \div 5.0$ and $\delta \approx 6.0 \div 6.2$ ppm accordingly characteristic for methylene and methine protons of vinyl groups, this proves that hydrosilylation reaction proceeds at the double bond of a bicyclic fragment with a low degree. Multiplet signal with chemical shift δ≈5.45÷5.55 ppm corresponds to the methine protons of -CH=CH- fragment. A signal with chemical shift δ≈4.1 ppm characteristic for ≡Si-H bond shows that full hydrosilylation does not occur and that various linked oligomers are obtained.

^{13}C and C,H-correlation NMR spectra of oligomer II (Figure 6) are in good correlation with ^{1}H and H,H-COSY NMR spectra and proves that reaction proceeds according to all three directions: hydride addition by Markovnikov and anti-Markovnikov rule, also hydrosilylation by participation of double bond of bicyclic fragment. In the ^{13}C NMR spectra one can observe signal δ≈0.013 characteristic for carbon atoms in ≡Si-Me and -$SiMe_3$ fragments. Signal δ≈14.04 ppm characteristic for methylene carbon of ≡Si-$\underline{CH_2}$- fragment and signal δ≈28.92 ppm corresponding to methylene carbon in C_6H_9-$\underline{CH_2}$- proves the proceeding of hydrosilylation according to the anti-Markov- nikov rule.

Signals δ≈22.98 and δ≈10.96 ppm correspond to the methine and methyl carbon atoms in =CH-CH_3 fragment accordingly and show that reaction proceeds by the Markovnikov rule. Signal δ≈113.12 ppm corresponds to methylene carbon of the vinyl fragment and signal δ≈143.76 ppm corresponds to the methine carbon of the same group. These two signals indicate that hydrosilylation also proceeds with a double bond of the bicyclic fragment.

To determine the ratio of the fragments obtained from the anti-Markovnikov and Markovnikov rules, also from hydride addition towards a double bond of the cyclic fragment, is not available, because the spectra have complicated character.

We have carried out wide-angle X-ray diffractometry analysis for the synthesized oligomers. Figure 7 show that the oligomer II is one-phase amorphous system. Diffraction patterns display two maxima $2\Theta=9.5^0$ and $2\Theta=16.5^0$.

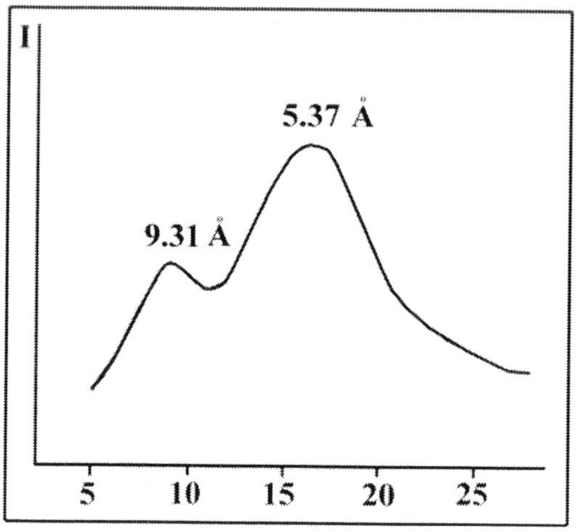

Figure 7. X-ray patterns for oligomer II.

The main one on the left side corresponds to the maximum of the interchain distance d_I=9.31 Å, while the second one corresponds to d_2=5.37 Å, which characterizes both intra-molecular and interchain interactions [19]. Here we want to mention that changing of the catalyst or reaction conditions does not influence the values of the interchain distances.

The synthesized oligomers were studied by GPC. Figure 8 shows the molecular weight distribution of oligomer II, from which it is evident that oligomer II has a bimodal molecular weight distribution. We have found the number average molecular mass $\overline{M}_n \approx 7.7x10^3$ and the weight average molecular mass $\overline{M}_\omega \approx 2.97x10^3$. Thus, the polydispersity index is equal to 3.8.

The average molecular mass of oligomer II in case of full hydrosilylation of PMHS by 5-vinyl-2-norbornene is equal to 6462, but we found that the average molecular mass 4.6 times exceed theoretical values of calculated molecular masses in case of full hydrosilylation. It indicates that during hydride addition reactions take place during macromolecular interchain hydrosilylation due to obtaining branched polymers.

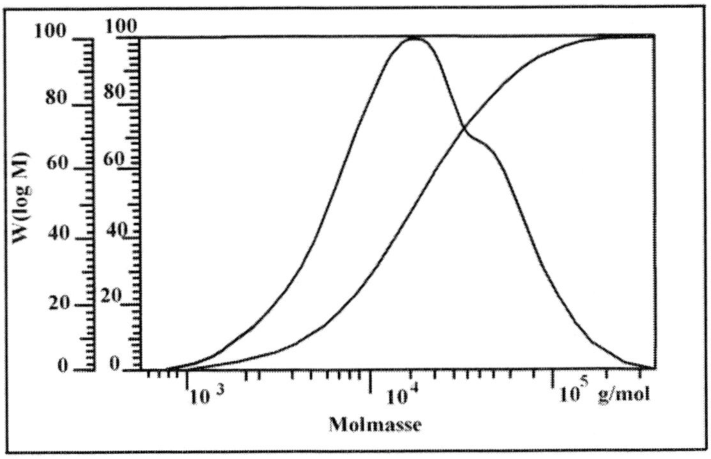

Figure 8. Gel-permeation chromatographic curves of oligomer II.

4. Quantum-Chemical Calculations

Semi empirical quantum methods are simplified versions of the Hartree-Fock approach using empirical corrections derived from experimental data. These methods are usually referred to through acronyms encoding some of the underlying theoretical assumptions. We apply one of the most frequently used AM1 methods. It is based on the neglect of differential diatomic overlap (NDDO) integral approximation. This approach belongs to the class of zero differential overlap (ZDO) methods, in which all two-electron integrals involving two-center charge distributions are neglected. A number of additional approximations are made to speed up calculations and a number of parameterized corrections are made in order to correct for the approximate quantum mechanical model. For AM1 the parameterization is performed so that we obtain enthalpies of formation H^{form} instead of total enthalpies [20] as a function of the distance $R_{C\text{-}Si}$. The calculations also provide us with $P_{x\text{-}y}$ values which represent bond orders.

The purpose of the quantum-chemical calculations has been the development of the method for the prediction of hydrosilylation reactions' directions. Chemists can use this method as an aid to experiment, particularly in studies of reaction mechanisms. A major problem in studying reactions by any current theoretical model is the lack of experimental data; because of this we have performed calculations parallel with the experiment.

We have performed calculations using a semi empirical AM1 method for the modelling reaction between methyldimethoxysilane [$Me(MeO)_2SiH$] and 5-vinyl-2-norbornene for full characterization of hydride addition of polymethylhydridesiloxane to 5-vinyl-2-norbornene using the software Chem3D Ultra 9.0 from Cambridge Soft.

Using a quantum-chemical half empiric AM1 method we have calculated for all initial, intermediate and final products, in the modelling reaction of hydrosilylation of methyldimethoxysilane with 5-vinyl-2-norbornene, the heats of formations (ΔH_f), changes of system energy (ΔH) depending on the change of distance ($R_{C\text{-}Si}$) between ≡C-Si≡ bonds, as well as the charge values (q) on the atoms, dipole moments (μ) and bonds orders (Pij).

We have discussed the hydrosilylation reaction of model compound - methyldimethoxysilane with 5-vinyl-2-norbornene according to the anti-Markovnikov and Markovnikov rules with a substituted vinyl group and also a hydride addition toward a double bond of the bicyclic fragment.

First we have considered the hydride addition of [$Me(MeO)_2SiH$] with 5-vinyl-2-norbornene according to the anti-Markovnikov rule. The model reaction proceeds according to the following scheme by obtaining compound VI:

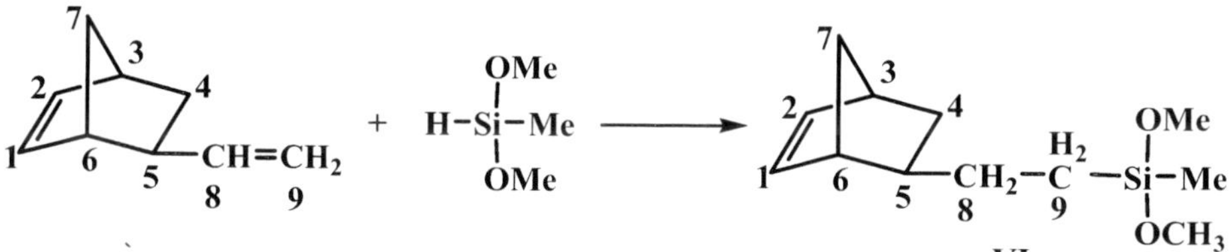

VI

The distance between the double bonded carbon atom and silicon atom was assumed 1.0 Å longer than the bond length expected in the product. The change of the distance $R_{C\text{-}Si}$ between the atom of silicon and the double bonded carbon atom was 0.05 Å. The change of energy (ΔH) calculated from AM1 as a function of the distance $R_{C\text{-}Si}$ is presented in Figure 9.

We have considered the hydrosilylation reaction of the modelling compound methyldimethoxysilane [$Me(MeO)_2SiH$] with 5-vinyl-2-norbornene according to the Markovnikov rule. The modelling reaction proceeds according to the following scheme with formation of compound VII:

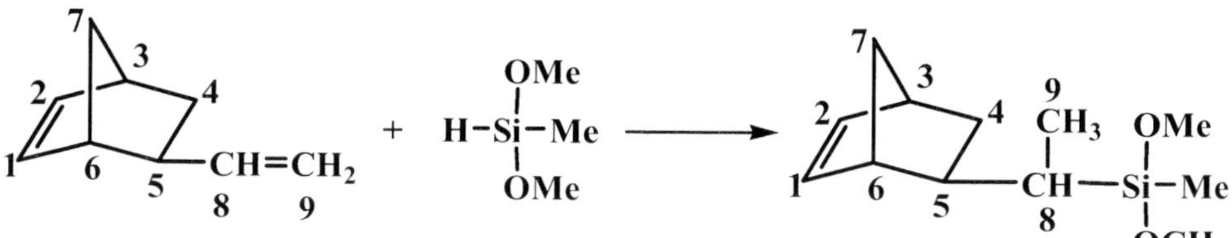

VII

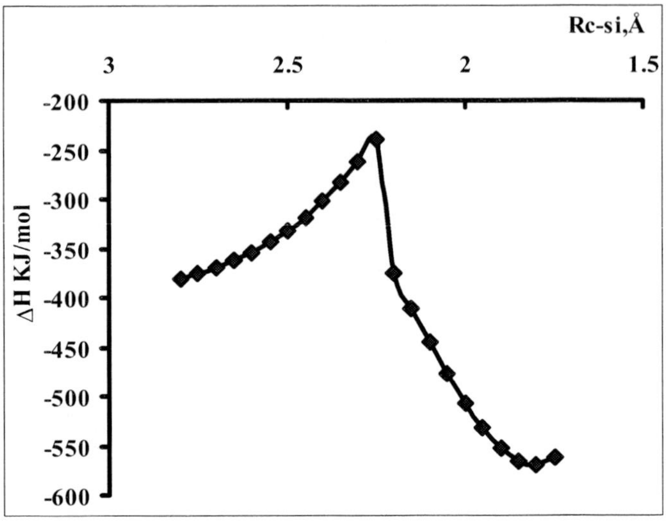

Figure 9. The change of energy (ΔH) as a function of the distance $R_{C\text{-}Si}$ between silicon and carbon atoms from modelling reaction of methyldimethoxysilane with 5-vinyl-2-norbornene according to anti-Markovnikov rule.

Here also the distance between the double bonded carbon atom and silicon was assumed 1.0 Å longer than the bond length expected in the product. The change of distance $R_{C\text{-}Si}$ between the atom of silicon and the double bonded carbon atom was assumed 0.05 Å. The change of energy (ΔH) calculated from AM1 as a function of the distance $R_{C\text{-}Si}$ is presented in Figure 10.

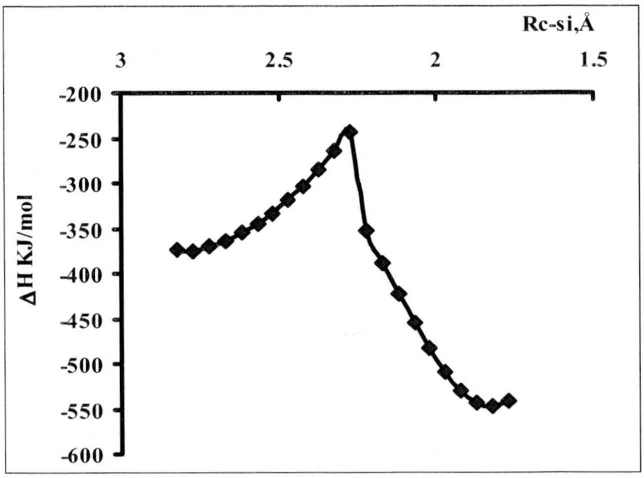

Figure 10. The change of energy (ΔH) as a function of the distance $R_{C\text{-}Si}$ between silicon and carbon atoms from modelling reaction of methyldimethoxysilane with 5-vinyl-2-norbornene according to Markovnikov rule.

We have also considered the hydrosilylation reaction of modelling compound methyldimethoxysilane [$Me(MeO)_2SiH$] with 5-vinyl-2-norbornene towards –CH=CH– double bond in the bicyclic fragment. The model reaction proceeds according to the following scheme with formation of compound VIII:

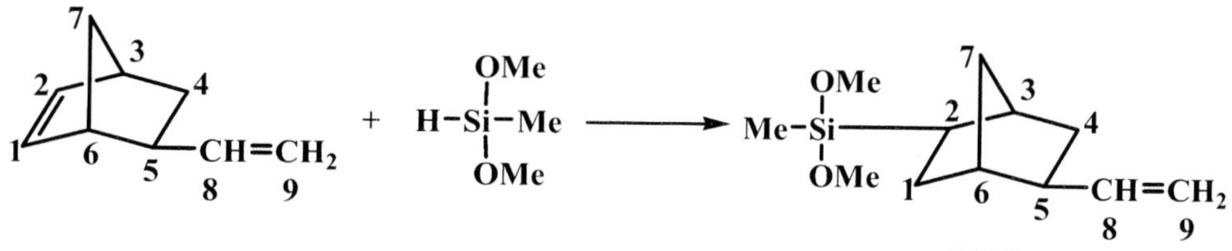

VIII

The curve of the change of energy (ΔH) calculated from AM1 as a function of the distance $R_{C\text{-}Si}$ was done according to above considered method and it is presented in Figure 11.

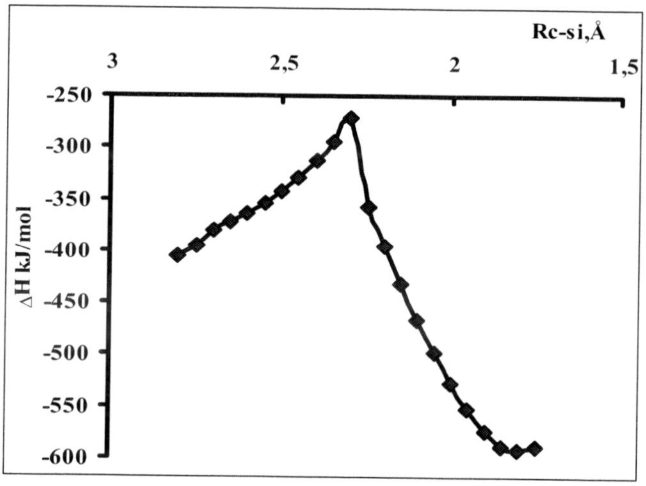

Figure 11. The change of energy (ΔH) as a function of the distance $R_{C\text{-}Si}$ between silicon and carbon atoms from modelling reaction of methyldimethoxysilane with 5-vinyl-2-norbornene towards double bond in the bicyclic fragment.

Using the quantum-chemical half empiric AM1 method we have calculated activation energies E_{act} and the heats of formation (ΔH) of the hydrosilylation reactions and their products. From the Figures it is obvious that ΔH monotonously decreases with decreasing distance while strengthening of recently formed bonds occurs. Activation energies for the modelling reactions were calculated from the difference of the maxima and the initial points from Figure 9, 10 and 11. In the following Table 2 activation energies and heats of formations are given for all three directions.

Table 2. Heats of formations and activation energies for modelling hydrosilylation reaction of methyldimethoxysilane with 5-vinyl-2-norbornene.

Reaction Direction	Heat of formation ΔH kJ/mol	Activation energy E_{act} kJ/mol
anti-Markovnikov rule	-188.64	142.71
Markovnikov rule	-173.104	130.84
Hydrosilylation towards the double bond of bicyclic fragment	-187,73	134.68

Taking into account activation energies and heats of formations, we can conclude that the reaction can proceed according to all three directions: anti-Markovnikov addition, Markovnikov addition and addition with a double bond of the norbornene fragment. The results of calculations are close to each other and it is difficult to conclude which direction is energetically most favourable. The results of theoretical calculations are in good agreement with NMR spectral data, which show proceeding of all four directions. Thus this method proves validity of the method for prediction of the hydrosilylation reaction direction as a rough approximation.

REFERENCES

[1] P. Cardiano, S. Sergi, M. Lazzari, P. Piraino. *Polymer* 2002, 43, 6635.
[2] C. Guizard, A. Bac, M. Barboiu, N. Hovnanian. *Sep Purif. Tech.* 2001, 25, 167.
[3] M.L. Sforca, I.V.P. Yoshida, S.P. Nunes. *J. Membr. Sci.* 1999, 159, 197.
[4] R.A. Zoppi, I.V.P. Yoshida, S.P. Nunes. *Polymer* 1997, 39, 1309.
[5] H.R. Kricheldorf, (Ed.), *Silicon in Polymer Synthesis*, Springer, Berlin, 1996.
[6] R.G. Jones, W. Ando, J. Chojnowski. *Silicon-containing Polymers*, Kluwer Academic Publishers, Dordrecht, 2000.
[7] B. Marciniec, J. Gulinski, H. Maciejewski.Hydrosilylation, in: I.T. Horvath (Ed.), *Encyclopedia of Catalysis*, vol. 4, Wiley, New York, 2003.
[8] M.P. McGrath, E. D. Sall, S. Tremontm. *J. Chem. Rev.* 1995, 95, 381.
[9] J.M. Buriak. *Chem. Rev.* 2002, 102, 1271.
[10] M.A. Brook, *Silicon in Organic, Organometallic, and Polymer Chemistry,* Wiley, New York, 2000.
[11] J.L. Speier, J.A. Webster, G.H. Barnes. *J. Am. Chem. Soc.* 1957, 79, 974.
[12] B. Marciniec (Ed.), *Comprehensive Handbook on Hydrosilylation,* Pergamon Press, New York, 1992.
[13] T. Iwahara, M. Kusakabe, M. Chiba, K. Yonezawa. *J. Polym. Sci.* A 1993, 31, 2617.
[14] J. Beckmann, D. Dakternieks, A. Duthie, S.L. Floate, R.C. Foitzik, C.H. Schiesser. *J. Organomet. Chem.* 2004, 689, 909.
[15] T. Hayashi, Y. Uozumi. *Pure Appl. Chem.* 1992, 64, 1911.
[16] T. Hayashi. *Acc. Chem. Res.* 2000, 33, 354.

[17] G. Titvinidze, Hydride addition reactions of α,ω-bis(trimethylsiloxy)methylhydrosiloxan with some unsaturated bonds containing compounds; PhD thesis, Ivane Javakhishvili University, Tbilisi, 2003.
[18] O. Mukbaniani, T. Tatrishvili, G. Titvinidze, N. Mukbaniani, L. Lezhava N., Gogesashvili. *J. Appl. Polym. Sci.,* 2006, 100, 2511.
[19] O.V. Mukbaniani, G.E. Zaikov, Cyclolinear Organosilicon Copolymers: Synthesis, Properties, Application; VSP, Utrecht–Boston, 2003.
[20] M.I.S. Dewar, E.G. Zeobish, E.F. Healy, J.S. Stewart. *J. Am. Chem. Soc.* 1985, 107, 3902.

In: Modification Reactions of Oligomethylhydridesiloxanes
Editors: O.Mukbaniani, et al. pp. 219-228
ISBN: 1-60021-365-0

Chapter 14

HYDROSILYLATION REACTION OF METHYLHYDRIDESILOXANE TO PHENYLACETYLENE

O. Mukbaniani, T. Tatrishvili, G. Titvinidze,and N. Mukbaniani
I. Javakhishvili Tbilisi State University,
I. Chavchavadze Ave., 1, 380028 Tbilisi, Georgia

ABSTRACT

The hydrosilylation reaction of α,ω–bis(trimethylsiloxy)methylhydridesiloxane to phe-nylacetylene in the presence of catalyst – platinum hydrochloric acid (0,1 M solution in tetrahydrofuran) at 1:35 ratio of initial compounds, at various temperature (40-60^{0}C) was investigated and methyl-siloxane oligomers with phenethenyl substituted groups in the side chain have been obtained. It was shown that complete hydrosilylation of all active ≡Si-H groups does not take place. The hydrosilylation reaction order, activation energy and rate constants were found. The synthesized oligomers were characterized by ^{1}H, ^{13}C NMR and IR spectra data. Gel-permeation chromatography, differential scanning calorimetry, thermogravimetric and wide-angle X-ray investigations of synthesized oligomers were carried out.

Key words: methylhydridesiloxane, phenylacetylene, hydrosilylation, hydride addition.

INTRODUCTION

The properties of organosilicon polymers depend on the structure of macromolecular chain and on the nature of organic groups surrounding the silicon atom [1]. Last time great attention was paid to copolymers, where different organic substituted groups are bonded on methylsiloxane hydrophobic matrix. By variation of these substituted groups utilization of the above-mentioned copolymers is possible in a wide range.

The hydrosilylation continues to be much studied and reviewed and the role of transition metal catalysts represents a key aspect of these studies. The results of such investigations revealed that product distributions vary considerably with the nature of the acetylene, the silane structure, the catalyst and the reaction conditions [2].

In the literature, the hydride addition reactions of silanes to phenylacetylene are known, using H_2PtCl_6 as a catalyst which results in the predominant formation of the *trans*-β-addition products via *cis*-addition, t-$R_3SiCH{=}CHPh$, with minor formation of the α-products $R_3Si(Ph)C{=}CH_2$. Few *cis*-β-products derived from *trans*-addition are observed using this catalytic system. The *cis*-addition occurs with re-tention of configuration at Si atom [3-7]:

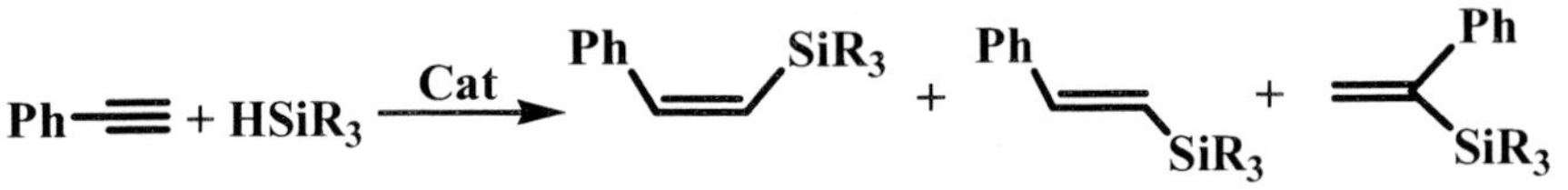

cis-β–product **trans–β–product** **α–product**

There are known hydride addition reactions of Ph_2MeSiH and $PhMe_2SiH$ to phenyl-acetylene in the presence of H_2PtCl_6/CO catalytic system. It was shown, that ratio between the obtained products in the case of $PhMe_2SiH$ is α: *trans*-β=66:34 (in the absence of CO it is equal to 25:75). In the case of Ph_2MeSiH the ratio between obtained products is α: *trans*-β=58:42 (in the absence of CO it is equal to 18:82). The observed increases of α-product formation demonstrate that by changing the atmosphere in which the reaction is performed the nature of the platinum hydrochloric acid system can be modified simply and it can play a significant role on the direction of reaction [8].

The reaction between phenylacetylene and transition metal-substituted silane $(\eta^5$-$C_5H_5)Fe(CO)_2SiPh_2H$ in the presence of platinum hydrochloric acid resulted in significantly greater amounts of the α-addition product than normally found when this catalyst is used (α:β=80:20) [9]:

$$\mathbf{FpSiPh_2H + PhC{\equiv}CH \longrightarrow t\text{-}FpSiPh_2CH{=}CHPh + FpSiPh_2(Ph)C{=}CH_2}$$

trans–β–product **α–product**

During hydride addition reactions of trichlorosilane and methyldichlorosilane to phenylacetylene in the presence of Pt/C and $H_2PtCl_6{\cdot}6H_2O$, obtained silyl-substituted styrene can attach to the second molecule of hydride silane [2]:

$$C_6H_5C{\equiv}CH + H\text{-}Si{\equiv} \xrightarrow{Cat} C_6H_5CH{=}CH{-}Si{\equiv}$$

$$C_6H_5CH{=}CH{-}Si{\equiv} + H\text{-}Si{\equiv} \xrightarrow{Cat} C_6H_5CH_2{-}CH(Si{\equiv})_2$$

EXPERIMENTAL PART

The starting materials for the synthesis of methylsiloxane oligomers with reactionable double bonded phenethenyl substituted groups in the side chain were α,ω–bis(trimethylsiloxy)methylhydridesiloxane and phenylacetylene.

The initial α,ω–bis(trimethylsiloxy)methylhydridesiloxane with the degree of polymerization n≈35 was received from Fluka. The organic solvents were cleaned by drying and distillation. Phenylacetylene was received from Fluka.

The IR spectra of all samples were taken on an UR-20 spectrophotometer from KBr pellets, while the ^1H NMR spectra was taken on a "Perkin-Elmer" instrument at operating frequency of 250 MHz. All spectra were obtained with the use of $CDCl_3$ as a solvent and internal standard. The Perkin –Elmer DSC-7 differential scanning calorimeter was used to determine TGA and the thermal transitions temperatures (T_g) were read at the maximum of the endothermic or exothermic peaks. Heating and cooling scanning rates were 10^0C/min.

Using Waters Model 6000A chromatograph with an R 401 differential refractometer detector gel-permeation chromatography investigation was carried out. The column set comprised 10^3 and 10^4 Å Ultra-styragel columns. Sample concentration was approximately 3% by weight in toluene and typical injection volume for the siloxane was 5 μL. standardization of the GPC was accomplished by using of styrene or polydimethylsiloxane standards with the known molecular weight.

Wide-angle X-ray diffractograms were taken on a "DRON-2" instrument. A-CuK_α was measured without a filter, the angular velocity of the motor ω≈2 deg/min.

Hydrosilylation Reaction of α,ω–Bis(Trimethylsiloxy)Methylhydridesiloxane to Phenylacetylene

Hydrosilylation reaction was carried out in a three-necked flask equipped with a tube for catalyst inlet, reflux condenser and inert gas carrier tube. The reaction products 1.4699 g (0.7491 mmole) α,ω–bis-(trimethylsiloxy)methylhydridesiloxane and 2.2958 g (22.473 mmole) phenylacetylene and 5.4 ml dry toluene were placed into the flask and thermostated in an oil bath until constant temperature was set up. Then the catalyst 0.1 M solution of platinum hydrochloric acid in tetrahydrofuran ($5 \div 9 \times 10^{-5}$ g per 1.0 g of starting substance) was introduced. After the reaction had been finished the solvent was eliminated.

RESULT AND DISCUSSION

From the literature it is known, that insertion of various heteroatom or cyclic fragments in the linear dimethylsiloxane backbone breaks the spiral structure of the polydimethylsiloxane chain and changes the physical-chemical indices and thermal-oxidative stability of copolymers [10]. The insertion of more rigid phenyl fragments in the main polydimethylsiloxane (PDMS) backbone hinder the depolymerization processes of the polymeric chain which is accompanied by release of cyclic D_n type dimethylcyclo-siloxanes $[Me_2SiO]_n$. Such types of polymers are characterized by higher thermal-oxidative stability.

Increased thermal-oxidative stability of copolymers may be explained with the presence of high resis-tance phenyl groups to oxidation and with their inhibiting effect on oxidation of methyl groups [11, 12].

For synthesis of new siliconorganic oligomers with double bonded phenethenyl groups in the side chain the hydrosilylation reaction of α,ω–bis(trimethylsiloxy)methylhydridesiloxane to phenylacetylene, in the presence of catalyst 0.1 M solution of platinum hydrochloric acid in tetrahydrofuran at 1:35 ratio of initial compounds has been investigated.

Preliminary heating of initial compounds in the temperature range of $40\div60^0$C in the presence of a catalyst, showed that in these conditions polymerization of phenylacetylene, or other changes due to the break in the siloxane backbone, or elimination of methane from methylhydridesiloxane does not take place.

Besides, there are no changes in the NMR and IR spectra of initial compounds. By gas-liquid chromatography it was established that the polymerization of phenylacetylene in this condition does not occur.

During the hydrosilylation reaction, the decrease of active ≡Si-H groups on the time was observed. The hydrosilylation reaction was investigated in a dry toluene solution ($C\approx6.8x10^{-2}$ mole/l). It was established that during the reaction all active ≡Si-H groups does not completely take place in the hydrosilylation reaction. In Figure 1 one can observe dependence of changes of active ≡Si-H groups on the time during the hydrosilylation reaction of α,ω–bis(trimethylsiloxy)methylhydridesiloxane to phenylacetylene.

From Figure 1 it is evident that at 40^0C the hydrosilylation reaction proceeds about ~80% conversion of active ≡Si-H groups, while at 60^0C the hydrosilylation reaction proceeds about 92% conversion. Consequently, with the rise of temperature the depth of the hydrosilylation reaction increases.

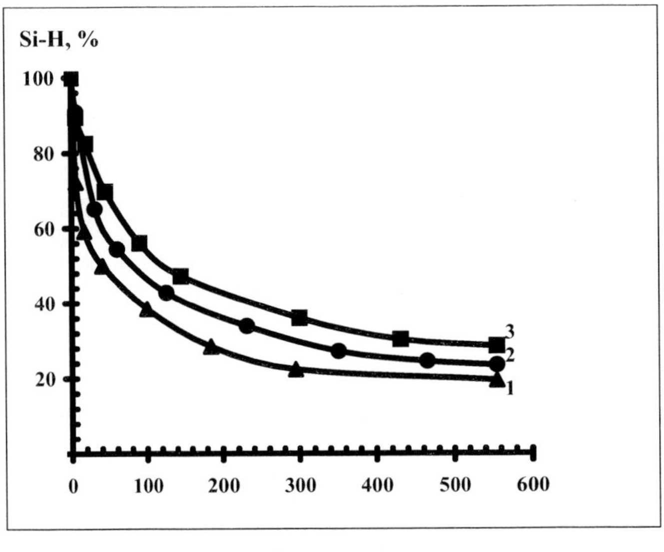

Figure 1. Dependence of changes of concentration of active ≡Si-H group on the time, upon hydrosilylation reaction of α,ω–bis(trimethylsiloxy)methylhydridesiloxane to phenylacetylene, where curve 1 is at 60^0C, curve 2 at 50^0C and curve 3 is at 40^0C.

So, the hydrosilylation reaction of α,ω–bis(trimethylsiloxy)methylhydridesiloxane to phenylacetylene in general proceeds according to the scheme 1 in which [(a)+(b)+(c)](d)=m≈35; 60^0C – I, 50^0C – I^1, 40^0C – I^2.

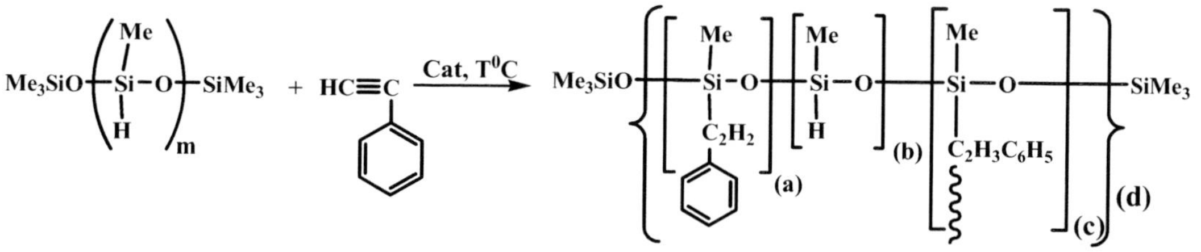

Scheme 1. Hydrosilylation reactions of α,ω–bis(trimethylsiloxy)methylhydridesiloxane to Phenylacetylene.

Hydrosilylation proceeds rapidly in the first hour, and then the reactions proceed at a low rate. Synthesized oligomers are orange liquid and well soluble in ordinary aromatic type organic solvents with $\eta_{sp}\approx 0.07\div0.08$. Structure and composition of oligomers were proven by elementary analysis, deter-mination of molecular masses, IR, ^{1}H and ^{13}C NMR spectral data. Some physical-chemical properties of synthesized oligomers are presented in table 1.

Table 1. Some physical-chemical properties of synthesized oligomers.

№ Oligo-mers	Yield, %	η_{sp} in 1% toluene solu-tion at 25^{0}C	Reaction temperature, ^{0}C	Conversion of active ≡Si-H % group	T_g, ^{0}C	10 % mass losses, ^{0}C	$\bar{M}_\omega$x10^{-4}
I	92	0.08	60	83	-45	350	4.14
I^1	91	0.07	50	76	-	-	-
I^2	91	0.07	40	71	-	-	-

After elimination of the solvent was carried out NMR and IR spectra investigations were performed.

In the IR spectra of oligomers one can observe absorption bands at 1025 cm^{-1} characteristic of asymmetric valence oscillation of linear Si-O-Si bonds. In the region 840, 1410 and 1275 cm^{-1}, one can observe absorption bands characteristic for –$SiMe_3$, ≡Si-CH= and ≡Si-Me bonds respectively. In IR spectra there was no signal at 2160 cm^{-1} characteristic for ≡Si-H bonds, which indicate that during elimination of the solvent, full intermolecular hydrosilylation takes place and branched systems were obtained.

In the ^{1}H NMR spectra of oligomer I (Figure 3), one can observe singlet signals for protons in ≡Si-Me, –$SiMe_3$ and Me group with chemical shift in the range $\delta\approx0.05\div0.44$ ppm. In the spectra one can observe a signal for protons of the methine group in the fragment ≡Si-CH=, with centre of chemical shift $\delta\approx5.8$ ppm, also a signal with a centre of chemical shift $\delta\approx6.2$ ppm, which corresponds to protons of the methine group in Ph-CH= fragment. In the spectra one can observe a multiplet signal characteristic for protons of the phenyl group in the range $\delta\approx6.7\div7.4$ ppm.

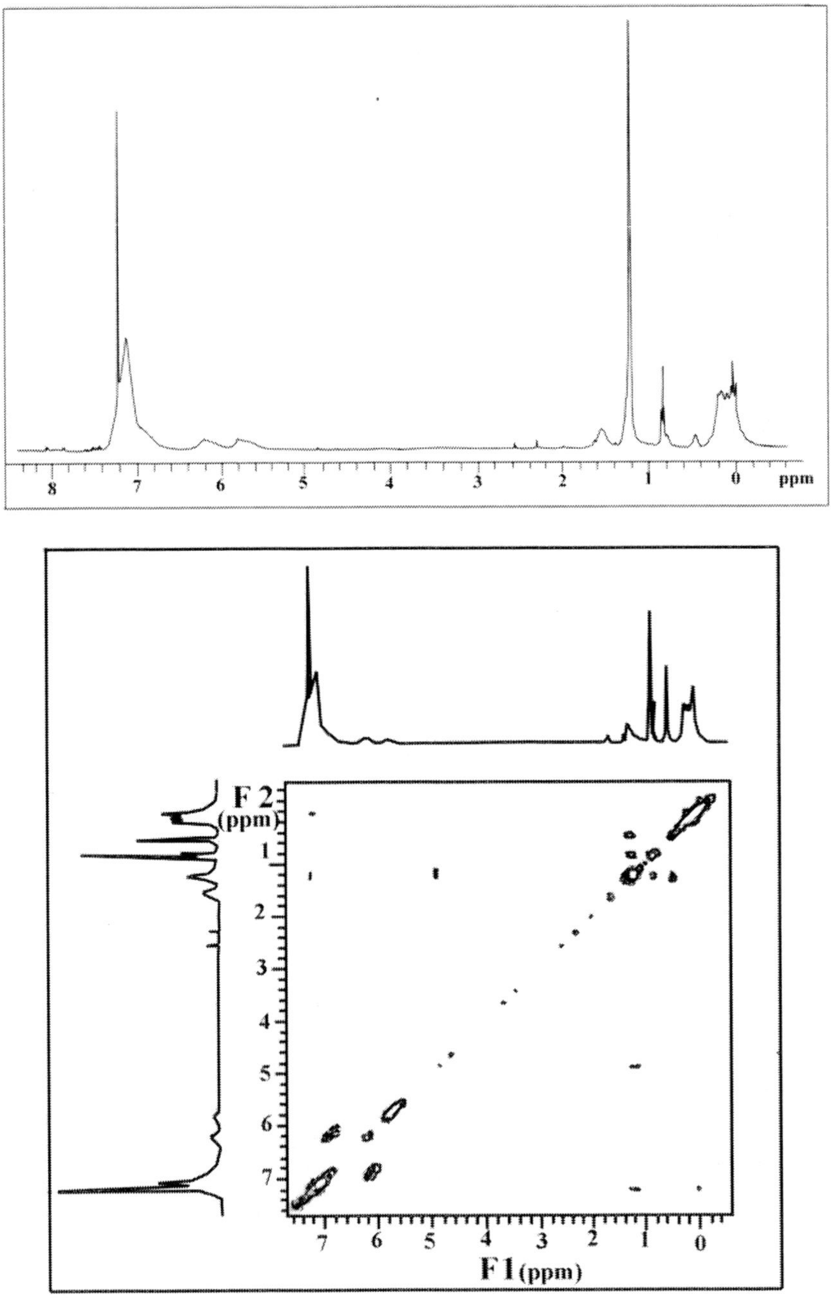

Figure 2. The ^{1}H NMR and COSY spectra of oligomer I.

One can also observe two signals with a centre of chemical shifts in the range 1.22 ppm and 1.57 ppm, which corresponds to methylene protons; triplet signal with a centre of chemical shift at 0.82 ppm corresponds to protons in the methine fragment. Formation of these complementary signals may be explained by the passing of branching processes after the elimination of the solvent.

^{13}C NMR spectra of oligomer I is in complete conformity with ^{1}H NMR spectra data. In the spectra one can observe double tuned signals for a carbon-silicon fragment with chemical shift $\delta \approx 125.3$ ppm, for C-C_{ar} double tuned signals with chemical shift $\delta \approx 125.5$ ppm. It shows that there are two isomeric products *cis*- and *trans*-β-vinylsilane. In the spectra one can observe the resonance signal for a carbon nucleus of an aromatic fragment with chemical shift $\delta \approx 126.1$-129.0 ppm.

Figure 3 shows dependence of reverse concentration on the time during the hydrosilylation reaction of methylhydridesiloxane to phenylacetylene. One can see that at the initial stages the hydrosilylation reaction is of second order.

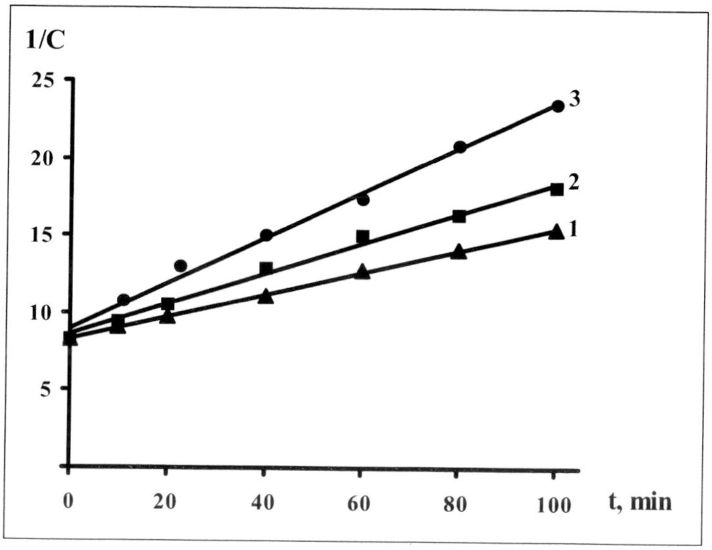

Figure 3. Dependence of inverse concentration on the time during the hydrosilylation reaction of methylhydridesiloxane to phenylacetylene. Where curve 1 is at 60^0C, curve – 2 at 50^0C and curve – 3 at 40^0C.

The reaction rate constants of hydrosilylation reactions of methylhydridesiloxane to phenylacetylene at various temperatures were determined, which values are: $k_{40°C} \approx 0.9457 \times 10^{-2}$, $k_{50°C} \approx 1.4998$ and $k_{60°C} \approx 2.3571$ l/mol·s. The reaction temperature coefficient is equal to $\gamma \approx 1.6$.

From the dependence of the hydrosilylation reaction rate constants logarithm on the reverse temperature, the activation energy of the hydrosilylation reaction was calculated, which is equal to: $E_{act.} \approx 54.6$ kJ/mol.

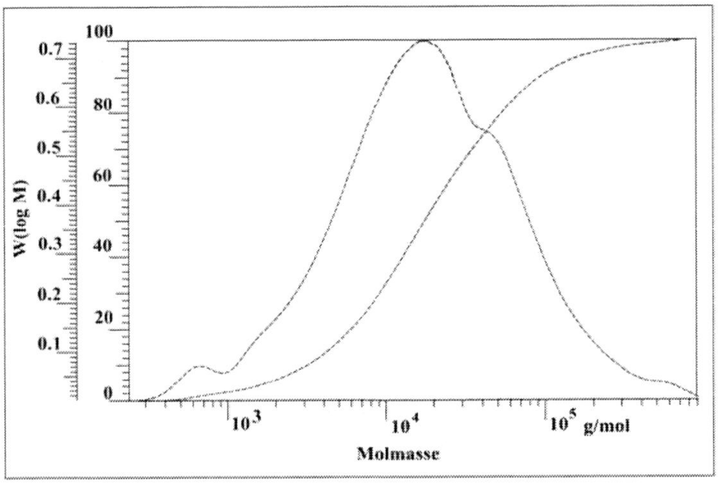

Figure 4. Gel-permeation chromatographic curves of oligomer I.

Figure 4 shows the molecular weight distribution of oligomer I. From the figure it is evident that oligomer I has trimodal molecular weight distribution. Its molecular weight is equal to: $\bar{M}_n \approx 6.94 x 10^3$, $\bar{M}_\omega \approx 4.14 x 10^4$, and polydispersity - D≈ 5.97.

As it is obvious the average molecular masses of synthesized oligomers exceed theoretical values of calculated molecular masses in case of full hydrosilylation several times. It indicates that during the hydrosilylation reaction the branching process also takes place according to the next general scheme 2.

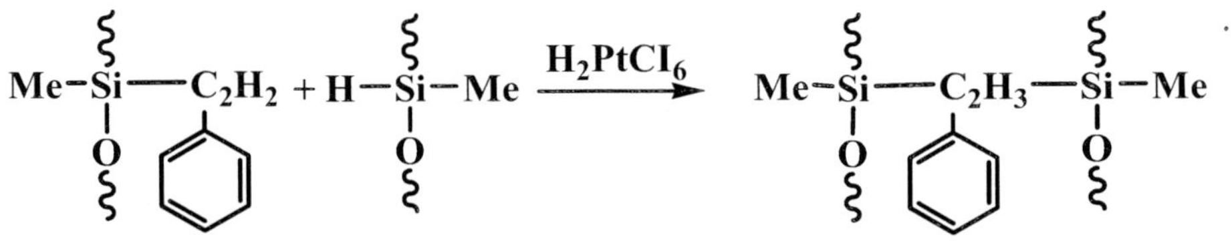

Scheme 2. Possible reaction of intermolecular macromolecular hydrosilylation.

Increased molecular masses proved that during the intermolecular hydrosilylation reaction according to Markovnikov and Farmer rule branched oligomers are obtained. Obtaining of branching systems also may proceed with participation of residual ≡Si-H groups in intermolecular cross-linking reactions with obtaining of ≡Si-O-Si≡ and ≡Si-Si≡ bonds, which were earlier discussed [13,14].

Differential scanning calorimetric investigation for synthesized oligomers has been carried out (Figure 5). As it is seen from DSC curves for oligomer I only one endothermic peak observed, with one temperature transition, which corresponds to the glass transition temperature $Tg \approx -45^0C$.

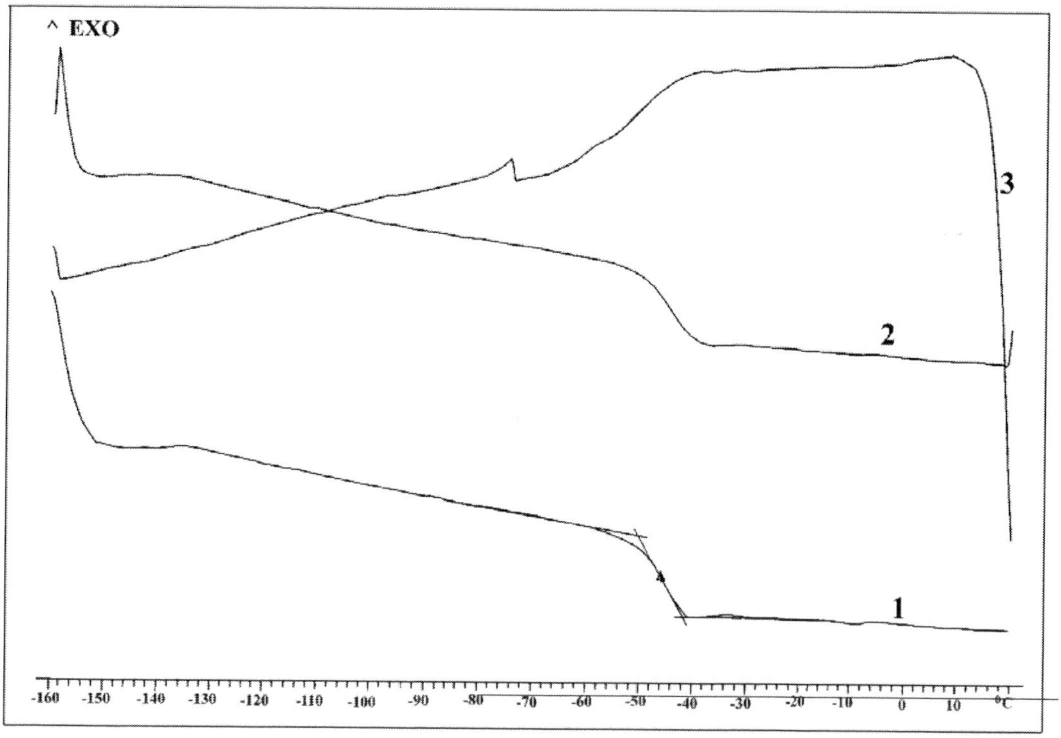

Figure 5. DSC curves of oligomer I. Where curves 1 and 2 correspond to first and second heating scanning, curve 3 – cooling scanning.

Thermogravimetric investigations of synthesized oligomers were carried out. For oligomers, the mass losses at 250^{0}C temperature are not observed, while the unblocked polydimethylsiloxane polymer is completely destroyed in the 270-280^{0}C-temperature range [15]. 10 % mass losses are observed at ~380^{0}C and the main destruction process proceeds from 400^{0}C. It was shown that by insertion of aryl fragments in the dimethylsiloxane chain thermal-oxidative stability of oligomers increased.

References

[1] O.V. Mukbaniani and L.M. Khananashvili *.Intern. J. Polymeric Materials*, 1994, 27, 31.

[2] Marciniec B., *Comprehensive Handbook on Hydrosilylation,* Pergamon Press, 1994.

[3] L.N. Lews, K.G. Sy, G.L. Bryant and P.E. Donahue. *Organometallics,* 1991,10, 3750.

[4] L.N. Lews *J. Am. Chem. Soc.*, 1990,112, 5998.

[5] E. Lukevics, R.Ya. Sturkovich and O.A. Pudova. *J. Organomet. Chem.*, 1985, 292, 151.

[6] I. Lovel, Yu.Sh. Goldberg, M.V. Shymanska, and E. Lukevics. *Organometallics,* 1987, 6, 1410.

[7] A.G. Brook, K.H. Pannell and D.G. Anderson. *J. Am. Chem. Soc.*, 1968, 90, 4374.

[8] M. Rivera-Claudio, J. Rozell, E. Ramirez-Oliva, J. Cervantes, K. H. Parmell. *Journal of Organomet. Chem.*, 1996, 521, 267.

[9] K.H. Pannell, J.M. Rozell, J. Lii and S.Y. Tien Mayr. *Organometallics,* 1988, 7, 2524.

[10] V.P. Mileshkevich. Kauchuk and Rezina, 1978, 4. (in Russian).

[11] T.H. Thomas, T.C. Kendrick. *J. Polymer Sci.*, 1969, **7**, 537.

[12] A.F.Bulkina, V.S. Papkov, A.A. Zhdanov, K.A. Andrianov. *Vysokomol. Soed.*, 1978, 20A, 70 (in Russian).

[13] T.A.C. Flipsen, R. Derks, H. Van Der Vegt, A.J. Pennings, G. J. Handzzionnou. *Polymer Science*, 1997, 35, 41.

[14] B.A. Trofimov, L.N. Parshina, M.I. Khilko, L.A. Oparina. *Thesis of the All-Russian Conference "Organosilicon Compounds: Synthesis, Characteristics and Application"*, Moscow, 2000, L-36. (in Russian).

[15] V.P. Yuriev, I.M. Salimgareva. "Hydrosilylation Reactions of Olefins", *Nauka,* 1982, 223 (in Russian).

INDEX

D

E

F

G

H

I

K

L

M

N

O

P

Q

R

U

V

W

X

Y

Z